# Advances in Intelligent Sensors and IoT Solutions

# Advances in Intelligent Sensors and IoT Solutions

Editors

**Behnam Mobaraki**
**Jose Turmo**

MDPI

Basel • Beijing • Wuhan • Barcelona • Belgrade • Novi Sad • Cluj • Manchester

*Editors*

Behnam Mobaraki
Universitat Internacional de Catalunya (UIC Barcelona)
Barcelona
Spain

Jose Turmo
Universitat Politècnica de Catalunya (UPC)
Barcelona
Spain

*Editorial Office*
MDPI
St. Alban-Anlage 66
4052 Basel, Switzerland

This is a reprint of articles from the Special Issue published online in the open access journal *Sensors* (ISSN 1424-8220) (available at: https://www.mdpi.com/journal/sensors/special_issues/735SR2H7C7).

For citation purposes, cite each article independently as indicated on the article page online and as indicated below:

Lastname, A.A.; Lastname, B.B. Article Title. *Journal Name* **Year**, *Volume Number*, Page Range.

**ISBN 978-3-7258-1323-0 (Hbk)**
**ISBN 978-3-7258-1324-7 (PDF)**
**doi.org/10.3390/books978-3-7258-1324-7**

# Contents

# About the Editors

**Behnam Mobaraki**

Behnam Mobaraki has previously worked in a research capacity at Minho University in Portugal from 2016 to 2017. Following this, he took on a research position at the Universitat Politècnica de Catalunya (UPC), in Spain, from 2017 to 2019. He holds a PhD with the distinction of cum laude in civil and building engineering from the Universidad de Castilla-La Mancha (UCLM), in Spain, which he completed between 2019 and 2023. During his doctoral studies, he developed a low-cost Internet of things (IoT) monitoring system for the thermal characterization of building envelopes. Throughout his academic career, he has secured no less than four competitive grants with public funding in Spain, both at the UCLM and UPC. Furthermore, he was honored with the golden medal at the Third International Invention and Innovation Competition of IFIA in Geneva, Switzerland, in 2023. His expertise lies in areas such as the finite element method (FEM), building monitoring, low-cost sensors, the Internet of things (IoT), and microcontrollers. He currently (2024) holds a position as an Assistant Professor in the Department of Basic Science at the Universitat Internacional de Catalunya (UIC Barcelona), in Spain.

**Jose Turmo**

Jose Turmo (Spain, 1974) received his 6-year program degree in civil engineering (1998) from the University of Cantabria (Santander, Spain) and his PhD (2003) from the Technical University of Catalonia BarcelonaTech—UPC (Barcelona, Spain). At the moment, he is a Professor at the School of Civil Engineering in Barcelona, BarcelonaTech (Spain), where he teaches construction engineering and bridges. His area of expertise is concrete bridges and structures. He has completed several research stays as a postdoc, being appointed as Visiting Faculty at the Indian Institute of Technology, Madras (2005), Fulbright Scholar at the University of California, San Diego, USA (2006), and Kwang-Hua Visiting Professor (2010) as well as High End Foreign Expert (2014–2016) at the Department of Bridge Engineering, Tongji University, Shanghai, China. He has authored around fifty SCI papers and one hundred conference papers. He is an international expert of the CTI (Quality Assurance Agency for the Schools of Engineering in France) and collaborates regularly with the ESTP Paris (École Supérieure de Travaux Publics), France. His technology transfer focuses on the design, maintenance, and monitoring of bridges and tunnels.

sensors

MDPI

*Article*

# Embedded Wireless Sensor for In Situ Concrete Internal Relative Humidity Monitoring

**Tai Ikumi [1,2], Ignasi Cairó [3], Jan Groeneveld [3], Antonio Aguado [1,4] and Albert de la Fuente [4,*]**

1 Smart Engineering Ltd., 08006 Barcelona, Spain; tai.ikumi@smarteng.es (T.I.); antonio.aguado@upc.edu (A.A.)
2 Department of Project and Construction Engineering, Technical University of Catalonia-BarcelonaTech, 08034 Barcelona, Spain
3 WitekLab, 08223 Terrassa, Spain; ignasi@witeklab.com (I.C.); jgroeneveld@chatu-tech.com (J.G.)
4 Department of Civil and Environmental Engineering, Universitat Politècnica de Catalunya, 08034 Barcelona, Spain
* Correspondence: albert.de.la.fuente@upc.edu

**Abstract:** The moisture content within the concrete pore network significantly influences the mechanical, thermal, and durability characteristics of concrete structures. This paper introduces a novel fully embedded wireless temperature and relative humidity sensor connected to an automatic acquisition system designed for continuous concrete monitoring. Relative humidity measurements from this new sensor are compared with those obtained by a commercial system based on the borehole method at different depths (2.5 and 4.0 cm) and exposure conditions (oven drying and humid chamber). The results allow for proving that both systems provide consistent internal relative humidity measurements aligned with the exposure conditions and highlight the capability of fully embedded wireless sensors as a practical and reliable alternative to the conventional borehole method. Additionally, the continuous monitoring of the wireless cast-in sensor exhibits reliability during unintended temperature fluctuations, emphasizing the effectiveness of permanently installed sensors in promptly detecting unintended curing variations in real time. The continuous real-time information provided combined with the practicality of these sensors might assist construction managers to improve the quality control of the concrete curing process and shrinkage behavior, and ensure the integrity of concrete surface finishing.

**Keywords:** wireless sensor networks; temperature; relative humidity; monitoring; concrete

**Citation:** Ikumi, T.; Cairó, I.; Groeneveld, J.; Aguado, A.; de la Fuente, A. Embedded Wireless Sensor for In Situ Concrete Internal Relative Humidity Monitoring. *Sensors* **2024**, *24*, 1756. https://doi.org/10.3390/s24061756

Academic Editors: Behnam Mobaraki and Jose Turmo

Received: 7 February 2024
Revised: 1 March 2024
Accepted: 7 March 2024
Published: 8 March 2024



## 1. Introduction

The amount and state of water in the concrete pore network influence the mechanical, thermal, and durability characteristics of concrete elements and structures. Proper monitoring of the water content in concrete is essential to address the following:

- *Control of the curing process.* Effective curing must ensure keeping the material under specific conditions of temperature and moisture to allow for the correct cement hydration and the development of its mechanical properties. Loss of moisture during curing can significantly slow down cement hydration as drying coarsens the pore structure of the cement paste matrix [1,2]. In fact, it has been observed that the hydration of cement paste might even stop when the relative humidity drops below about 80% [3]. This is particularly relevant in hot-weather concreting [4] and high-performance concrete applications, where low water-to-binder ratios and high binder contents are used.
- *Corrosion prevention.* Carbonation and chloride ingress rates are moisture-dependent. The progression of carbonation is the highest at intermediate moisture contents (between dry and saturated state) [5], while the diffusion of chlorides needs water in the pores to diffuse.
- *Freeze-Thaw.* The rising damp and alkali–silica reactions are aggravated by high moisture contents [6]. The prevention of these phenomena may involve the limitation

of water availability, and therefore, moisture monitoring in concrete is essential to assess the effectiveness of the corrective actions.

- *Shrinkage control*. Concrete shrinks as its internal moisture content decreases, either through exchange with the environment (evaporation) or through self-desiccation (hydration of cement particles), with its magnitude proportional to the amount of moisture lost [4,7,8]. The restraint of shrinkage-induced strains caused by moisture gradients is one of the most common causes of early cracking in concrete elements. This phenomenon is particularly relevant in concrete structures with large surface areas, where water evaporation is magnified and might lead to significant shrinkage strain gradients with concrete stresses superior to its early cracking strength.
- *Concrete surface finishing quality*. When concrete is poured in slabs, water vapor migrates from the bottom to the surface to evaporate. Early applications of impermeable finishes might lead to damage associated with moisture retention such as delamination of the floor adhesive, blistering of the epoxy coating, re-emulsification of the adhesive and curling, cracking or bubbling of flooring materials. ASTM F2170-19a [9] recognizes the need to monitor relative humidity levels before the installation of floor coverings and coatings and define threshold levels [10] to avoid subsequent damages to the system [11].

The examination of the moisture condition of concrete usually involves characterization of the internal moisture content and/or the relative humidity. The first refers to the amount of water present in the concrete, typically expressed as a percentage of the total mass of the concrete. On the other hand, the latter represents the percentage of moisture in the air relative to the maximum amount it could hold at a given temperature [12]. The relationship between these two critical indicators is established by the water sorption isotherms of cement pastes, as this link relative humidity with saturation degrees [13].

The measurement of moisture content is usually established by gravimetric methods. The average moisture content (free water) is assessed by the overall weight changes on test specimens of a particular shape and size. The moisture content distribution with the depth (referred to as moisture profile) is usually assessed by sampling from a series of depths. A review of sampling techniques can be found in [14]. Despite this being a straightforward method, the sample preparation is labor-intensive and the final measured values are often affected by variability related to the slicing process [7,15]. Moreover, this method cannot measure changes in the moisture condition produced by hydration of the cement or by changes in the capillary structure of the concrete [15]. Surface-based methods are not discussed here as these are not suitable for internal humidity assessments.

Internal relative humidity measurements are a relatively cheap and reliable alternative for in situ assessment of the internal moisture state of concrete [7,11]. Nowadays, the moisture condition is usually assessed by measuring the relative humidity with an electronic RH-probe placed in a closed volume of air in contact with the material [6,8,16–20]. This closed volume of air can be created during casting by placing a PVC tube or sleeve into a concrete structure on site with an open end at an intended distance from the exposed surface and a rubber cork in the other end. The other alternative is to drill a hole in the hardened material and seal it with a plastic tube with an open end and a rubber cork. Readings can be obtained once hygrometric equilibrium conditions are reached in the enclosed air volume and the open concrete surface. The time required depends on the geometry, the properties of the material and the properties of the RH-probe [14]. This standardized procedure is formally outlined in ASTM F2170-19a [9], with several commercial systems available in the market, mainly oriented to the pavement construction industry (Vaisala SHM40, Proceq hygropin, Wagner Meter RH L6, amongst others).

Despite its widespread implementation, this procedure presents several limitations when it comes to the continuous long-term monitoring of concrete internal humidity. Firstly, there is a recurrent need for sealing/unsealing the closed air cavity in contact with the concrete at each measurement, which might induce disturbances in the internal humidity distribution [7]. Secondly, probe installation and placement (either drilled hole or cast-in tube) affect the concrete surface at multiple locations and constitute external objects that might interfere with the construction practices on site. Finally, the readings are manual and time-consuming (about 30 min per reading), which is not compatible with continuous monitoring.

Nowadays, there is a growing interest in full autonomous wireless embedded solutions for internal relative humidity monitoring in concrete to overcome these limitations, most of them being experimental prototypes not currently commercialized [16,17,21,22]. These solutions are based on the same principles as the aforementioned ones, as the measurements are also conducted with the use of sensors in a closed air volume in open contact with the concrete pore network. The closed air volume is integrated in the sensor body, which is kept completely embedded inside the concrete element since casting. This air cavity is protected from fresh concrete ingress through the use of a water-impermeable fabric which is permeable to vapor. This way, the air cavity is continuously sealed, guaranteeing no disruption of the humidity distribution. The humidity measurements from the sensor are localized and reflect only the depth at which the sensor is positioned. To precisely measure the moisture gradient in a concrete section, various sensors should be placed at different depths at a considerable distance from each other (to avoid disturbance of the moisture flow).

The data in such fully embedded solutions are not only securely held on the sensor inside the concrete, but data collection is also simpler and faster as there are no external units that need to be maintained. In fact, these mostly incorporate wireless automatic acquisition systems (by means of various communication systems such as LoRa, Zigbee, Bluetooth Low Energy, ISM or Wifi [23]). This allows for the autonomous and continuous recording of the relative humidity following casting, which allows for the monitoring of the progressive reductions in the internal moisture associated with concrete self-desiccation at early ages [7]. Such a high operating complexity combined with the total embedment of the sensors in concrete significantly limits the battery lifespan of this technology, being unsuitable for long-term monitoring.

Alternatively, some experimental prototypes implement passive radio frequency identification technology (RFID) oriented to reach longer service lives (required for structural health monitoring), and to reduce the complexity and economic cost of the solution [21,22]. These systems are able to communicate with the interrogator on a zero-powered backscatter mechanism but have significant limitations regarding the wireless sensing range between the tag and the reader. As a result, data collection is often compromised or impractical for continuous live concrete monitoring. Commercialized fully embedded wireless concrete sensors for the monitoring of temperature (such as [24–28]), distance ([29,30]), or other properties are not considered here.

Despite the recognized importance of internal moisture monitoring in concrete, there is still a lack of real experimental data openly available to support progress in both the measurement methods and the prediction of moisture in buildings [6]. This is evidenced by the lack of established general criteria among practitioners regarding common concrete drying rates and even a basic understanding of what moisture and relative humidity in concrete really represent.

The objective of this paper is to introduce a new fully embedded wireless temperature and relative humidity sensor for continuous concrete monitoring. Relative humidity measurements from this new sensor at various exposure conditions and depths are compared with those obtained from a commercial sensor based on the borehole method according to ASTM F2170-19a [9].

## 2. Materials and Methods

### 2.1. Mortar Cubes

A total of 16 cubes of 10 cm × 10 cm × 10 cm were cast to assess the internal relative humidity of mortars at 2 different depths (2.5 and 4 cm) during varying wet–dry exposure. The mortar mix adopted is specified in Table 1, which intends to simulate a typical composition of the mortar that surrounds the coarse aggregate in a conventional concrete applied in pavements. Table 2 describes the particle size distribution of the aggregates used.

**Table 1.** Mortar mix evaluated [31].

| Material | [kg/m$^3$] |
|---|---|
| CEM II/A-L 42.5 N (Promsa, Barcelona, Spain) | 330 |
| Coarse agg. 4/10 mm (Promsa, Barcelona, Spain) | 260 |
| Sand 0/4 mm (Promsa, Barcelona, Spain) | 1543 |
| Water | 165 |
| Water/cement ratio | 0.50 |
| Master Ease 3850 Superplasticiser | 0.90% by cement weight |
| Master Pozzolith 7003 Plasticiser | 0.18% by cement weight |

**Table 2.** Aggregate grading [31].

| Sieve Size [mm] | 4–10 | 0–4 |
|---|---|---|
| | [% Passing] | |
| 40 | 100.0 | 100.0 |
| 20 | 100.0 | 100.0 |
| 10 | 96.1 | 100.0 |
| 4 | 0.8 | 99.9 |
| 2 | 0.40 | 83.3 |
| 1 | 0.4 | 52.1 |
| 0.5 | 0.4 | 33.7 |
| 0.25 | 0.4 | 22.4 |
| 0.125 | 0.4 | 17.6 |
| 0.063 | 0.4 | 14.9 |

The mixing procedure defined in UNE-EN 196-1:2005 [32] was adopted, where all solid components (cement, sand, and gravel) were initially dry-mixed. At the end of mixing process, the fresh mortar was poured into 10 cm × 10 cm × 10 cm molds.

### 2.2. *Internal Relative Humidity Monitoring*

#### 2.2.1. Borehole Method (Conforms to ASTM F2170-19a [9])

The Vaisala Structural Humidity Measurement Kit SHM40 (Vaisala, Vantaa, Finland) is adopted here to obtain reference internal relative humidity measurements of the mortar specimens, which has been designed for use with the borehole method. It uses the humidity and temperature probe HMP110 with a rugged polyurethane filled stainless steel body. The measurement range is [0, 100%RH] and [−40, +80 °C]. Reported measurement accuracy at temperatures between 0 and 40 °C is ±1.5%RH (0–90%RH) and ±2.5%RH (90–100%RH). Vaisala reports [33] errors between 5–6%RH when there is a difference of ±1 °C between the measured object and the probe at temperatures between 20 and 40 °C.

Plastic tubes (Ø17.4 mm and 120 mm length, Vaisala, Vantaa, Finland) were cast-in in 8 mortar cubes in such a way that the ends of the tubes had an open concrete surface at an intended distance from the exposed surface. Four specimens were prepared for each depth evaluated (2.5 and 4 cm). Long paper plugs were placed inside the tube to prevent the fresh mortar from blocking it. Due to the presence of aggregates, trowelling was performed to ensure a flat surface around the tube.

After 24 h, the specimens were demolded and paper plugs were removed. Then, the bottoms of the tubes were cracked with a flat-head screwdriver to help the air in the tube reach equilibrium with the humidity in the concrete. The crack between the tube and the hole was sealed with a thermo-silicon gun (Taurus Group, Oliana, Spain). Finally, the mortar dust at the bottom of the tube was cleaned with a vacuum cleaner and all tubes were sealed with a rubber plug.

Humidity measurements were initiated 3 days after the tube installation to allow for the airspace humidity to reach humidity equilibrium with the mortar, as indicated by the producer. Humidity measurements were performed by inserting the probe into the tube. Then, the tube was sealed with the rubber plug on the cable of the probe for 30 min to stabilize before starting the measurements.

2.2.2. Wireless Totally Embedded Sensor

The new temperature and relative humidity sensor Monsec, developed by ChatuTech (Terrassa, Spain) and Smart Engineering (Barcelona, Spain), is presented and used here as a practical alternative for concrete internal relative humidity monitoring. Monsec incorporates a silicon-based integrated circuit (IC) sensible to both temperature and relative humidity with a size of 1.5 mm × 1.5 mm × 0.5 mm in contact with an airspace integrated in the body. The sensing element consists of a mixed signal application-specific integrated circuit (ASIC) that provides measurement information through the IC (also possible SPI) digital serial interface to the local microcontroller unit. Such a sensing element consists of a polymer dielectric planar structure, capable of detecting relative humidity and is manufactured using a dedicated silicon process. The digitalization of the humidity sensor is carried out in the ASIC in a digital signal processing unit (DSP). Figure 1 represents a block diagram of the sensor unit.

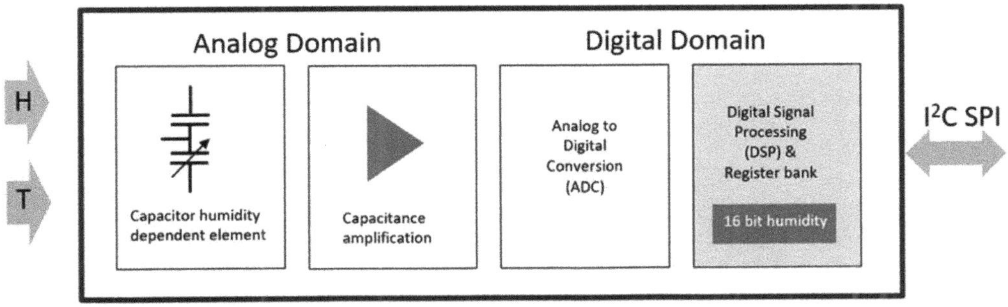


**Figure 1.** Block diagram of the sensor unit.

The sensor unit is integrated with the rest of the electronic system, taking special precautions to ensure that the embodiment of the sensor IC minimizes the differences between the humidity and temperature conditions of the environment under test conditions and those that represent the conditions around the sensor area. It is also important to consider the influence of heat generated by other devices close to the sensing area or due to the heating of the sensor itself. Changes in temperature are critical because these will also determine relative humidity deviations and, consequently, a slower response of the system. In this case, to improve the thermal decoupling of the sensor from the system, milling slits were created, and all unnecessary metals from the PCB around the sensor were etched. Figure 2 depicts a schematic diagram of the humidity sensor system integration.

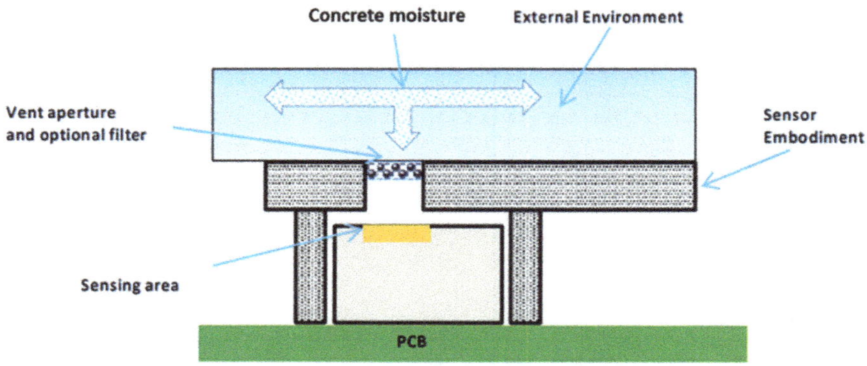


**Figure 2.** Humidity sensor system integration.

Considering the small window of the IC device exposed to the external environment and in order to obtain reliable and consistent measurements, the design is optimized to maximize sensor exposure to the external environment of concrete. This ensures a faster time response in terms of humidity and temperature. Additionally, it is crucial to guarantee that the environmental conditions match the sensing area conditions, not only in the steady state (static conditions) but also under dynamic conditions. As depicted in Figures 2 and 3, the impermeable fabric located on top protects this air cavity from fresh concrete ingress while being permeable to vapor exchange with the surrounding concrete. The sensor is fully functional in condensing environments and provides an operating range of [0, 100%RH] and [−40, +125 °C]. Relative humidity and temperature accuracies reported are ±[1.5, 1.8%RH] (30–70%RH) and ±0.1 °C, respectively. For RH > 70%, the accuracy tolerance is ±[2, 3%RH].

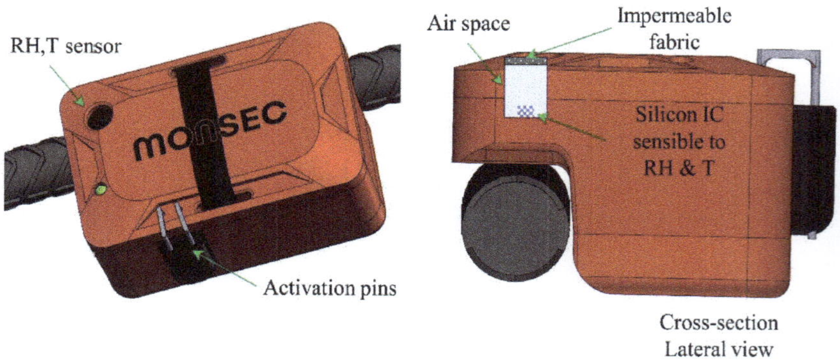


**Figure 3.** Details of the RH sensor incorporated in Monsec.

Besides the sensor element, which performs local measurements of temperature and relative humidity (Figure 3), the Monsec solution is composed of a station (a single receiver operating on the same ISM band), which wirelessly receives the recorded data and transmits it immediately to the cloud and a webApp from which measured data can be accessed. This system is entirely automated, and the receiver station can function independently using batteries and solar panels, making it a fully autonomous system without relying on the grid. The communication protocol between sensors is proprietary, secure, and exclusively compatible with Monsec sensors. Figure 4 illustrates a schematic representation of all the components involved, each necessary to achieve the wireless and remote continuous monitoring capabilities of the solution.

**Figure 4.** Components of the Monsec solution.

The wireless communication technology adopted by this sensor and the lack of accessibility to replace batteries due to complete embedment in concrete increase the cost and compromise the long-term performance of this solution. By configuring a measuring frequency of 15 min, the sensor battery lasts three months. This duration is sufficient for relative humidity monitoring during concrete setting but does not allow for the long-term monitoring of structures due to the limited battery lifespan. Additionally, the advantages in terms of practicality in installation and data collection often compensate for the higher initial cost.

The wireless sensors were placed in 8 of the freshly cast mortar specimens at the desired depths (4 sensors at 2.5 and 4 sensors at 4 cm). Measurements were initiated prior to the installation at intervals of 10 min. Figure 5 shows the final layout of the 16 specimens used, with the measuring depth for the wireless sensors and the borehole method highlighted.

**Figure 5.** Layout of all specimens used.

### 2.3. *Exposure Conditions*

After demolding, the samples were exposed to dry–wet cycles to induce rapid changes in the relative humidity of concrete. The drying method adopted consisted of hot air exposure in the oven. In this method, the heat evaporates the water in the specimens and increases its vapor pressure while lowering the relative humidity of the air in the oven [14]. The constant temperature used was 47 ± 2 °C to minimize any potential damage in the pores and/or cement paste degradation [34]. The wetting cycle was performed in a humidity chamber, where all specimens were kept at 100%RH and 20 ± 2 °C. Figure 6 shows the exposed ambient temperature and relative humidity over the test duration.

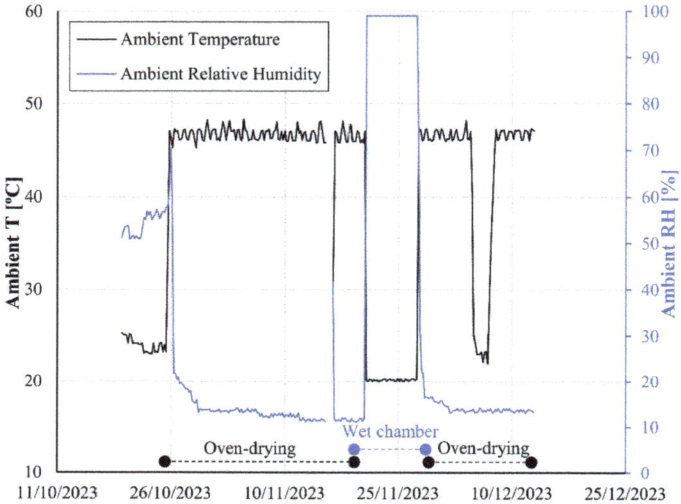


**Figure 6.** Ambient temperature over the test duration.

## 3. Results

### 3.1. *RH Measurements from the Vaisala Borehole Method*

Figure 7 depicts the ambient relative humidity and the internal relative humidity of the mortar specimens obtained through the borehole method (ASTM F2170-19a [9]) at 2.5 and 4 cm over the test duration. Relative humidities registered by the Vaisala system are presented as averages of the 4 samples evaluated and with error bars corresponding to ±1 standard deviation. As expected, mortar measurements show a gradual reduction in internal relative humidity during the oven-drying phase and a rapid increase during the wet chamber phase. Notably, the readings show minimal dispersion, with maximum coefficients of variation of only 3% among different replicates throughout the experiment.

Figure 7 illustrates that the internal moisture content of the mortar exhibits relatively minor variation across the two depths under evaluation. The average difference during the drying phase is merely 2.2%RH, and during the wetting phase, the difference is virtually negligible (<0.1%RH). The technical literature features a scarcity of studies documenting concrete relative humidity gradients in depth during the drying conditions [1,16,35–38]. Some studies indicate marginal variations in relative humidity at depths within 2.5–4 cm from the external surface in uncracked specimens [16,36], while others reveal substantial differences [1,35]. This diversity of results underscores the multitude of factors influencing moisture transport in concrete elements.

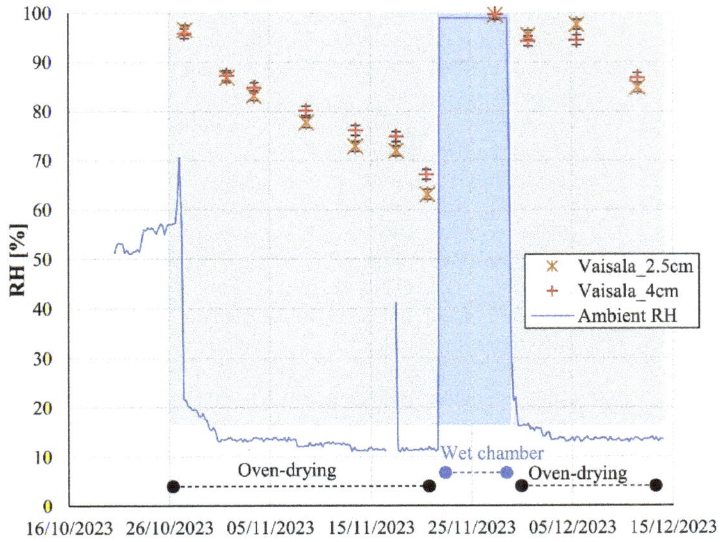


**Figure 7.** Concrete internal relative humidity at 2.5 and 4 cm measured with the borehole method.

## 3.2. RH Measurements from the Wireless Embedded Sensors

Figure 8 shows the evolution of the mortar internal relative humidity at 2.5 and 4 cm obtained by the embedded wireless system (Monsec) and the borehole method (Vaisala). The Monsec sensors, permanently installed in the mortar, record data every 10 min. For these sensors, each data series is presented separately, as these could not be grouped under statistical criteria due to their slightly different time references.

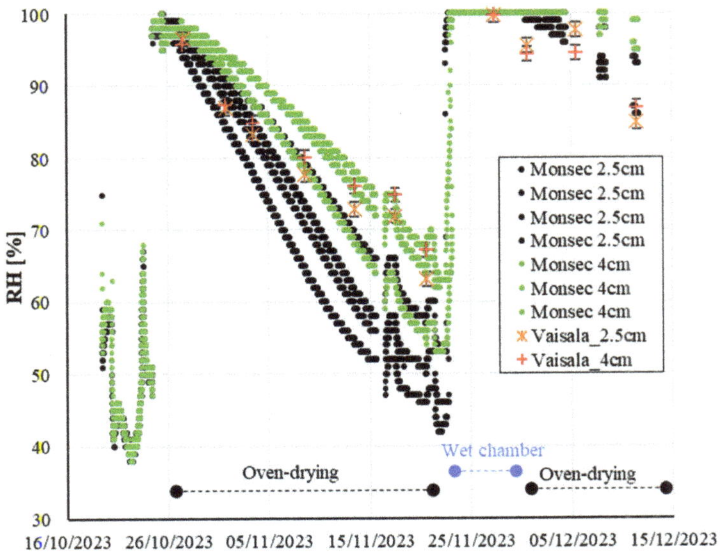


**Figure 8.** Internal relative humidity evolution in the concrete specimens at 2.5 and 4 cm.

In the pre-concreting phase, the Monsec sensors detect relative humidities ranging from 40% to 60%, aligning with laboratory conditions. Upon contact with the mortar, the humidity swiftly elevates to nearly 100%RH. Subsequently, the recorded humidity values begin a gradual decline attributed to oven-drying. As anticipated, sensors positioned closer to the surface (2.5 cm) exhibit faster drying rates compared to those at greater depths (4 cm). Approximately two weeks later, the relative humidity initiates a stabilization phase, with values between 40% and 60% RH. During this period, the cast-in sensors show a significant dispersion within specimens of the same depth (2.5 and 4 cm). This might be explained by slight sensor position variations in the vertical direction (significant in regions with large moisture gradients) [39] and/or differences on the characteristics of the material (components and pore network) around the sensor location.

Following one month of oven-drying, the samples undergo transfer to a wet chamber, triggering a rapid increase in the internal humidity of the mortar as detected by all sensors, reaching values close to 100%RH. Subsequently, after a 4-day interval, the samples are reintroduced to the oven under identical drying conditions as the initial phase. At this stage, a gradual reduction in the relative humidity of the mortar is once again observed, with more pronounced effects noted in the sensors situated at the outermost layers. Unfortunately, complete drying curves could not be completed due to limitations in the sensor's internal battery. With a measurement frequency of 10 min, the sensor signals were lost approximately 50 days post-activation.

### 3.3. RH Measurements from the Borehole Method vs. Wireless Cast-In Sensors

As depicted in Figure 8, both the cast-in sensors and the borehole method yield internal relative humidity measurements of the mortar that align with the adopted exposure conditions. Throughout the initial drying cycle, the average difference between measurements from both systems at a depth of 4 cm amounts to 3.4%RH. In the more superficial layers (2.5 cm), the average difference increases to 6.7%RH. These disparities cannot be directly ascribed to inaccuracies in the precision of either system's readings, as potential variations in the exact positioning of the sensors/tubes among the various test specimens considered might have played a significant role.

Additionally, the discrepancy observed between the two systems might not be attributed to the difference in the size of the air chamber where the relative humidity of the concrete is measured (significantly larger in the borehole method). Previous studies reported negligible influence in the size of the macro-pore inside which the embedded RH sensor was inserted [7]. This is attributed to the quantity of water that is necessary to shift the humidity of the macro-pore being very low in comparison to the quantity of water that the cementitious matrix can release upon drying [7]. However, it is essential to acknowledge that the studies have not assessed air chambers as small like the one in the Monsec system, making it impossible to dismiss its potential influence.

Another potential factor that could account for the reported measurement discrepancies relates to the disturbance of moisture flow around each system [14,39]. The distinct sizes and shapes of the two systems may lead to variations in moisture conditions in close proximity to the sensors. Nilsson and Fredlund [39] quantified this phenomenon through experimental and numerical analysis for various geometries and orientations, reporting differences of up to 10%RH in larger width probes. Additional research is needed to assess the influence of both sensor bodies on the moisture flow and how this might affect the reliability of the measurements.

### 3.4. Temperature Variation Effects

Figure 8 shows RH peaks at some specific dates that deviate from the overall trends previously described. These deviations are directly linked to variations in the exposure temperature conditions, as temperature plays a key role in moisture measurement [40,41]. In Figure 9, the internal temperature of the concrete, as recorded by the same sensors, reveals instances or intervals where the power supply from the laboratory was interrupted. This interruption led to a shift in exposure temperature, as the oven temperature equilibrated with the laboratory's ambient temperature (approximately 20 °C). Upon resumption of power, the furnace was reactivated, initiating a progressive heating of the specimens until the previous temperature was reestablished.

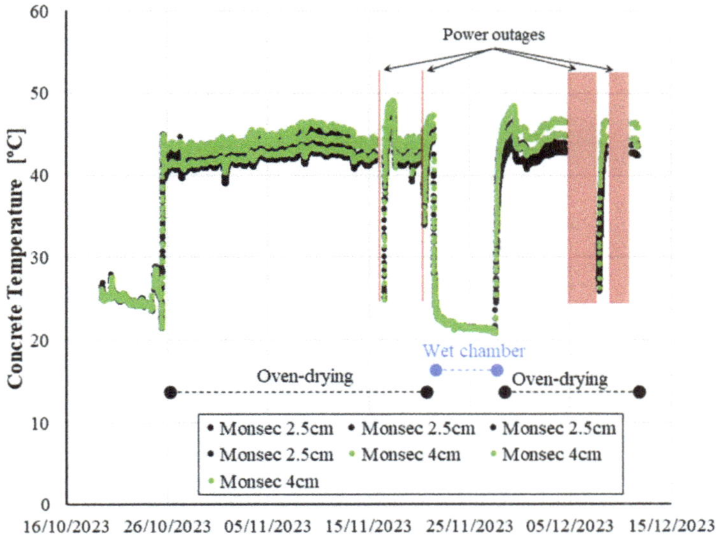


**Figure 9.** Internal concrete temperature evolution measured by Monsec sensors.

This inadvertent thermal fluctuation accounts for the atypical relative humidity readings recorded within the specimens on 17 and 21 November. These peaks align precisely with the furnace reactivation and the ensuing temperature increase, which underscores the efficacy of continuous monitoring sensors in promptly detecting unintended temperature variations in real time. Previous studies also reported reliable measurement results during temperature variations for cast-in sensors as this method ensures no temperature difference between the sensor and concrete [14]. Jensen and Hansen [42] proposed that a temperature difference of 1 °C between the sensor and the concrete material might introduce an error of approximately 6%RH.

Furthermore, examining the relative humidity evolution during these periods of furnace interruption/reactivation reveals interesting trends. It can be observed that a positive correlation between the rise in concrete temperature and the increase in measured humidity within the concrete. This contradicts the inverse correlation established between relative humidity and ambient temperature in open-air conditions, where an increase in temperature results in a decrease in relative humidity. This distinctive behavior in concrete aligns with findings reported by other researchers, indicating the impact of temperature on the vaporization of capillary water bound in the pores and the resulting water vapor transfer in capillary passages between the pores [14,17].

## 4. Conclusions

This paper presents a novel fully embedded wireless temperature and relative humidity sensor for continuous concrete monitoring. Relative humidity measurements from this new sensor at various exposure conditions and depths are compared with those obtained from a commercial sensor based on the borehole method. The following outlines can be concluded:

- The comparison at various depths and exposure conditions indicates that both systems yield consistent internal relative humidity measurements aligned with the adopted conditions. These results highlight the capability of fully embedded wireless sensors as a practical and reliable alternative to conventional methods.
- The wireless cast-in sensor method has given reliable relative humidity measurements during unintended temperature variations, as this method ensures no temperature difference between the sensor and the concrete. This emphasizes the efficacy of perma-

nently installed sensors over discrete monitoring in promptly detecting unintended curing variations in real time.

- Further research is needed to assess the influence of moisture flow disturbance around the cast-in sensor body and how this might affect the reliability of the measurements.

**Author Contributions:** Conceptualization, T.I., I.C., A.A. and A.d.l.F.; methodology, T.I., I.C. and A.A.; software, I.C. and J.G.; validation, T.I. and A.A.; formal analysis, T.I., A.d.l.F., J.G. and A.A.; investigation, T.I., J.G. and I.C.; resources, I.C. and A.d.l.F.; data curation, T.I.; writing—original draft preparation, T.I.; writing—review and editing, T.I., A.A., A.d.l.F.; visualization, T.I.; supervision, A.A., I.C. and A.d.l.F. All authors have read and agreed to the published version of the manuscript.

**Funding:** Support from the INNOSUP Horizon 2020 Metabuilding 1st GROW/HARVEST CALL through the research project MONSEC is greatly acknowledged (Agreement Number: 370).

**Institutional Review Board Statement:** Not applicable.

**Informed Consent Statement:** Not applicable.

**Data Availability Statement:** The data presented in this study are available on request from the corresponding author.

**Acknowledgments:** We would like to acknowledge AntennaLAB Group from Politechnical University of Catalonia (UPC), and in particular Lluis Jofre and Jordi Romeu.

**Conflicts of Interest:** Authors Tai Ikumi and Antonio Aguado were employed by the company Smart Engineering Ltd. Authors Ignasi Cairó and Jan Groeneveld were employed by the company WitekLab. The remaining authors declare that the research was conducted in the absence of any commercial or financial relationships that could be construed as a potential conflict of interest.

## References

1. Parrot, L.J. Moisture profiles in drying concrete. *Adv. Cem. Res.* **1988**, *1*, 164–170. [CrossRef]
2. Parrot, L.J. Factors influencing relative humidity in concrete. *Mag. Concr. Res.* **1991**, *43*, 45–52. [CrossRef]
3. Flatt, R.J.; Scherer, G.W.; Bullard, J.W. Why alite stops hydrating below 80% relative humidity. *Cem. Concr. Res.* **2011**, *41*, 987–992. [CrossRef]
4. Mehta, P.K.; Monteiro, P.J. *Concrete: Microstructure, Properties, and Materials*, 4th ed.; McGraw-Hill Education: New York, NY, USA, 2014.
5. Venuat, M. Carbonation—Technical Commission 16-C. *Matér. Constr.* **1978**, *11*, 142–146. [CrossRef]
6. Andrade, C.; Sarría, J.; Alonso, C. Relative humidity in the interior of concrete exposed to natural and artificial weathering. *Cem. Concr. Res.* **1999**, *29*, 1249–1259. [CrossRef]
7. Granja, J.L.; Azenha, M.; De Sousa, C.; Faria, R.; Barros, J. Hygrometric assessment of internal relative humidity in concrete: Practical application issues. *J. Adv. Concr. Technol.* **2014**, *12*, 250–265. [CrossRef]
8. Zhang, J.; Huang, Y.; Qi, K.; Gao, Y. Interior Relative Humidity of Normal- and High-Strength Concrete at Early Age. *J. Mater. Civ. Eng.* **2012**, *24*, 615–622. [CrossRef]
9. *ASTM F2170-19a;* Standard Test Method for Determining Relative Humidity in Concrete Floor Slabs Using In Situ Probes. ASTM International: West Conshohocken, PA, USA, 2019.
10. *SisäRYL 2000 Code of Building Practice, RT 14-10668;* Rakennustieto Oy: Helsinki, Finland, 2000; p. 318. ISBN 951-682-506-0.
11. Åhs, M. Moisture Redistribution in Screeded Concrete Slabs. Ph.D. Thesis, Lund Institute of Technology, Lund University, Lund, Sweden, 2007.
12. Baroghel-Bouny, V.; Mainguy, M.; Lassabatere, T.; Coussy, O. Characterization and identification of equilibrium and transfer moisture properties for ordinary and high-performance cementitious materials. *Cem. Concr. Res.* **1999**, *29*, 1225–1238. [CrossRef]
13. Wang, J.; Yio, M.H.N.; Zhou, T.; Wong, H.S.; Davie, C.T.; Masoero, E. Water sorption isotherms and hysteresis of cement paste at moderately high temperature, up to 80 °C. Cem. *Concr. Res.* **2023**, *165*, 107076. [CrossRef]
14. RILEM Technical Committee 248-MMB. *Methods of Measuring Moisture in Building Materials and Structures, State of the Art Report;* Nilsson, L.O., Ed.; Springer: Berlin/Heidelberg, Germany, 2018.
15. Gause, G.R.; Tucker, J., Jr. Method For Determining the Moisture Condition in Hardened Concrete. *J. Res. Natl. Bur. Stand.* **1940**, *25*, 403–416. [CrossRef]
16. Zhou, J.K.; Chen, X.; Zhang, J.; Wang, Y. Internal relative humidity distribution in concrete considering self-desiccation at early ages. *Int. J. Phys. Sci.* **2011**, *6*, 1604–1610.
17. Chang, C.Y.; Hung, S.S. Implementing RFIC and sensor technology to measure temperature and humidity inside concrete structures. *Constr. Build. Mater.* **2012**, *26*, 628–637. [CrossRef]
18. Grasley, Z.C.; Lange, D.A.; D'Ambrosia, M.D. Internal relative humidity and drying stress gradients in concrete. *Mater. Struct.* **2006**, *39*, 901–909. [CrossRef]

19. Jiang, Z.; Sun, Z.; Wang, P. Autogenous relative humidity change and autogenous shrinkage of high-performance cement pastes. *Cem. Concr. Res.* **2005**, *35*, 1539–1545. [CrossRef]
20. Shen, D.; Liu, C.; Wang, M.; Jin, X.; Tang, H. Prediction model for internal relative humidity in early-age concrete under different curing humidity conditions. *Constr. Build. Mater.* **2020**, *265*, 119987. [CrossRef]
21. Zhou, S.; Deng, F.; Yu, L.; Li, B.; Wu, X.; Yin, B. A Novel Passive Wireless Sensor for Concrete Humidity Monitoring. *Sensors* **2016**, *16*, 1535. [CrossRef] [PubMed]
22. Strangfeld, C.; Johann, S.; Bartholmai, M. Smart RFID Sensors Embedded in Building Structures for Early Damage Detection and Long-Term Monitoring. *Sensors* **2019**, *19*, 5514. [CrossRef] [PubMed]
23. Cabezas, J.; Sánchez-Rodríguez, T.; Gómez-Galán, J.A.; Cifuentes, H.; Carvajal, R.G. Compact Embedded Wireless Sensor-Based Monitoring of Concrete Curing. *Sensors* **2018**, *18*, 876. [CrossRef] [PubMed]
24. Converge. Available online: https://converge.io/ (accessed on 1 March 2024).
25. WAKE Inc. Available online: https://www.wakeinc.com/ (accessed on 1 March 2024).
26. AOMS Technologies. Available online: https://lumicon.io/ (accessed on 1 March 2024).
27. Giatec. Available online: https://www.giatecscientific.com/ (accessed on 1 March 2024).
28. Quadrel. Available online: https://vorb.io/ (accessed on 1 March 2024).
29. Liu, Y.; Bao, Y. Real-time remote measurement of distance using ultra-wideband (UWB) sensors. *Autom. Constr.* **2023**, *150*, 104849. [CrossRef]
30. Liu, Y.; Bao, Y. Review of electromagnetic waves-based distance measurement technologies for remote monitoring of civil engineering structures. *Measurement* **2021**, *176*, 10919. [CrossRef]
31. Ikumi, T.; Monserrat López, A.; Aidarov, S.; Aguado, A.; de la Fuente, A. Design-oriented method for concrete pavements with volumetric stability admixtures: An integrated experimental and analytical approach. *Case Stud. Constr. Mater.* **2023**, *19*, e02583. [CrossRef]
32. *UNE-EN 196-1:2005*; Métodos de Ensayo de Cementos. Parte 1: Determinación de Resistencias Mecánicas. AENOR: Madrid, Spain, 2005.
33. Vaisala. *Vaisala Structural Humidity Measurement Kit SHM40. User's Guide*; Vaisala Oyj: Helsinki, Finland, 2015.
34. *ISO 12570:2000*; Hygrothermal Performance of Building Materials and Products: Determination of Moisture Content by Drying at Elevated Temperature. International Organization for Standardization: Geneva, Switzerland, 2000.
35. Kim, J.K.; Lee, C.S. Moisture diffusion of concrete considering self-desiccation at early ages. *Cem. Concr. Res.* **1999**, *29*, 1921–1927. [CrossRef]
36. Ryu, D.W.; Ko, J.W.; Noguchi, T. Effects of simulated environmental conditions on the internal relative humidity and relative moisture content distribution of exposed concrete. *Cem. Concr. Compos.* **2011**, *33*, 142–153. [CrossRef]
37. Oxfall, M.; Johansson, P.; Hassanzadeh, M. Long-term hygrothermal performance of nuclear reactor concrete containments–Laboratory evaluations of measurement setup, in situ sampling, and moisture flux calculations. *Cem. Concr. Compos.* **2016**, *65*, 128–138. [CrossRef]
38. Oxfall, M.; Hassanzadeh, M.; Johansson, P. Moisture levels and drying potential of the concrete in Swedish reactor containments. *Eur. Phys. J. Web Conf.* **2013**, *56*, 03002. [CrossRef]
39. Nilsson, L.O.; Fredlund, P. *Cast-in Probe for Measuring Drying of Concrete—Simulation of the Effect of the Size and Orientation of the Sensor*; Sensobyg D4. Report TVBM-7198; Laboratory of Building Materials, University of Lund: Lund, Sweden, 2009. (In Swedish)
40. Fredin, H.; Skoog, H. Temperature effects and corrections on relative humidity measurements. In Proceedings of the 7th Symposium on Building Physics in the Nordic Countries, Reykjavik, Iceland, 13–15 June 2005; pp. 1–3.
41. Paroll, H.; Nykanen, E. Measurement of relative humidity and temperature in a new concrete bridge vs. laboratory samples. *Nord. Concr. Res.-Publ.* **1998**, *21*, 103–119.
42. Jensen, O.M.; Hansen, P.F. Influence of temperature on autogenous deformation and relative humidity change in hardening cement paste. *Cem. Concr. Res.* **1999**, *29*, 567–575. [CrossRef]

sensors

MDPI

*Article*

# IoT-Based Framework for Digital Twins in the Industry 5.0 Era

**Ahmed Awouda, Emiliano Traini, Giulia Bruno * and Paolo Chiabert**

Department of Management and Production Engineering, Politecnico di Torino, 10129 Turin, Italy; ahmed.awouda@polito.it (A.A.); emiliano.traini@polito.it (E.T.); paolo.chiabert@polito.it (P.C.)
* Correspondence: giulia.bruno@polito.it

**Abstract:** Digital twins are considered the next step in IoT-based cyber–physical systems; they allow for the real-time monitoring of assets, and they provide a comprehensive understanding of a system behavior, allowing for data-driven insights and informed choices. However, no comprehensive framework exists for the development of IoT-based digital twins. Moreover, the existing frameworks do not consider the aspects introduced by the Industry 5.0 paradigm, such as sustainability, human-centricity, and resilience. This paper proposes a framework based on the one defined as the outcome of a project funded by the European Union between 2010 and 2013 called the IoT Architectural Reference Model (IoT-A or IoT-ARM), with the aim of the development and implementation of a standard IoT framework that includes digital twins. This framework establishes and implements a standardized collection of architectural instruments for modeling IoT systems in the 5.0 era, serving as a benchmark for the design and implementation of an IoT architecture focused on digital twins and enabling the sustainability, resilience, and human-centricity of the information system. Furthermore, a proof of concept of a monitoring digital twin for a vertical farming system has been developed to test the validity of the framework, and a discussion of applications in the manufacturing and service sectors is presented.

**Keywords:** digital twin; Internet of Things; Industry 5.0; aeroponics; IoT-ARM

**Citation:** Awouda, A.; Traini, E.; Bruno, G.; Chiabert, P. IoT-Based Framework for Digital Twins in the Industry 5.0 Era. *Sensors* **2024**, *24*, 594. https://doi.org/10.3390/s24020594

Academic Editors: Behnam Mobaraki and Jose Turmo

Received: 25 December 2023
Revised: 13 January 2024
Accepted: 16 January 2024
Published: 17 January 2024



## 1. Introduction

In recent years, after undergoing a transformative paradigm shift called Industry 4.0, the manufacturing landscape is now deeply immersed in the Fourth Industrial Revolution. This evolution involves the seamless integration of digital technologies, data analytics, and automation into conventional manufacturing processes, giving birth to what is commonly referred to as "smart factories". Within these cutting-edge environments, cyber–physical systems (CPS), cloud computing, and the Internet of Things (IoT) collaboratively optimize efficiency, elevate productivity, and catalyze innovation [1]. At its core, the IoT interconnects devices and systems in a synergistic endeavor to augment industrial processes, extending its influence beyond typical computing devices [2]. This extensive network encompasses physical devices embedded with sensors and actuators, forming a real-time visibility network across the entire manufacturing ecosystem.

Digital twins (DTs) emerge as another pivotal element in this revolutionary wave. Serving as virtual replicas of physical objects or processes, digital twins facilitate data-driven decision-making, intricate systems monitoring, and product validation [3]. In the realm of cyber–physical systems, particularly in manufacturing, digital twins function as adaptive models that seamlessly integrate data between physical and virtual machines, holding transformative potential [4,5]. Digital twins, with the characteristics of ultra-high synchronization and fidelity, and convergence between physical and virtual products, have many potential applications in product design, product manufacturing, and product service [6]. One of the main benefits of digital twins is their ability to decouple physical flows from planning and control [7]; they create virtual replicas of physical assets and processes, allowing for simulation, real-time monitoring, and remote management. This decouples planning and control processes from physical flows by enabling efficient decision-making, optimization, and scenario planning in a virtual environment before implementation in the real world.

The symbiotic relationship between digital twins and the Internet of Things is evident in how IoT contributes data, and digital twins leverage these data to create dynamic, real-time models of their physical counterparts. This process involves integrating IoT data with virtual models, enabling the real-time monitoring, analysis, and optimization of physical entities [8].

Keeping well in mind the importance of designing information management systems that can work with human knowledge cognitive processes, in this paper, it is given as a hypothesis that such IoT systems should be able to manage industrial digital twin and complex knowledge management systems that are usually described as Industrial IoT (IIoT) systems [9].

Such complex Knowledge-Based Systems, well explained in [10,11], are considered to be based on hybrid modeling techniques, i.e., hybrid modeling is the process of making use of two or more modeling techniques belonging to different philosophies or methodologies and then synthesizing the results into a single score or spread. Hybrid modeling is conceptually similar to ensemble modeling but, while the second refers to the unification of different models of the same family, hybrid modeling refers to completely different approaches of modeling with a consequent increase in terms of the complexity of managing heterogeneous characteristics [12]. Moreover, hybrid knowledge in Industry 5.0 requires information systems with adequate information management platforms to digitize, manage, and use data involved in the information system: a more detailed description of the relationships between data, information, and knowledge is given in [11] through the DIKW method [13].

Despite the technological prowess of Industry 4.0 and its array of technologies, it is often perceived as a techno-economic vision. This vision outlines how broader technological advancements, originating outside industrial contexts, will influence industrial value chains and redefine the industry's economic landscape [11]. However, Industry 4.0 falls short of effectively addressing the urgent environmental, social, and sustainability challenges of our contemporary world. Its emphasis on digitalization and AI-driven technologies for efficiency tends to overshadow the original principles of social fairness and sustainability. In response to these limitations, institutions and policymakers are shifting their focus toward human-centered design and ethical, responsible innovation in the factories of the future. This shift gives rise to the concept of Industry 5.0, introduced by the European Commission in 2020. Industry 5.0 aims to make production more sustainable, human-centric, and resilient, moving beyond shareholder value to embrace stakeholder value for all parties involved [14,15]. This vision is realized by acknowledging the capacity of the industry to attain societal objectives beyond mere employment and economic expansion. The aim is for the industry to transform into a robust source of prosperity by ensuring that production adheres to the limits of our planet and prioritizes the well-being of its workers throughout the production process. Furthermore, the focus is on utilizing technology not solely for economic gains, but rather for the advantage and convenience of all stakeholders involved [16].

The principles of Industry 5.0—human-centricity, sustainability, and resilience—build upon the technological foundations of Industry 4.0, placing the essential needs and interests of humans at the center of the production process [14,15,17]. This approach advocates for adapting technology to the needs of workers and ensuring that new technologies respect fundamental rights [18]. To achieve sustainability, Industry 5.0 promotes circular processes that rejuvenate, repurpose, and recycle natural resources, thereby reducing waste and environmental impacts [19]; this involves implementing innovative strategies and technologies. Rejuvenation refers to restoring resources through sustainable practices such as afforestation, soil restoration, and water conservation. Repurposing entails finding new applications or uses for materials, extending their lifecycle and reducing overall demand.

Recycling involves the systematic collection and processing of materials to manufacture new products, minimizing the need for fresh raw materials. Resilience involves cultivating robust industrial production capable of withstanding disruptions and supporting critical infrastructure during crises [18].

Motivated by the glaring absence of IoT frameworks for digital twin development and the inadequacy of existing frameworks in meeting the requisites of Industry 5.0, this work proposes a framework rooted in the IoT Architectural Reference Model (IoT-A or IoT-ARM), i.e., a project funded by the European Union between 2010 and 2013. The proposed framework utilizes the IoT-A reference architecture for the base of structuring the IoT system as based on three sub-models: the domain, the information, and the functional model, to deal with the information systems managed by the IoT-A data and knowledge. Another main characteristic taken from the IoT-A reference is the formalization of the digital twin component acting in the system as a virtual entity. This concept is further elaborated in the following sections. Basing the proposed framework on the IoT-A one, there are many benefits like achieving interoperability, developing system roadmaps, managing product life cycles, and benchmarking in the IoT domain. This type of analysis is more deeply discussed by [20], from which the most important benefits described are briefly introduced below.

- Cognitive aid: IoT-ARM serves as a common language, facilitating discussions and communication among stakeholders involved in product development.
- As a reference model for common grounding, it establishes a shared understanding by defining IoT entities and their basic interactions, providing a common foundation for collaboration.
- The generation of architectures according to guidelines for translating the ARM into concrete designs.
- Identifying differences in derived architectures when using the IoT ARM, can be attributed to specific use cases and related design choices, providing insights into variations.

The proposed framework serves as a standardized collection of architectural instruments for modeling IoT systems, aligning seamlessly with the Industry 5.0 vision [21]. The significance of having a standardized set of architectural instruments ensures interoperability, compatibility, and scalability across diverse IoT ecosystems, fostering seamless integration and communication among heterogeneous devices and platforms. This standardization not only streamlines the development process but also facilitates data exchange, enhancing the reliability and efficiency of digital twin implementations. Additionally, it promotes a common language and framework, enabling cross-domain collaboration and accelerating innovation in the rapidly evolving landscape of IoT-based digital twins. Ultimately, a standardized collection of architectural elements serves as the linchpin for achieving cohesive, reliable, and scalable IoT architectures that underpin the robust development and deployment of digital twins across various domains [22].

Unfolding in subsequent sections, this article elucidates the state of the art and delineates the methodology employed for the literature review in Section 2. Section 3 delves into the intricacies of the developed framework, while Section 4 scrutinizes the case study in which the framework has been applied. Finally, Section 5 encapsulates the primary findings and conclusions drawn from this endeavor.

## 2. State of the Art

The literature review begins with the outcomes of the European Lighthouse Integrated Project, IoT-A, that aimed to address the IoT interoperability and scalability challenges alongside challenges faced when modeling IoT systems. These outcomes are summarized in the book "Enabling Things to Talk: Designing IoT solutions with the IoT Architectural Reference Model" [20].

The first step of the literature review method was a forward citation process that was conducted on Scopus to find all the articles that cited IoT-A to understand how the outcomes of this project contributed to the scholarly conversation. The main components of this process are presented below, and the review methodology is highlighted in Figure 1.

a. Initially, 237 documents were found to have cited the book and 88 more to have cited specific chapters of it, leading to 325 documents.
b. After that, manual filtering based on the article titles was done to include relevant articles dealing with IoT-A and to exclude any that were far from topic, resulting in 109 documents without duplication. These documents are composed of 32 journal articles, 52 conference papers, 3 review articles, 12 book chapters, 2 books, and 1 survey paper.
c. The following step in the literature review process was abstract-based filtering. This involved going through all the abstracts and extracting articles that have a specific focus or contribution to IoT-ARM. Moreover, since the objective of this work is to provide a framework enabling digital twins that takes into consideration social and human aspects addressed by Industry 5.0, another extraction criterion was included in the abstract search. Specifically, any articles touching on digital twins, Industry 5.0, sustainability, human-centricity, and resilience were extracted. This process resulted in 28 articles that fit the extraction criteria. These contain 7 journal articles, 3 book chapters, 17 conference papers, and 1 book.
d. The final step was a deep review of these documents focusing on contributions to IoT-ARM and keywords related to Industry 5.0 and the digital twin concept. It was found that 5 documents mention the digital twin concept while only 2 of them provide meaningful contributions inherent to the scope. More than this, 11 articles introduce and delve into IoT-ARM with only 3 of them providing considerable contributions. And finally, 11 documents touch on concepts related to Industry 5.0; however, none of them explicitly mention Industry 5.0, but only related concepts such as sustainability (2) and human-centricity. Direct contributions to resilience were not found. The analyzed manuscripts, the ones that are the results of the research method described in the figure, fall within three main themes: comparison between various IoT architecture reference models, discussion of applications in specific domains, and detailed contributions to the IoT-A Framework.

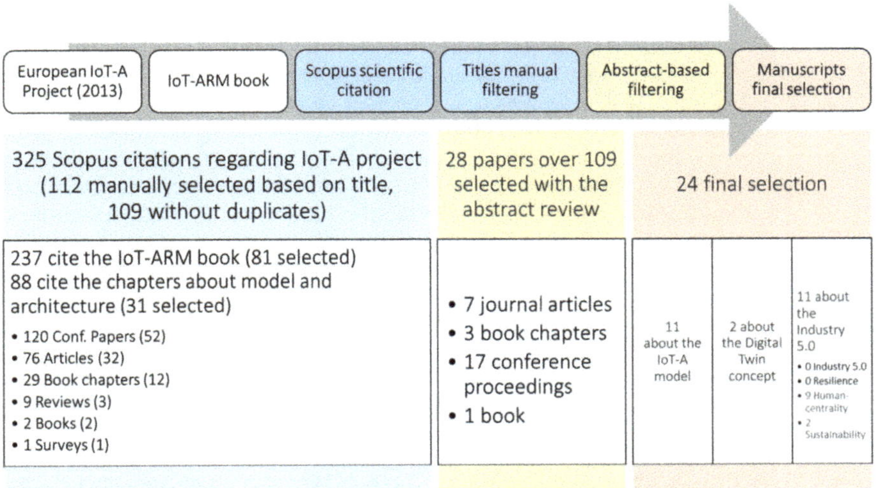


**Figure 1.** Summary of the literature review process and results from the forward citations research: manuscripts citing IoT-A on Scopus. European IoT-A Project (2013) [23].

The authors of [24] compared various IoT reference architectures to the IoT-ARM, with a focus on reference architectures that enable IoT integration with cloud computing or fog and edge computing; they also highlighted the importance of socio-technical perspectives when dealing with IoT and highlighted the importance of associate actors and processes which IoT interacts with. Ref. [25] also compared IoT reference architectures using a quantitative approach; they provided a set of criteria that can be modified based on the requirements of the application and the end users. Ref. [26] proposed a strategy to integrate the methods and architecture of IoT with cyber–physical systems and multi agent systems. The authors of [27] demonstrated the application of IoT-ARM in generating system architectures for IoT platforms in the smart city sector with a focus on security privacy, and scalability. The developed platform provides architectural artifacts for efficient and scalable security and user-centric privacy.

Ref. [28] introduced a conceptual framework for designing and implementing digital twins, tailored towards the farming sector; their framework consists of a control model based on a general systems approach and an implementation model based on the functional model of IoT-A. The work of [29] also proposed a reference architecture based on IoT-A for the development and implementation of digital twins in the agricultural sector. They built on the work of [28] by defining a set of architectural views that address different stakeholder concerns. More specifically, they focused on the functional decomposition and deployment views of IoT-A but tailored these towards digital twin development. The authors of [30] introduced different model-based approaches used in the lifecycle of IoT application development. They highlighted the benefit of IoT-ARM in the fact that it not only deals with the physical layer of an IoT system but also on the digital twin and the application-specific components.

The work conducted by [31] proposed a new IoT communication reference model based on the IoT-A communication model. According to them, the original communication model does not suitably handle the security and quality of service features; to this end, they introduced a security and a quality-of-service (QoS) layer. Ref. [32] proposed a framework that encompasses a model-driven approach for applying QoS attributes in the development of the IoT systems. They emphasize the importance of introducing quality attributes such as confidentiality, scalability, reliability, and fault tolerance into the IoT-A modeling process. Another contribution of their work is the extension of the domain and information models of IoT-ARM by adding a perspective metamodel and linking it to each of them.

The usage of IoT-A to achieve Industry 5.0 objectives is not highly emphasized in scientific literature; out of all the literature that cites IoT-A, there is a very limited number of contributions to sustainability, human-centricity, and resilience. In [33], the authors underscored the pivotal role of the IoT in fostering a sustainable world and enhancing human well-being. Their work delves into both hardware and software approaches aimed at minimizing energy consumption in IoT-based services. By providing insights into strategies deployed at different levels of IoT-based services, the authors contribute to the overarching goal of reducing power consumption and fostering a more sustainable environment. The comprehensive approach presented in their work aligns with the imperative of creating IoT solutions that not only advance technological innovation but also prioritize energy efficiency for long-term environmental and human well-being benefits.

With regard to human inclusion in IoT reference architectures, Ref. [34] introduced a human–computer interaction quality evaluation method based on ubiquitous computing software measures, while employing IoT-ARM as the reference architecture for their implementation. More specifically, they introduced the human actor as part of the IoT-A domain model and provided insight on how to link the human model to the machine and platform model considering them as multiple agents. They also highlighted the importance of providing the human worker with context-based information reduction to minimize information overload. The paper written by [35] presented an architecture based on IoT-A and subject-oriented process representations which allows one to encode and adapt to human properties on different levels of an organizational control process; they highlighted the importance of continuously adapting manufacturing system behavior to worker needs by incorporating human-aware modeling. Refs. [36,37] also highlighted this aspect by emphasizing the vital role of Humans in Production Industries and the need for modeling human interactions in such systems. The concept of humanized IoT was introduced by [38] by applying Fiskes's Four Elementary Forms of sociality to IoT, they developed a personal and communal IoT platform that allows users, with different profiles and skills, to share information, interact and build on top of IoT applications. Considering the role of humans in cyber physical systems a hierarchy of design level for socio-technical systems was introduced by [39] and was developed according to the IoT-A domain model.

One of the more recent works was a review conducted on the role of human digital twins (HDTs) in the context of Industry 5.0 [40]. The authors conducted a comprehensive survey and proposed a conceptual framework and architecture for HDTs that emphasizes interactions, responsibilities, and collaborations between components. They also highlighted the importance of IoT as an enabler for human DTs due to its ability to integrate all the actors in the manufacturing ecosystem. Another review carried out by [41] analyzed the application of digital twins in the context of Industry 5.0. Their review found that DTs can generate value, shorten the time to market, enhance plant machinery and final product performance, and offer unique insight not found in any other solution.

## 3. IoT-DT Framework

### 3.1. Overview and First Considerations

The proposed framework aims for a 5.0 characterization by deciding the structure of the information model, functional model, and information view in alignment with the IoT-A reference architecture. Additionally, a key focus is placed on designing the IoT system around digital twins, emphasizing the connection between physical entities and their corresponding virtual representations. To achieve these goals, specific requirements are addressed in the subsequent section, aligning with the contextual demands of the 5.0 framework.

The proposed approach starts by defining the generic requirements of industrial IoT systems [42], then, these are complemented by Industry 5.0 requirements that focus on human-centricity, sustainability, and resilience that form the basis for the KPIs developed. Furthermore, a methodology for implementing these requirements into an IoT-a-based domain model is defined [43], along with the sub-models that complement it, namely an information, functional, and communication model. Based on this model and frameworks found in the literature, a system architecture is proposed. Figure 2 shows the general steps followed throughout the framework and the derivations of every step.

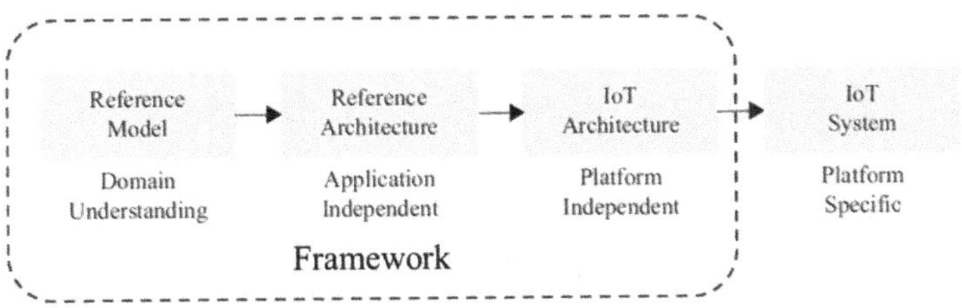


**Figure 2.** Framework application in IoT development process, adopted from [24].

In accordance with the IoT Architectural Reference Model (IoT-ARM) [20], the prototypical IoT scenario delineates a generic user interfacing with a physical entity (PE), which may be spatially distant, within the tangible domain, as depicted in Figure 3. In the physical realm, direct interactions, exemplified by the manual relocation of an assembled part from point X to Y, are operationally viable. Nevertheless, the core tenet of IoT lies in enabling indirect engagements through a third-party intermediary. This is manifested through the invocation of a Service capable of imparting information about the physical entity or effectuating a specified action upon it. Thus, the framework is categorically denoted as service-oriented, underscoring this inherent dynamism. In practical instantiation, a human user gains entry to a service through a service client—a software entity featuring an accessible user interface. This access modality aligns seamlessly with the service-oriented paradigm of the framework, accentuating the capacity to interact with Physical Entities through services facilitated by software interfaces. The limitations of interacting with physical entities through services facilitated by software interfaces may arise from various technological, practical, and ethical factors. The main limitations are the hardware dependencies, since interacting with physical entities often requires compatible hardware components, and connectivity issues, since poor or unstable connections can lead to communication failures or delays, impacting the effectiveness of remote control and monitoring. Other limitations can be the latency and response time, the security concerns, the privacy of the data, and the energy and resource constraints. It is possible to address these limitations by using a combination of technology advancements, robust security measures, and thoughtful design, to ensure safe and effective interactions with physical entities through software interfaces. The orchestration of such interactions embodies the intrinsic technicality and sophistication integral to the engineering and scientific underpinnings of this IoT framework.

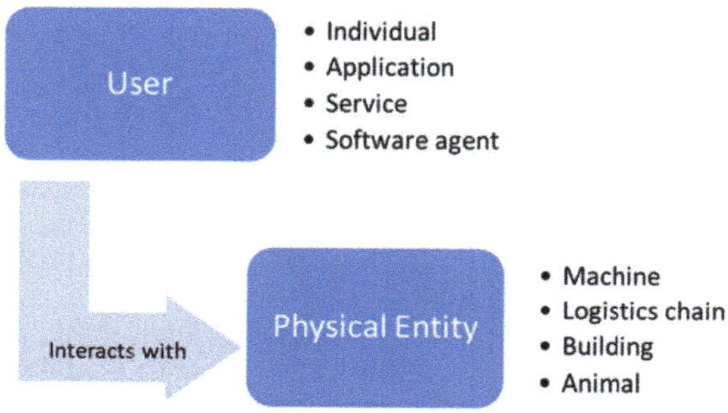


**Figure 3.** Basic IoT interaction as defined by [20].

It is noteworthy to mention that the main component of the IoT-A models is the virtual entity; this is intuitive and in line with the notion of "things" in IoT since physical entities are the "things" and virtual entities are representations of these physical entities in the information system. However, this definition of a virtual entity is tightly coupled with other concepts in the IoT realm such as digital twins, digital shadows, and digital models. According to [44], a digital twin is a collection of adaptive models designed to mimic the behavior of a physical system within a virtual environment, continuously updating in real-time throughout the system's life cycle. This digital twin replicates the intricacies of the physical system, enabling the anticipation of potential failures and identification of opportunities for change. Through this emulation, the digital twin can prescribe real-time actions, optimizing and/or mitigating unexpected events by keenly observing and evaluating the operational profile of the system. The essence lies in the synchronized, dynamic relationship between the physical and virtual representations, fostering proactive decision-making and enhancing the overall system performance.

Considering this definition, a DT is a virtual entity with specific characteristics and highly complex interactions. On the other hand, a digital shadow is a virtual model that represents the physical model only, with one-way data flow [45]. A digital shadow is characterized by a unidirectional data flow, meaning data can only go from one location to the other. It is possible to make digital shadows of digital twins because they can capture and simplify the multitude of information that passes through a twin; however, a digital shadow cannot influence its physical counterpart or conduct an action upon it (actuation), unlike the digital twin. We highlight the difference between these terms in Figure 4. IoT-A goes further and defines the augmented entity as a combination of a virtual entity and its physical counterpart. However, we have omitted this entity as it contains no additional information about the physical entity than ones contained in the virtual entity, so it does not really provide any additional functionalities found in a digital twin. Furthermore, since the concept of digital twins was fairly unknown when IoT-A was developed, we argue that the augmented entity was an attempt to model digital twins in IoT systems and since virtual entities can fit this description of DTs, we felt no added value was provided by including this entity in the domain model.

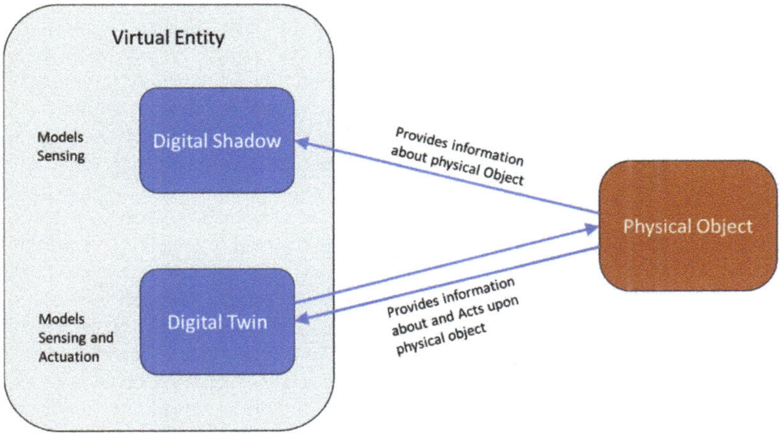


**Figure 4.** Virtual entities in relation to digital twins and digital shadows.

To capture the various complexities of an IoT system, a single model will not suffice; on the contrary, multiple models that capture different aspects of the system are required. In the present work, three models are considered, namely the domain, information, and functional. The virtual entity can be considered the interaction point for these models, since the domain model considers all elements of the IoT domain. The information model can be considered a breakdown of the virtual entity since all the information about the "thing" is contained in the virtual entity. Thirdly, the functional model identifies groups of functionalities that build on each other following the relations identified in the domain model. The functionalities of these functional groups that manage information use the IoT information model as the basis for structuring their information, and therefore are closely linked to the virtual entity concept. These models will be explained in more detail throughout this section.

It is noteworthy to mention that IoT-A includes a security, trust, and privacy model that interacts with the functional model at the security layer (as shown in Section 3.5). This model focuses on trust at the application level, by defining a trust mechanism that provides data integrity and confidentiality, and endpoint authentication between any two system entities that interact with each other. Moreover, it defines a security reference model that includes application security, service security, and communication security. Furthermore, it includes a privacy sub-model that describes the privacy measures that stop a subject's (either user's or entity's) data from being misused, such as access controls, encryption/decryption methods, and credential-based security measures.

While security is a very important aspect and has high implications for the success and failure of IoT deployments [46], it is beyond the scope of this article to delve into the intricacies of this deep discussion; however, Ref. [23] provides a broader discussion about the security, trust, and privacy model.

### 3.2. Industry 5.0 Requirements

System requirements are usually categorized in terms of functional and non-functional requirements. Functional requirements outline what a system should do by specifying its features and capabilities from a user's perspective, while non-functional requirements are not directly observable by end-users. While functional requirements focus on the users and system interaction and they are closer to the user interface (the ability for users to transfer funds between accounts, for instance, in a banking system), non-functional requirements define how the system should perform, emphasizing qualities like performance, security reliability, and scalability. When talking about IoT systems specifically, several general functional requirements are usually defined for the developed system, which can include real-time data acquisition from sensors, the reliable transmission of data, and device management, to name a few. These requirements are implemented as features by the developer. Several features may need to be present to implement a single requirement, and vice versa, a single feature can satisfy multiple functional requirements.

The emphasis of this work is not on supplementing functional requirements for IoT systems; instead, the focus lies on non-functional requirements, specifically those essential for Industry 5.0 compliance. Considering non-functional requirements as expressions of how functional requirements are achieved, the goal is to formalize the analysis and management of how the IoT system should operate during development, ensuring alignment with functional requirements. Given the core tenets of Industry 5.0—sustainability, human-centricity, and resilience—the formalization of I5.0 requirements must inherently incorporate these pillars. The central discussion revolves around translating the key attributes of human-centricity, resilience, and sustainability into system requirements. The identified requirements, derived from the Industry 5.0 vision, are intended to be integrated into the framework. The framework is expected not only to facilitate an assessment of system conformity to these requirements but also to serve as the foundation for developing indicators and metrics for evaluating Industry 5.0 conformity.

### 3.2.1. Human-Centricity

Human-centricity in IoT systems refers to the prioritization of human needs, motivations, and beliefs in the design and implementation of these systems [47]. This approach is usually applied in the context of Internet of Medical Things (IoMT) systems, where it can lead to improved user satisfaction and system functionality [48]; however, it should not be limited to this sector but rather to all IoT systems that provide services to humans or have any form of human interaction.

Human-centricity requirements synthesized from [49] pivot on safeguarding human potential and prioritizing humanity over the exploitation of vulnerabilities. These requirements, rooted in a commitment to placing human welfare at the forefront, can be summarized as follows:

- Emphasize the role of both the designer and the human user in shaping the overall experience with the result of active human involvement, where the synergy between human creativity and technological design contributes to a richer experiential landscape;
- Consider the human user as belonging to a society, and this society, restricted to the corporate system and not, builds the experience and thus the wisdom that constitutes the know-how of humans at the basis of a system, in this case, of a production system;
- Promote an environment that empowers individuals to exercise their self-direction and creativity (not individualism, but the enhancement of the individual);
- Value people beyond their role as "users," recognizing the needs and contributions of the individual, transcending conventional labels and standardization to ensure a holistic and respectful approach to human interactions with technology.

### 3.2.2. Sustainability

In terms of sustainability, the Brundtland report [50] introduced the principles of environmental, social, and economic sustainability as the three dimensions of sustainability and sustainable development. The sustainability of IoT systems is a critical concern, given their potential to contribute to energy consumption, toxic pollution, and electrical waste. The proliferation of IoT devices contributes to increased energy consumption, often relying on non-renewable sources, leading to a higher carbon footprint. Additionally, the manufacturing, use, and disposal of these devices can result in toxic pollution due to the presence of hazardous materials. The rapid pace of technological advancements and device obsolescence further contributes to electrical waste, exacerbating the global e-waste problem. Several studies have explored the concept of "green IoT" to address these challenges. Refs. [51,52] both emphasize the importance of green sensing and communication in IoT systems, with the latter proposing a framework for minimizing carbon footprints and promoting the use of green IoT.

- Sustainability, defined in [53], is a concept based on three main aspects:
- The environmental aspect, with environmental KPIs (e.g., carbon footprint, energy consumed per message);
- The economic sustainability, with the indicator for the cost of information (e.g., cost of sending a message);
- The social impact, especially on users outside the system boundaries.

#### 3.2.3. Resilience

Resilience in IoT systems refers to their ability to withstand and recover from malicious activities, disruptions, and failures. This is particularly important given the complex and dynamic nature of these systems [54,55]. The concept of resilience is crucial in the design and operation of IoT systems; however, the practical implementation of resilience mechanisms in IoT systems can be challenging, as they impose constraints on developers [56]. Despite these challenges, the need for resilient IoT systems is clear, given their increasing integration into critical infrastructures. An example of a resilient IoT system is a smart energy grid that incorporates self-repairing capabilities [57]. Through the real-time monitoring and analysis of grid performance, the system can quickly identify faults or disruptions. Autonomous responses, such as rerouting energy flows or isolating affected areas, ensure the continuous and reliable supply of electricity. This resilient IoT application enhances the stability of the energy grid, minimizing the impact of failures and contributing to the overall resilience of the critical infrastructure. Other examples include intelligent transportation networks with adaptive traffic management that can self-adapt to mitigate congestions [58]. Other examples in smart manufacturing facilities include systems with predictive maintenance strategies, forecasting potential equipment failures and optimizing operational efficiency [11,59,60]. Additionally, these systems excel in fault diagnosis by promptly identifying and diagnosing anomalies in real-time data streams [61–63].

- Resilience, i.e., being future-proof and able to adapt to disturbances is considered as the following:
- The ability to adapt to disturbances;
- A sufficient scalability level;
- The estimated quality level of the data sent.

### 3.3. *Domain Model*

The IoT domain model defines the fundamental components of the IoT. Additionally, it describes the fundamental features of these concepts, such as the name and identifier, as well as the relationships between concepts [64]. The primary objective of a domain model is to develop a shared knowledge of the target domain and to capture the key concepts and relationships that are pertinent to IoT stakeholders. The domain model defines various types of elements that make up its basic building blocks and allow for the modeling of various characteristics and systems within an IoT environment.

The first is the physical entity, an observable component of the physical world relevant to the user's objective. Physical entities may be practically any object or environment, including persons, animals, automobiles, retail or supply chain products, computers, electrical equipment, jewelry, and clothing.

Virtual entities are representations of physical entities in the digital world. There are many kinds of digital representations of physical entities: 3D models, database entries, objects (or instances of a class in an object-oriented programming language). Active or passive classifications can be applied to virtual entities, Active Digital Artifacts (ADA) are software programs, agents, or services that have access to other services or resources. Passive Digital Artifacts (PDA) are inert software components, such as database records, that can serve as digital representations of physical entities. For an element to be considered a virtual entity it must satisfy two fundamental properties:

a. Virtual entities are coupled with a single physical entity, which the virtual entity itself represents; however, a single PE can be represented by multiple unique VEs, providing a distinct representation for each application domain;
b. Synchronization, i.e., VEs must provide synchronized representations of a specified set of features (or qualities) of the physical entity, i.e., a change in the physical world must be reflected in the physical entity and vice versa.

Devices, as technological artifacts, serve as interfaces bridging the gap between the digital and physical realms, establishing a connection between virtual entities and physical entities. Consequently, these devices must exhibit functionality in both the physical and digital domains. However, the IoT domain model predominantly focuses on its capability to facilitate observation and the modification of the physical environment from the digital sphere. If other attributes of a device are significant, the device itself would be modeled as an entity. In line with the IoT-ARM specifications, three primary types of devices are defined: sensors, actuators, and tags.

Resources, within the IoT context, refer to software components that either provide data sourced from physical entities or are integral to their operational processes. On-Device Resources are discernible from Network Resources. As the name implies, On-Device Resources reside locally on devices, comprising software installed directly on the device connected to the physical entity. These resources encompass executable code designed for accessing, analyzing, and storing sensor data, as well as code for controlling actuators. In contrast, Network Resources refer to resources accessible over the network, such as cloud-based databases. This distinction underscores the essential role that both On-Device and Network Resources play in the overall architecture of the IoT ecosystem.

Services, as defined by [65], serve as the mechanism for aligning requirements with capabilities. Within the IoT framework, services are confined to technical services delivered through software. They function as the interface linking the IoT components of a system with other, non-IoT-specific elements within an information system, such as enterprise systems. The orchestration of both IoT-related services and non-IoT services allows for the comprehensive construction of a system. Unlike heterogeneous resources, which may heavily depend on the underlying hardware of the device for their implementation, a service provides an open and standardized interface encompassing all the essential functionalities required for managing resources and devices associated with Physical Entities. Within the service hierarchy, low-level services play a pivotal role by directly interacting with resources and residing closest to the actual hardware of the device. These low-level services may be further invoked by other services to deliver higher-level functionalities, such as the execution of a business process activity. This hierarchical structure highlights the layered nature of services within the IoT spectrum, where low-level services form the foundation, facilitating interaction with device hardware, and higher-level services build upon this foundation to enable complex functionalities and business processes.

Figure 5 shows the IoT domain model which highlights the relationships between the above-mentioned elements. The model shows a physical entity that has a device attached to it; we use the term smart device (in contrast to the original device term used by IoT-A), basing our definition on [66], to indicate an electronic component that is able to connect, share, and interact with its user and other smart devices; furthermore, this definition of a device can encompass sub-device classes such as sensors that can directly interact with IoT systems without the need for an interface (such as a PLC or a gateway), which is the case in many Industry 4.0 environments found today. The smart device can contain a sensor or an actuator that monitors or acts on the physical entity. The smart device hosts on device resources and can utilize Network Resources. These resources are exposed to the user via a service that the user can invoke or subscribe to. The physical entity is represented in the digital realm using a virtual entity, which in certain cases, becomes the digital twin entity. The virtual entity is associated with a specific set of resources that are exposed to it via a service. The relationship between virtual entities, resources, and services will be elaborated more in the information model section.

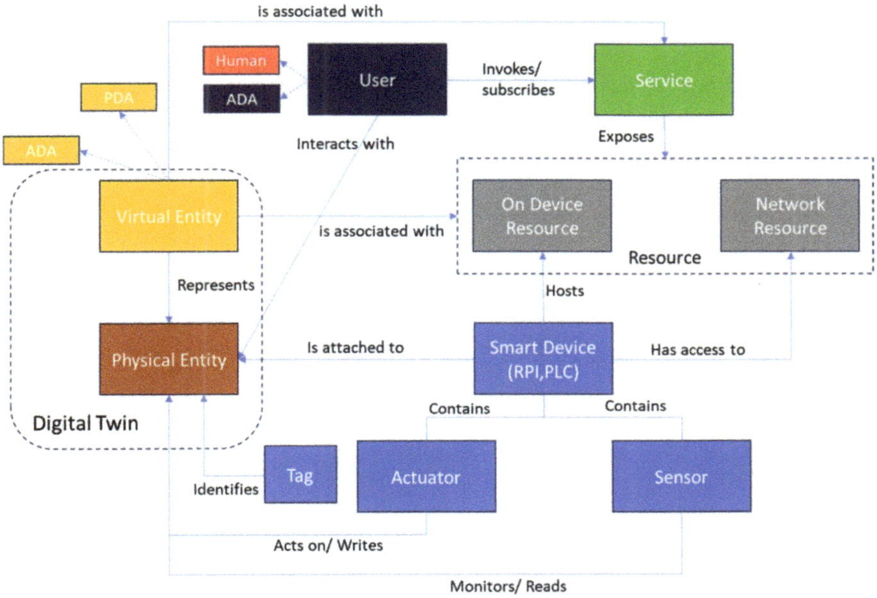


**Figure 5.** IoT domain model.

The significance of the domain model stems from its incorporation of fundamental abstract concepts, outlining their roles and interconnections. Concerning granularity, the domain model should distinguish between relatively constant elements and those subject to variation. For instance, in the IoT domain, the concept of a device is likely to endure, even as the specific types of devices evolve over time or differ based on the application context. As a result, the model refrains from detailing specific technologies, instead focusing on abstract representations.

Moreover, due to the abstract nature of the domain model, its applicability in various industries becomes part of the core nature of the framework. In the current work, the framework is applied to vertical farming, but it can be expanded beyond this sector to other sectors beyond it such as manufacturing and service management. Furthermore, the core elements of an IoT system are shared among the various sectors, allowing for cross-sector applications of the technology. These elements include perception for data collection and interfacing the physical and digital worlds, networks and communication for connecting the IoT ecosystem, middleware (platform layer) for data processing and analytics, and finally, an application layer for decision support and integration with business systems.

### 3.4. Information Model

On a conceptual level, the IoT information model describes the structure (e.g., relationships, characteristics, and services) of all the information for virtual entities. The virtual entity is the key concept of any IoT system, as it models the physical entity or the thing that is the real element of interest. The word information is used in conjunction with the definitions of the DIKW hierarchy [11,67], where data are defined as values devoid of context that is meaningful or useful. Information provides context for data and answers to common queries such as who, what, and when.

The IoT information model provides information on the modeling of a virtual entity. As shown in Figure 6, the virtual entity (virtual entity) contains attributes with a name, a type, and one or more values that can relate to metadata (MetaData). Metainformation includes, for instance, the unit of measurement, the time at which and the location where a value is digitized, or the quality of this digitization process, i.e., the quality, for example, in terms of the veracity of the measure. The virtual entity is interfaced to the remainder of the system (sources of information) by its ability to access a service through the description of the service itself that describes how a service serves information.

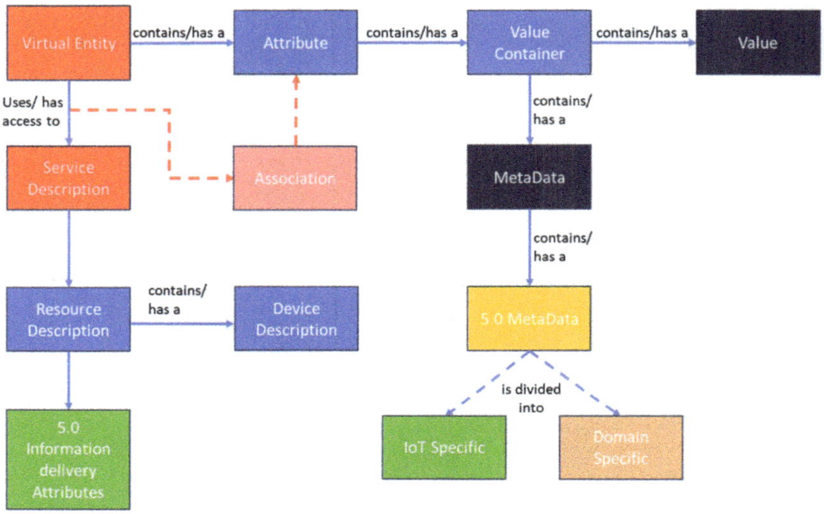


**Figure 6.** IoT information model, adopted from [20].

Our contribution to the IoT-A information model is the modification of the virtual entity to make it compliant with Industry 5.0. This is done by formalizing a description of how the service, resource, and device descriptions handle information related to Industry 5.0, and by defining a structure for the MetaData component that will contain the metainformation pertaining to the Industry 5.0 requirements described previously.

The association between a virtual entity and a service is specified by a particular attribute of the virtual entity. The type of the service can be decided to be either information or actuation, depending on whether it offers value to be read or written.

A service description describes the relevant aspects of a service, including its interface. Additionally, it may contain one (or more) resource description(s) describing a resource whose functionality is exposed by the service. The resource description in turn may contain information about the device on which the resource is hosted.

With regard to the Industry 5.0 information delivered to the virtual entity, it is eventually all encompassed in the service description in some form of container; when linking these attributes to the virtual entity, the same method of association described above is used, i.e., through an association, the connection between an attribute of a virtual entity and the service description is modeled. However, once they are part of the virtual entity, they will not be modeled as one or more attributes but as MetaData, since they fit the definition of metadata or meta information as being "information about the value of a piece of information" [68], to this end, the block Industry 5.0 metadata are added.

To summarize, the IoT information model is a metamodel that provides a structure for the information being handled by IoT systems. This structure provides the basis for all aspects of the system that deal with the representation, gathering, processing, storage, and retrieval of information, and as such, is used as a basis for defining the functional interfaces of the IoT system.

### 3.5. Functional Model

Functional Decomposition (FD) refers to the process by which the different Functional Components (FCs) that make up the IoT ARM are identified and related to one another [69]. The primary objective of Functional Decomposition is, on the one hand, to reduce the complexity of an IoT ARM-compliant system into smaller, more manageable components, and, on the other hand, to comprehend and show their relationships. IoT-A defines the functional model as "an abstract framework for understanding the main Functionality Groups (FG) and their interactions". This framework defines the common semantics of the main functionalities and will be used for the development of IoT-A-compliant Functional Views.

The IoT functional model used in this framework is adopted from [28], which is based on the original IoT-A functional model and its modifications suggested by the same authors in [70]. The main addition provided by this "improved functional model" is that the virtual entity management is replaced with digital twin management.

As shown in Figure 7, the functional model addresses eight layers (functional groups), ranging from a device layer, which is attached to the physical entities, to an application layer, which allows various user interactions. The device layer provides the hardware components that are attached to and directly interact with physical objects as defined by the IoT domain model entity of device; this includes sensors and actuators.

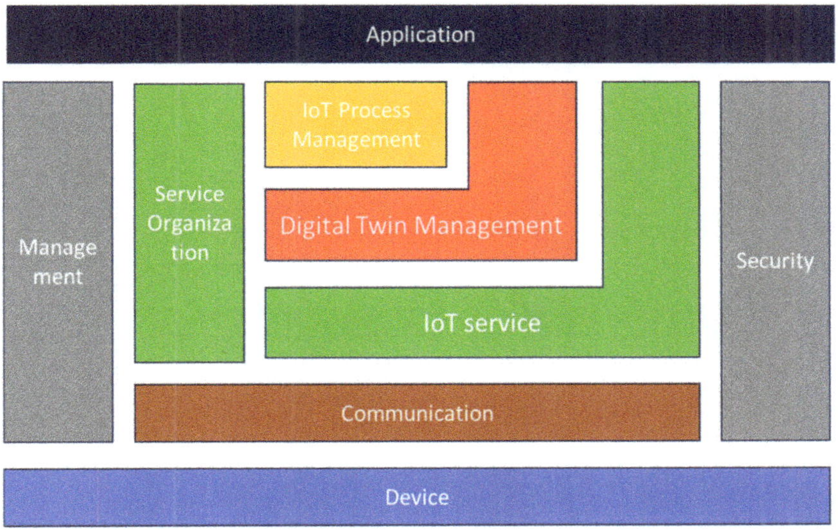


**Figure 7.** IoT functional model, adopted from [28].

The communication layer regulates the relationships between the various components and facilitates communication from the devices to the IoT services. It facilitates end-to-end communication across diverse networking contexts by providing networking, connection, and data transport capabilities. Moreover, it abstracts the heterogeneous interaction schemes stemming from the myriad technologies associated with IoT systems, encapsulated within the device Function Group (device FG), and establishes a unified interface to the IoT Service FG. This abstraction serves to simplify the instantiation and management of high-level information flow. Specifically, it addresses various aspects of the ISO/OSI communication model, encompassing considerations related to data representation, end-to-end path information, and network management. This unified interface ensures a cohesive and standardized approach to handling communication intricacies within the IoT ecosystem, providing a streamlined mechanism for the efficient instantiation and management of information flows at a higher level of abstraction [71].

The IoT service layer includes services and features for IoT service discovery, lookup, and name resolution. It can be used to retrieve data from a sensor device or send data to control actuator devices. As mentioned previously, sensors and actuators capture and facilitate the change in certain aspects of the physical world. The resources associated with the sensors and actuators are exposed as IoT services on the IoT service layer. Examples of applications and IoT system interactions in this layer can be "Give me the value of Sensor 123-55" or "Set Actuator 123-31 to On".

The digital twin Management layer contains functions for interacting with the IoT system based on virtual entities. It can give access to all the information about the digital twin, from sensor devices, databases, or applications. Furthermore, it contains all the functionality needed for managing associations with physical objects and monitoring their validity. The relationship between the virtual entity FG and the IoT service layer is close one, this is because this layer models higher-level aspects: examples are "Give me the outdoor temperature of Car 123" or "Set lock of Car 123 to locked".

The IoT Process Management FG is concerned with the integration of process management activities given by the business with the IoT system. Its overarching objective is to furnish the functional concepts requisite for seamlessly integrating the peculiarities of the IoT realm into conventional (business) processes. This layer establishes an environment conducive to the modeling and execution of processes that are cognizant of IoT considerations. The deployment of process models to execution environments is facilitated through the utilization of IoT Services. It is crucial to note that this layer is inherently proximate to enterprise systems. Therefore, the modeling of processes within this context must not only account for the nuances of the IoT domain but also incorporate the specificities of the underlying business domain. This dual consideration ensures a comprehensive and coherent approach to process management, acknowledging both the intricacies of IoT functionalities and the broader context of enterprise operations.

The Security layer is responsible for the security and privacy of the systems and its users. It handles the initial registration of a client to the system in a secure manner and protects the private data of users. It also ensures legitimate interactions between peers that are authorized to interact with each other, and it manages secure communications.

The service organization layer acts as a central hub between the other functional groups; this is important since the final objective is to achieve a service-oriented model or architecture. The IoT Process Management FG relies on Service Organization to map the abstract process definitions to more concrete service invocations. Moreover, it enables the translation of high-level requests to specific IoT services, therefore also linking the digital twin management to the IoT service layer.

Finally, the Management layer indeed plays a crucial role in overseeing and controlling the following aspects of the system according Industry 5.0 guidelines: system setup and configuration (for example, adding devices and communication protocols), error reporting (monitoring activity for maintaining system reliability and performance, i.e., to ensure resilience), system health monitoring (sustainability real-time assessment), strategic decision supporting and functionalities execution management (defining and implementing strategies to guarantee that the functions and actions of the resulting IoT system contribute to achieve the 5.0 goals in terms of human-centricity, sustainability, and resilience.

The human-centricity aspect of the Management layer, for example, is guaranteed by a management system of all the functions given by the proposed functional model (summarized by Figure 7 between the applications and devices of the IoT system) in order to pursue measurable goals thanks to several human-centric functions of the whole information system. The following are examples of functions, or groups of functions (services), to be made sustainable through the design of an IoT system to support the management of the information system in which users (humans) and resources operate.

- Flexible work arrangements, allowing for smart (flexible and remote) working hours and a personal level of responsibility for processes.

- Encourage open communication between employees to share their ideas, concerns, and feedback.
- Provide learning opportunities to support continuous learning and to encourage employees to pursue projects or initiatives that align with their personal and professional development goals.
- Recognition and rewards for creativity, implementing a recognition system that rewards employees for innovative ideas and creative problem-solving (impossible to consider during the IoT design).
- Diversity and inclusion towards a diverse and inclusive environment for humans from various backgrounds (without digital barriers of IT platforms).
- Supportive leadership by cultivating all human leadership styles, following clear guidance, and acting as a facilitator rather than an authority.
- Experimentation and risk-taking, promoting the culture of learning from failures, creating a safe space.
- Provide tools and resources (access to tools and technologies that are needed to pursue human ideas and projects).

## 4. Case Study

### 4.1. Overview

The global population is projected to reach an estimated 8.5 billion by 2030 and is anticipated to further increase to 9.7 billion by 2050 [72]. This substantial population growth necessitates a significant boost in agricultural production to ensure food security. However, the expansion of such production is constrained by environmental crises and the adverse effects of conventional open-field agricultural practices [73,74]. To overcome these challenges, smart farming techniques, such as aeroponics, offer a promising avenue for optimizing resource utilization. In accordance with [28], a smart farming system can be conceptualized as a cyber–physical control cycle, seamlessly integrating sensing and monitoring, smart analyses and planning, and the intelligent control of farm operations across all pertinent processes. However, the successful implementation of these modern agricultural techniques relies heavily on reliable and up-to-date information about farm operations. This is precisely where digital twins (DTs) come into play, providing a valuable tool to enhance and optimize agricultural practices by offering real-time insights and data for informed decision-making. In this context, DTs can help farmers get information on the fields and overall farm behavior, and thus decide better crop management strategies. By collecting data, it can also forecast yield prediction, growth stage, nutrient information, and weather reports.

The use case is about soilless farming in controlled agriculture environments (referred to as vertical farming) that are characterized by a high degree of technology usage. Lettuce crops are grown in a completely controlled environment where all parameters relevant to plant growth are artificially provided, such as growth medium, irrigation, nutrition, and climatic conditions. The plants are grown in a growth chamber that consists of a number of sensors and actuators that are connected to a controller; each growth chamber with its associated sensors, actuators, and controller is referred to as a module. Each module sends its data to a remote database for storage and further advanced analytics. Also, a production planning system is connected to the IoT system via a specific API to acquire plant data. The overall IoT system is shown in the context diagram in Figure 8.

A monitoring digital twin is developed at the plant level (crop digital twin) with the objective of optimizing the growth of the lettuce plants. By considering the module as a production system that has inputs (seeds, energy, water, etc.) which are processed into outputs (plants, $O_2$, inedible biomass), each plant stays in production from germination until maturity.

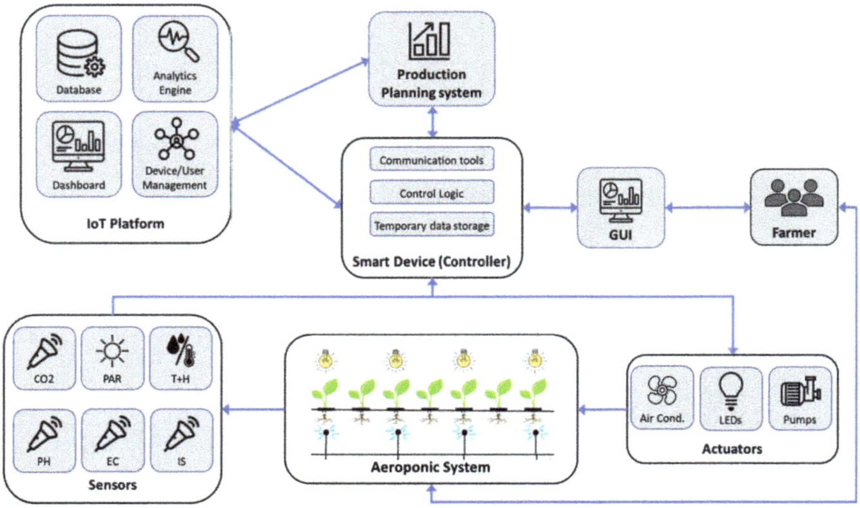


**Figure 8.** Context diagram of aeroponic system.

### *4.2. Aeroponic System Domain Model*

As mentioned in the previous section, the IoT domain model serves as the foundation for the other sub-models by establishing a shared taxonomy for the primary concepts of IoT and their interconnections. This taxonomy forms the fundamental structure of the information model for a particular (IoT) domain. By using the modeling tools described in the previous sections, a domain model is created for the vertical farm system, as shown in Figure 9. The starting point is the physical entity under study, which is a single lettuce plant This PE is identified by a specific tag that contains the plant ID. The plant is monitored by a set of sensors that collect information about the plant environment; these sensors feed their information to a Raspberry Pi [75]. The Raspberry Pi also controls various effectors (actuators) that act on the physical entity and its environment to provide the most suitable growth conditions. The physical entity is represented by a virtual entity and together they constitute the digital twin.

The virtual entity is associated with a set of services that either provide or acquire data to or from the virtual entity. In our model, we consider four types of services that are available to the virtual entity: The first is a monitoring service that acquires sensor data and updates the VE accordingly; the second is the control service that invokes a change in the physical environment by triggering an effector. A data storage service is used to store data in any data archive, whether it is a simple CSV file or a complex database; and finally, the data retrieval service acquires data from any data storage and uses it to update the VE.

All the above-mentioned services expose a set of resources; these resources can be on-device resources, i.e., resources hosted by the Raspberry Pi. These are sensor drivers used for data acquisition, control logic to trigger actuators, communication tools, and local data storage. It is noteworthy to mention that many other resources exist on the device (e.g., device operating system, GUI enabling software, etc.); however, our focus is on the high-level resources that interact directly with the VE services. Network Resources, on the other hand, are hosted on cloud or on some remote server with higher computational capabilities; these are the database management systems (DBMSs) that include the actual DB and the MQTT broker, which is responsible for IoT-related communications.

Two types of users are defined in this system: the first is a human who can be the farmer or a vertical farm operator; this user invokes the farm dashboard service, which in turn invokes the monitoring service to monitor the plant during its growth. This user also interacts directly with the PE for standard farm operations such as weeding, cleaning, and harvest, to name a few. The second user is the production planning system, responsible for managing plant production in the farm. This user subscribes to the monitoring service if it needs updated plant data, and it invokes the data storage and retrieval services in case it needs to acquire historical data or if it needs to update the DB.

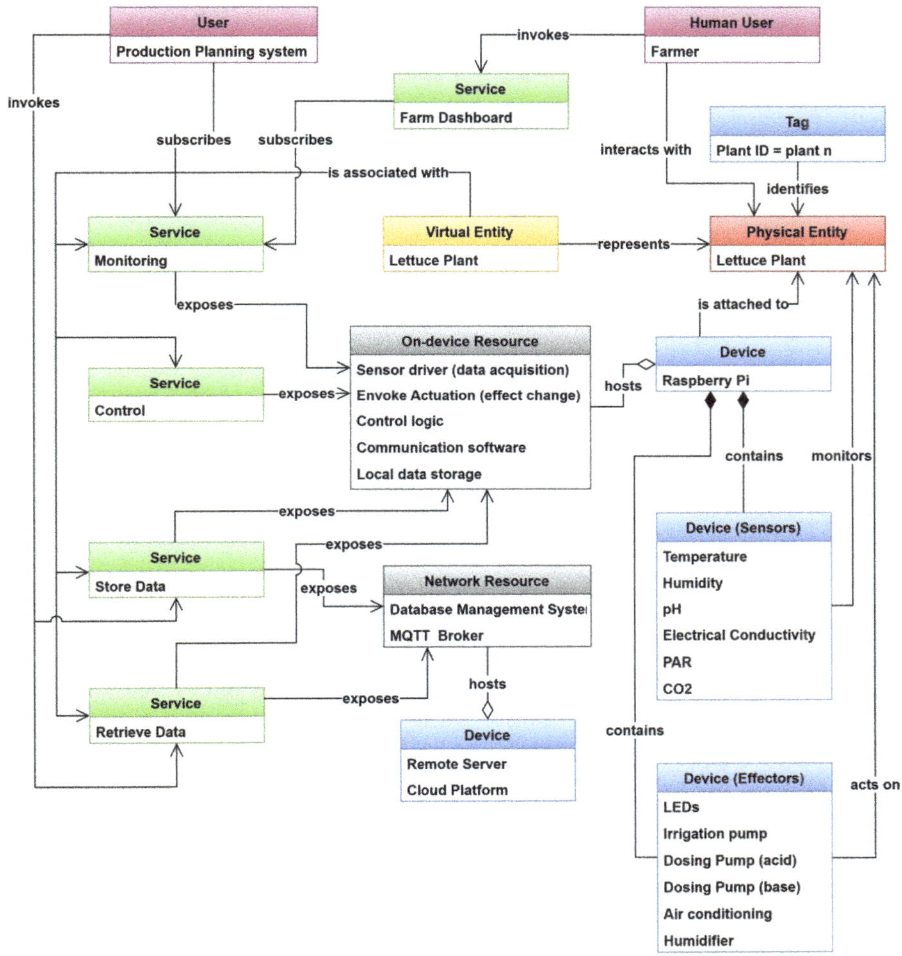


**Figure 9.** IoT domain model for lettuce plant in the aeroponics system.

### 4.3. Information View

The information model defines the structure, characteristics, and relationships of information managed by the virtual entity; it also defines the IoT services linked to the specific virtual entity. Based on the defined information model, we develop the information view that is specific to this use case.

Figure 10 shows the information structure of the digital twin. Starting from a digital twin of a single plant represented by a virtual entity "Plant n" that contains many attributes, these attributes hold the relevant information pertaining to the specific twin, as mentioned in the previous section. Information is provided or taken from the digital twin via services and the connection between the service and virtual entity is achieved through an association that maps the information of the service to a specific attribute. The type of information stored in the attribute can refer to the specific conditions of the physical entity such as the attribute "temperature", or they can refer to KPIs that provide qualitative indicators about the physical entity, such as the indicator "Leaf area". To this end, and by adopting the KPI formalization for aeroponic systems suggested by [76], the digital twin attributes are grouped as follows:

1. Environment attributes that provide information about the plant growth conditions (i.e., about the environment of the physical entity);
2. Productivity attributes which are KPIs that provide information about the overall biomass;
3. Quality attributes, which refer to vegetable quality;
4. Environmental sustainability attributes, that take into consideration the sustainability of the growth process.

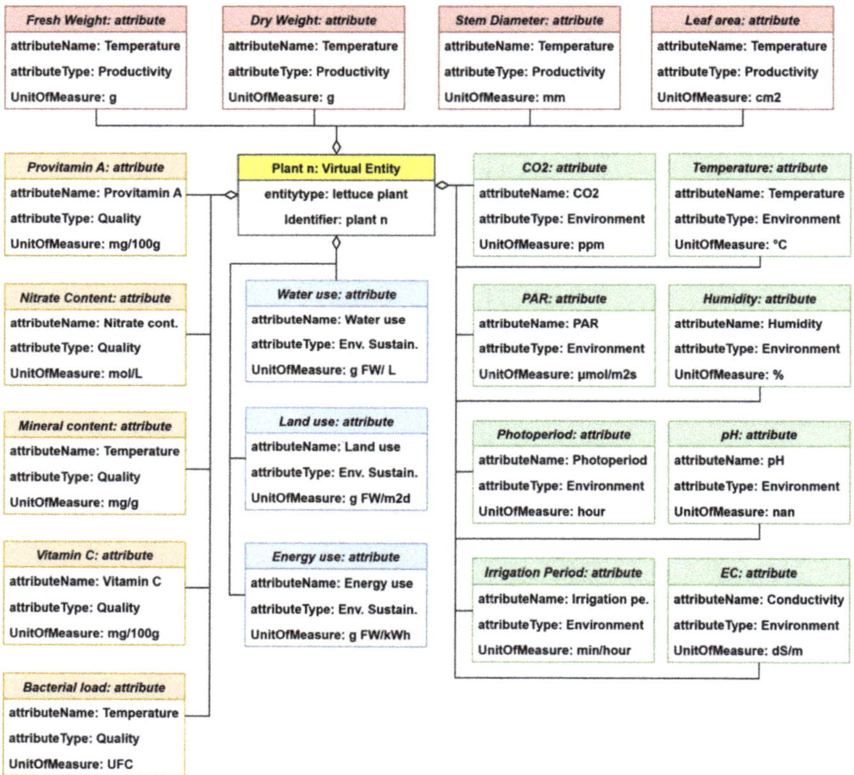


**Figure 10.** Information structure showing the virtual entity attributes (i.e., digital twin attributes).

The specific attributes associated with the VE are shown in Figure 10; each attribute contains the fields: attributeName, attributeType, which indicates the type of attribute in relation to the above-mentioned groups, and the unit of measurement. The productivity attributes Fresh weight, Dry weight, Stem diameter, and Leaf area are shown in red. Environmental attributes are in green, Quality in orange, and Environmental Sustainability in blue. Each of these attributes has a value container, which has a value and the metadata associated with it as defined by the IoT information model in Section 3. For the sake of simplicity, we will only describe the metainformation structure for a single attribute.

Considering the attribute "temperature", as shown in Figure 11, it has a value container categorizing one value and the zero-to-many information of the related value via means of metadata. The metadata are categorized into general metadata and Industry 5.0-specific metadata; the former contains fields such as the time stamp, the device issuing the data, and the resource used to obtain this information along with the service that delivered the information. Industry 5.0 metadata contains information about the sustainability, human-centricity, and resilience of the delivered information. In this example, we consider two metrics for each of these pillars. Namely, in terms of human-centricity, we observe the degree of human involvement and the impact of the delivered information on the human, for sustainability we use the metric average per-bit delivery cost (APBDC) to quantify economic sustainability and energy per bit for environmental sustainability. We do not consider social sustainability since it considers the social aspects outside of the system border and this is difficult to quantify within the scope of the work. For resilience, indicators of scalability and fault tolerance are used.

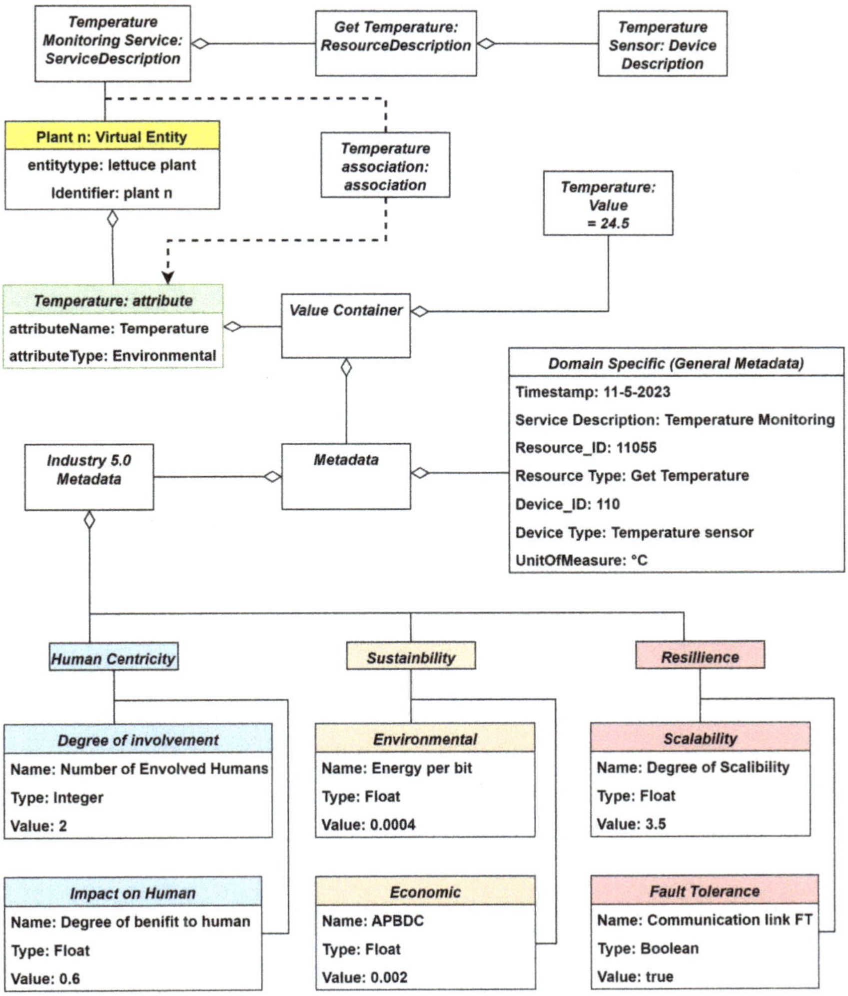


**Figure 11.** Information model for attribute "Temperature".

The information flow is initiated by a temperature sensor that sends the temperature value measured at the physical entity (the plant). This value is delivered to the attribute of the virtual entity associated with the temperature measurement through an association. This way, the measured value replaces the value in the temperature attribute of the virtual entity.

### 4.4. IoT System Architecture

Taking into consideration the elements described in the domain model and the information structure highlighted in the information view in the previous sections, the system architecture of the implemented aeroponic system is developed as shown in Figure 12; it consists of various sensors and effectors connected to a Raspberry Pi along with an IoT platform. The left portion of the diagram displays the sensors that function in two primary areas: the nutrient solution area, which is home to the pH, EC, and water level sensors; and the ambient environment of the plants, which includes the PAR, $CO_2$, and temperature and humidity sensors.

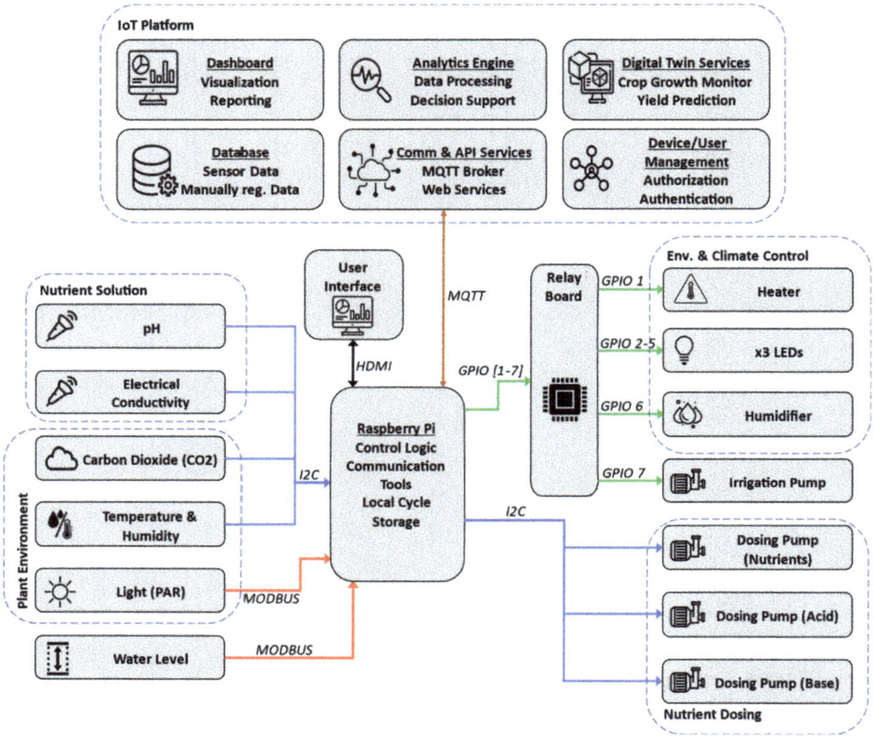


**Figure 12.** System architecture of developed aeroponics system.

The PAR and water level sensors transmit their data to the Raspberry using the Modbus protocol [77]. Modbus is an application layer communication protocol that has been considered the de facto standard for industrial automation for many years. Moreover, Modbus allows for a high degree of interoperability because it is an open standard. It is also a very cost-effective solution since the protocol can be supported by low-cost hardware due to its low complexity.

All the other sensors employ the I2C protocol to send their data. I2C, or Inter-Integrated Circuit, is a widely used serial communication protocol that facilitates communication between electronic devices [78]. With a straightforward two-wire design (SDA and SCL), I2C simplifies hardware implementation and reduces the pin count. Each device connected to the I2C bus has a unique address, allowing for the seamless integration of multiple components without conflicts.

The Raspberry Pi is considered the central hardware component in the system, as it is responsible for data acquisition, managing communications, control logic, and driving actuators depending on the outcome of the control logic. The main software component running in the Raspberry is Mycodo, which is an open-source environmental monitoring and automation system designed for home and industrial applications [79]. It is commonly used for tasks such as managing and controlling greenhouse environments, aquariums, and other settings where monitoring and automation are crucial. Moreover, Mycodo provides the tools for dashboard configuration and as a result, a graphical user interface is attached to the Raspberry Pi. The GUI shows sensor readings, trends, historical information, and notifications associated with the system.

Effectors (which include actuators and LEDs) are found on the right side of the diagram, similar to the sensors the effectors operate in the nutrient solution and the plant's ambient environment. A heater is used alongside a humidifier to regulate the growth climate. A heater has been used in place of an air conditioner due to the cold climate where the chamber is implemented. Three LEDs are used to provide artificial night and day cycles. All these actuators are connected to the Raspberry Pi through its GPIO pins since they are controlled by standard ON/OFF switching. A relay board is used to allow for logic switching between the Raspberry Pi (3.3 V) and the power supply of the actuator (e.g., 220 V). Three dosing pumps are connected to the system via I2c; specifically, two dosing pumps are used for pH regulation (acid and base) and one for EC regulation.

The communication between the aeroponic system and the IoT platform is implemented using the MQTT protocol [80]. MQTT, which stands for Message Queuing Telemetry Transport, is a lightweight and open messaging protocol designed for low-bandwidth, high-latency, or unreliable networks. It is commonly used in the Internet of Things (IoT) to facilitate communication between devices. MQTT operates on the publish/subscribe model, where devices can publish messages to specific topics, and other devices can subscribe to those topics to receive the messages.

The IoT platform consists of six main modules. The first module is the communication and API module responsible for providing communication functionality to and from the platform; it contains tools such as the MQTT broker and the tools used for web services to provide machine to machine interaction across the network. These data are stored in the database module, which is a robust storage system that efficiently manages time series sensor data and user input data, ensuring data integrity and accessibility for analytics and reporting purposes. The analytics engine employs machine learning algorithms to process sensor data, offering descriptive, diagnostic, predictive, and prescriptive analytics for actionable insights and optimized decision support. The dashboard module offers a user-friendly interface providing real-time insights and historical trends, allowing stakeholders to visualize and analyze data generated by IoT devices, enhancing decision-making processes. Digital twin services enabling the creation of virtual representations of physical assets are provided by the DT module, facilitating growth monitoring and accurate yield predictions through continuous synchronization with real-world data. Finally, is the component used for device and user management for authorization and authentication; the function of this module is to provide secure access through strong authentication methods, role-based authorization, and the efficient administration of IoT devices and user accounts.

The developed architecture tries to take into account several qualities and requirements that are essential to IoT systems. First, in terms of interoperability, the realm of IoT contains many types of devices with different technologies and varying communication protocols. Several tactics such as building variation points in software and employing open-source tools and protocols can be used to help in this issue; to this end, Raspberry Pi, MQTT, Modbus, and Mycodo are used due to their open-source nature.

With regard to scalability, numerous strategies used in standard information systems such as trying to predict future system expansions, modular architecture design, and using scalable communication protocols can be used. However, compared to traditional information systems, it is even harder to cope with in a highly distributed and fast-evolving scenario, as we have in IoT. In the developed architecture, a Raspberry has been used due to the availability of I/O expansion boards (I/O shields); moreover, Modbus and I2C both allow for a maximum number of 255 on its communication bus, which is quite satisfactory for a single aeroponic module.

## 5. Conclusions

### *5.1. Summary of Scientific Contributions*

The article introduces an IoT-DT framework, building upon the IoT-A reference architecture, to address the evolving requirements of Industry 5.0. This framework aims to describe IoT systems mainly focused on digital twin entities and capable of taking into account human-centricity, sustainability, and resilience. It recognizes and addresses the intricate interactions between virtual (digital twins) and physical entities, a core concept in digital twin technology. A pointwise description of the contributions of the research work presented by this manuscript is presented below.

- The proposed framework is built by focusing the IoT-A European reference architecture on the digital twin information holistic management (providing an integrated perspective on IoT systems and recognizing and addressing intricate interactions between virtual and physical entities) and by adding specific attributes and consideration regarding the increasingly popular European concept of Industry 5.0.
- The literature review section includes a systematic review of the the IoT-A concept to highlight existing research contributions and to synthesize scientific insights regarding the IoT-ARM project. The same section includes the state of the art regarding the technologies of digital twin and IoT systems appropriate in the 4.0 context and any research considering these technologies from a 5.0 perspective (scientific manuscripts regarding digital twin and human-centrality are mainly cited).
- The proposed framework provides a domain model that is robust within different domains of application (the new generation of smart farming and agriculture or Industry 5.0 environment for smart products and smart factories).
- The proposed functional model provides a further 5.0 characterization of the framework given by the management level of the model, i.e., the level that deals with types of decisions that are the most strategic and in common between different application areas.
- In the section dedicated to the case study, the information view referring to the information model for the vertical farming application is described through the textual contents of the subsection and Figure 11 regarding the information model for the attribute "Temperature". The case study is a concrete guide for applications in a smart farming scenario, specifically, aeroponic systems in vertical farming, with the general objective of optimizing growth conditions through monitoring and control services, and it demonstrates versatility by engaging various stakeholders like human users, production resources, and other various stakeholders, referring to dedicated digital components.

### *5.2. Future Improvements*

Looking ahead, we envision further enhancements and future directions that will fortify our framework's utility.

- One critical area of exploration involves refining and expanding the metrics for Industry 5.0 requirements, especially for aspects like social sustainability and resilience.
- Real-world validation across diverse IoT scenarios is a natural progression to strengthen the practicality of our framework and, for this reason, works in progress are aimed at applying this methodology to three use cases in manufacturing, agricultural, and service sectors, to validate a general framework and to highlight common factors and specificities.
- Dynamic adaptation is another avenue we are eager to explore, delving into methods that allow our digital twin framework to dynamically respond to evolving Industry 5.0 standards and emerging technologies, especially for ensuring the resilience of the system.

- Additionally, future improvements will include an intensified focus on security and privacy aspects within the framework to ensure the robust protection of user data and interactions, aligning with the ever-growing importance of secure and private IoT systems.

**Author Contributions:** Conceptualization, A.A. and E.T.; methodology, A.A., E.T. and G.B.; data curation, A.A. and E.T.; writing—original draft preparation, A.A. and E.T.; writing—review and editing, G.B. and P.C. All authors have read and agreed to the published version of the manuscript.

**Funding:** This research received no external funding.

**Institutional Review Board Statement:** Not applicable.

**Informed Consent Statement:** Not applicable.

**Data Availability Statement:** The data presented in this study are available in the manuscript text.

**Conflicts of Interest:** The authors declare no conflicts of interest.

## References

1. Yin, Y.; Stecke, K.E.; Li, D. The evolution of production systems from Industry 2.0 through Industry 4.0. *Int. J. Prod. Res.* **2018**, *56*, 848–861. [CrossRef]
2. Garrido-Hidalgo, C.; Olivares, T.; Ramirez, F.J.; Roda-Sanchez, L. An end-to-end Internet of Things solution for Reverse Supply Chain Management in Industry 4.0. *Comput. Ind.* **2019**, *112*, 103127. [CrossRef]
3. Botín-Sanabria, D.M.; Mihaita, A.-S.; Peimbert-García, R.E.; Ramírez-Moreno, M.A.; Ramírez-Mendoza, R.A.; Lozoya-Santos, J.D.J. Digital twin technology challenges and applications: A comprehensive review. *Remote Sens.* **2022**, *14*, 1335. [CrossRef]
4. Fuller, A.; Fan, Z.; Day, C.; Barlow, C. Digital Twin: Enabling Technologies, Challenges and Open Research. *IEEE Access* **2020**, *8*, 108952–108971. [CrossRef]
5. Rasheed, A.; San, O.; Kvamsdal, T. Digital twin: Values, challenges and enablers. *arXiv* **2019**, arXiv:1910.01719.
6. Tao, F.; Cheng, J.; Qi, Q.; Zhang, M.; Zhang, H.; Sui, F. Digital twin-driven product design, manufacturing and service with big data. *Int. J. Adv. Manuf. Technol.* **2018**, *94*, 3563–3576. [CrossRef]
7. Verdouw, C.; Kruize, J.W. Digital Twins in Farm Management: Illustrations from the FIWARE Accelerators SmartAgriFood and Fractals. In Proceedings of the 7th Asian-Australasian Conference on Precision Agriculture Digital, Hamilton, New Zealand, 16–18 October 2017; pp. 16–18.
8. Al-Ali, A.-R.; Gupta, R.; Batool, T.Z.; Landolsi, T.; Aloul, F.; Al Nabulsi, A. Digital twin conceptual model within the context of internet of things. *Future Internet* **2020**, *12*, 163. [CrossRef]
9. Sisinni, E.; Saifullah, A.; Han, S.; Jennehag, U.; Gidlund, M. Industrial Internet of Things: Challenges, Opportunities, and Directions. *IEEE Trans. Ind. Inf.* **2018**, *14*, 4724–4734. [CrossRef]
10. Sajja, P.S.; Akerkar, R. Knowledge-based systems for development. *Adv. Knowl. Based Syst. Model Appl. Res.* **2010**, *1*, 1–11.
11. Traini, E. Hybrid Modeling to Support the Smart Manufacturing: Concepts, Theoretic Contributions and Real-Case Applications about Hybrid and Wisdom-Based Systems. Ph.D. Thesis, Politecnico di Torino, Turin, Italy, 2022.
12. Traini, E.; Bruno, G.; Lombardi, F. Design of a Physics-Based and Data-Driven Hybrid Model for Predictive Maintenance. In *Advances in Production Management Systems. Artificial Intelligence for Sustainable and Resilient Production Systems*; Dolgui, A., Bernard, A., Lemoine, D., Von Cieminski, G., Romero, D., Eds.; IFIP Advances in Information and Communication Technology; Springer International Publishing: Cham, Switzerland, 2021; Volume 634, pp. 536–543. [CrossRef]
13. Kostek, B. Data, Information, Knowledge, Wisdom Pyramid Concept Revisited in the Context of Deep Learning. In *Intelligent Decision Technologies*; Czarnowski, I., Howlett, R.J., Jain, L.C., Eds.; Smart Innovation, Systems and Technologies; Springer Nature Singapore: Singapore, 2023; Volume 352, pp. 3–12. [CrossRef]
14. Müller, J. *Enabling Technologies for Industry 5.0*; European Commission: Brussels, Belgium, 2020; pp. 8–10.
15. Xu, X.; Lu, Y.; Vogel-Heuser, B.; Wang, L. Industry 4.0 and Industry 5.0—Inception, conception and perception. *J. Manuf. Syst.* **2021**, *61*, 530–535. [CrossRef]
16. European Commission. *Directorate General for Research and Innovation. Industry 5.0, a Transformative Vision for Europe: Governing Systemic Transformations towards a Sustainable Industry*; Publications Office: Luxembourg, 2021. [CrossRef]
17. Colella, M.; Barberio, M.; Figliola, A. The Big Vision: From Industry 4.0 to 5.0 for a New AEC Sector. In *Architecture and Design for Industry 4.0*; Barberio, M., Colella, M., Figliola, A., Battisti, A., Eds.; Lecture Notes in Mechanical Engineering; Springer International Publishing: Cham, Switzerland, 2024; pp. 3–17. [CrossRef]
18. Directorate-General for Research and Innovation (European Commission); Breque, M.; De Nul, L.; Petridis, A. *Industry 5.0: Towards a Sustainable, Human Centric and Resilient European Industry*; Publications Office of the European Union: Luxembourg, 2021. [CrossRef]
19. Aheleroff, S.; Huang, H.; Xu, X.; Zhong, R.Y. Toward sustainability and resilience with Industry 4.0 and Industry 5.0. *Front. Manuf. Technol.* **2022**, *2*, 951643. [CrossRef]

20. Bassi, A.; Bauer, M.; Fiedler, M.; Kramp, T.; Van Kranenburg, R.; Lange, S.; Meissner, S. (Eds.) *Enabling Things to Talk*; Springer Berlin/Heidelberg, Germany, 2013. [CrossRef]
21. Fortino, G.; Gravina, R.; Russo, W.; Savaglio, C. Modeling and Simulating Internet-of-Things Systems: A Hybrid Agent-Oriented Approach. *Comput. Sci. Eng.* **2017**, *19*, 68–76. [CrossRef]
22. Ferko, E.; Bucaioni, A.; Pelliccione, P.; Behnam, M. Standardisation in Digital Twin Architectures in Manufacturing. In Proceedings of the 2023 IEEE 20th International Conference on Software Architecture (ICSA), L'Aquila, Italy, 13–17 March 2023; pp. 70–81.
23. Bauer, M.; Boussard, M.; Bui, N.; Carrez, F.; Jardak, C.; De Loof, J.; Magerkurth, C.; Meissner, S.; Nettsträter, A.; Olivereau, A.; et al. *Internet of Things—Architecture IoT-A, Deliverable D1.5—Final Architectural Reference Model for the IoT v3.0*; European Commission: Brussels, Belgium, 2013.
24. Lynn, T.; Mooney, J.G.; Lee, B.; Endo, P.T. (Eds.) *The Cloud-to-Thing Continuum: Opportunities and Challenges in Cloud, Fog and Edge Computing*; Palgrave Studies in Digital Business & Enabling Technologies; Springer International Publishing: Cham, Switzerland, 2020. [CrossRef]
25. Boyanov, L.; Kisimov, V.; Christov, Y. Evaluating IoT Reference Architecture. In Proceedings of the 2020 International Conference Automatics and Informatics (ICAI), Varna, Bulgaria, 1–3 October 2020; pp. 1–5. [CrossRef]
26. Jost, J.; Kirks, T.; Mattig, B. Multi-agent systems for decentralized control and adaptive interaction between humans and machines for industrial environments. In Proceedings of the 2017 7th IEEE International Conference on System Engineering and Technology (ICSET), Shah Alam, Malaysia, 2–3 October 2017; pp. 95–100. [CrossRef]
27. Beltran, V.; Skarmeta, A.F.; Ruiz, P.M. An ARM-Compliant Architecture for User Privacy in Smart Cities: SMARTIE—Quality by Design in the IoT. *Wirel. Commun. Mob. Comput.* **2017**, *2017*, 3859836. [CrossRef]
28. Verdouw, C.; Tekinerdogan, B.; Beulens, A.; Wolfert, S. Digital twins in smart farming. *Agric. Syst.* **2021**, *189*, 103046. [CrossRef]
29. Spyrou, O.; Verdouw, C.; Hurst, W. A digital twin reference architecture for pharmaceutical cannabis production. *Int. J. Comput. Integr. Manuf.* **2023**, 1–21. [CrossRef]
30. Hirmer, P. *Model-Based Approaches to the Internet of Things*; Springer International Publishing: Cham, Switzerland, 2023. [CrossRef]
31. Alhamedi, A.H.; Snasel, V.; Aldosari, H.M.; Abraham, A. Internet of things communication reference model. In Proceedings of the 2014 6th International Conference on Computational Aspects of Social Networks, Porto, Portugal, 30 July–1 August 2014; pp. 61–66. [CrossRef]
32. Alves, M.P.; Delicato, F.C.; Pires, P.F. IoTA-MD: A model-driven approach for applying QoS attributes in the development of the IoT systems. In Proceedings of the Symposium on Applied Computing, Marrakech, Morocco, 3–7 April 2017; pp. 1773–1780. [CrossRef]
33. Albreem, M.A.; Sheikh, A.M.; Alsharif, M.H.; Jusoh, M.; Yasin, M.N.M. Green Internet of Things (GIoT): Applications, Practices, Awareness, and Challenges. *IEEE Access* **2021**, *9*, 38833–38858. [CrossRef]
34. Streitz, N.; Markopoulos, P. (Eds.) *Distributed, Ambient and Pervasive Interactions: 5th International Conference, DAPI 2017, Held as Part of HCI International 2017, Vancouver, BC, Canada, July 9–14, 2017, Proceedings*; Lecture Notes in Computer Science; Springer International Publishing: Cham, Switzerland, 2017; Volume 10291. [CrossRef]
35. Neubauer, M.; Krenn, F.; Majoe, D. Towards an Architecture for Human-aware Modeling and Execution of Production Processes. *IFAC-PapersOnLine* **2015**, *48*, 294–299. [CrossRef]
36. Stary, C.; Neubauer, M. Industrial Challenges. In *S-BPM in the Production Industry*; Neubauer, M., Stary, C., Eds.; Springer International Publishing: Cham, Switzerland, 2017; pp. 7–25. [CrossRef]
37. Gianotti, M.; Riccardi, F.; Cosentino, G.; Garzotto, F.; Matera, M. Modeling Interactive Smart Spaces. In *Conceptual Modeling*; Dobbie, G., Frank, U., Kappel, G., Liddle, S.W., Mayr, H.C., Eds.; Lecture Notes in Computer Science; Springer International Publishing: Cham, Switzerland, 2020; Volume 12400, pp. 403–417. [CrossRef]
38. Elicegui, I.; López, C.; Sánchez, L.; Lanza, J.; Muñoz, L.; Pintus, A.; Manchinu, A.; Serra, A. Design and Implementation of a Cloud-Based Platform for Unleashing the Personal and Communal Internet of Things. *Mob. Inf. Syst.* **2017**, *2017*, 2164072. [CrossRef]
39. Jost, J.; Kirks, T. Integrating CPS into Socio-Technical Systems. In Proceedings of the 2019 4th Asia-Pacific Conference on Intelligent Robot Systems (ACIRS), Nagoya, Japan, 13–15 July 2019; pp. 129–133. [CrossRef]
40. Wang, B.; Zhou, H.; Li, X.; Yang, G.; Zheng, P.; Song, C.; Yuan, Y.; Wuast, T.; Yang, H.; Wang, L. Human Digital Twin in the context of Industry 5.0. *Robot. Comput. Integr. Manuf.* **2024**, *85*, 102626. [CrossRef]
41. Lv, Z. Digital Twins in Industry 5.0. *Research* **2023**, *6*, 0071. [CrossRef]
42. Pattar, S.; Buyya, R.; Venugopal, K.R.; Iyengar, S.S.; Patnaik, L.M. Searching for the IoT Resources: Fundamentals, Requirements, Comprehensive Review, and Future Directions. *IEEE Commun. Surv. Tutor.* **2018**, *20*, 2101–2132. [CrossRef]
43. Bauer, M.; Bui, N.; Giacomin, P.; Gruschka, N.; Haller, S.; Ho, E.; Kernchen, R.; Lischka, M.; De Loof, J.; Magerkurth, C.; et al. *Internet of Things—Architecture IoT-A, Project Deliverable D1.2—Initial Architectural Reference Model for IoT*; European Commission: Brussels, Belgium, 2011.
44. Semeraro, C.; Lezoche, M.; Panetto, H.; Dassisti, M. Digital twin paradigm: A systematic literature review. *Comput. Ind.* **2021**, *130*, 103469. [CrossRef]
45. Errandonea, I.; Beltrán, S.; Arrizabalaga, S. Digital Twin for maintenance: A literature review. *Comput. Ind.* **2020**, *123*, 103316. [CrossRef]

46. Noor, B.M.; Hassan, W.H. Current research on Internet of Things (IoT) security: A survey. *Comput. Netw.* **2019**, *148*, 283–294. [CrossRef]
47. Abdelghani, W.; Zayani, C.A.; Amous, I.; Sèdes, F. User-centric IoT: Challenges and perspectives. In Proceedings of the Twelfth International Conference on Mobile Ubiquitous Computing, Systems, Services and Technologies (UBICOMM 2018), Athens, Greece, 18–22 November 2018; pp. 27–34.
48. Kotronis, C.; Routis, I.; Politi, E.; Nikolaidou, M.; Dimitrakopoulos, G.; Anagnostopoulos, D.; Amira, A.; Bensaali, F.; Djelouat, H. Evaluating Internet of Medical Things (IoMT)-based systems from a human-centric perspective. *Internet Things* **2019**, *8*, 100125. [CrossRef]
49. Ystgaard, F.; Atzori, L.; Palma, D.; Heegaard, P.E.; Bertheussen, L.E.; Jensen, M.R.; De Moor, K. Review of the theory, principles, and design requirements of human-centric Internet of Things (IoT). *J. Ambient Intell. Human. Comput.* **2023**, *14*, 2827–2859. [CrossRef]
50. Brundtland, H. Our common future—Call for action. *Environ. Conserv.* **1987**, *14*, 291–294. [CrossRef]
51. Gupta, V.; Tripathi, S.; De, S. Green sensing and communication: A step towards sustainable IoT systems. *J. Indian Inst. Sci.* **2020**, *100*, 383–398. [CrossRef]
52. Sharma, N.; Panwar, D. Green IoT: Advancements and sustainability with environment by 2050. In Proceedings of the 2020 8th International Conference on Reliability, Infocom Technologies and Optimization (Trends and Future Directions) (ICRITO), Noida, India, 4–5 June 2020; pp. 1127–1132.
53. Strezov, V.; Evans, A.; Evans, T.J. Assessment of the Economic, Social and Environmental Dimensions of the Indicators for Sustainable Development. *Sustain. Dev.* **2017**, *25*, 242–253. [CrossRef]
54. Fortino, G.; Messina, F.; Rosaci, D.; Sarnè, G.M.L. ResIoT: An IoT social framework resilient to malicious activities. *IEEE/CAA J. Autom. Sin.* **2020**, *7*, 1263–1278. [CrossRef]
55. Tsigkanos, C.; Nastic, S.; Dustdar, S. Towards resilient Internet of Things: Vision, challenges, and research roadmap. In Proceedings of the 2019 IEEE 39th International Conference on Distributed Computing Systems (ICDCS), Dallas, TX, USA, 7–10 July 2019; pp. 1754–1764.
56. Berger, C.; Eichhammer, P.; Reiser, H.P.; Domaschka, J.; Hauck, F.J.; Habiger, G. A survey on resilience in the IoT: Taxonomy, classification, and discussion of resilience mechanisms. *ACM Comput. Surv.* **2021**, *54*, 1–39. [CrossRef]
57. Song, Y.; Wan, C.; Hu, X.; Qin, H.; Lao, K. Resilient power grid for smart city. *iEnergy* **2022**, *1*, 325–340. [CrossRef]
58. Lilhore, U.K.; Imoize, A.L.; Li, C.T.; Simaiya, S.; Pani, S.K.; Goyal, N.; Kumar, A.; Lee, C.-C. Design and implementation of an ML and IoT based adaptive traffic-management system for smart cities. *Sensors* **2022**, *22*, 2908. [CrossRef] [PubMed]
59. Traini, E.; Bruno, G.; Lombardi, F. Tool condition monitoring framework for predictive maintenance: A case study on milling process. *Int. J. Prod. Res.* **2021**, *59*, 7179–7193. [CrossRef]
60. Natarajan, S.; Thangamuthu, M.; Gnanasekaran, S.; Rakkiyannan, J. Digital Twin-Driven Tool Condition Monitoring for the Milling Process. *Sensors* **2023**, *23*, 5431. [CrossRef] [PubMed]
61. Mian, T.; Choudhary, A.; Fatima, S.; Panigrahi, B.K. Artificial intelligence of things based approach for anomaly detection in rotating machines. *Comput. Electr. Eng.* **2023**, *109*, 108760. [CrossRef]
62. Lu, S.; Lu, J.; An, K.; Wang, X.; He, Q. Edge Computing on IoT for Machine Signal Processing and Fault Diagnosis: A Review. *IEEE Internet Things J.* **2023**, *10*, 11093–11116. [CrossRef]
63. Huang, M.; Liu, Z.; Tao, Y. Mechanical fault diagnosis and prediction in IoT based on multi-source sensing data fusion. *Simul. Model. Pract. Theory* **2020**, *102*, 101981. [CrossRef]
64. Serbanati, A.; Maria, C.; Biader, U. Building Blocks of the Internet of Things: State of the Art and Beyond. In *Deploying RFID—Challenges, Solutions, and Open Issues*; Turcu, C., Ed.; InTech: Milton, Australia, 2011. [CrossRef]
65. Desai, P.; Sheth, A.; Anantharam, P. Semantic Gateway as a Service Architecture for IoT Interoperability. In Proceedings of the 2015 IEEE International Conference on Mobile Services, New York City, NY, USA, 27 June–2 July 2015; pp. 313–319. [CrossRef]
66. Silverio-Fernández, M.; Renukappa, S.; Suresh, S. What is a smart device?—A conceptualisation within the paradigm of the internet of things. *Vis. Eng.* **2018**, *6*, 3. [CrossRef]
67. Rowley, J. The wisdom hierarchy: Representations of the DIKW hierarchy. *J. Inf. Sci.* **2007**, *33*, 163–180. [CrossRef]
68. Baca, M. *Introduction to Metadata*; Getty Publications: Los Angeles, CA, USA, 2016.
69. Van Eck, D.; McAdams, D.A.; Vermaas, P.E. Functional Decomposition in Engineering: A survey. In Proceedings of the International Design Engineering Technical Conferences and Computers and Information in Engineering Conference, Las Vegas, NV, USA, 4–7 September 2007; pp. 227–236.
70. Verdouw, C.; Wolfert, S.; Tekinerdogan, B. Internet of Things in agriculture. *CABI Rev.* **2016**, *2016*, 1–12. [CrossRef]
71. Alani, M. *Guide to OSI and TCP/IP Models*; Springer: Berlin/Heidelberg, Germany, 2014.
72. World Population Prospects 2022: Summary of Results | Population Division. Available online: https://www.un.org/development/desa/pd/content/World-Population-Prospects-2022 (accessed on 1 December 2023).
73. Malhi, S.; Kaur, M.; Kaushik, P. Impact of climate change on agriculture and its mitigation strategies: A review. *Sustainability* **2021**, *13*, 1318. [CrossRef]
74. Avgoustaki, D.D.; Xydis, G. Indoor vertical farming in the urban nexus context: Business growth and resource savings. *Sustainability* **2020**, *12*, 1965. [CrossRef]

75. Raspberry, P. Raspberry Pi 4 Model B Specifications. Available online: https://www.raspberrypi.com/products/raspberry-pi-4-model-b/specifications/ (accessed on 21 November 2023).
76. Grasso, N.; Fasciolo, B.; Bruno, G.; Lombardi, F. A Smart Vertical Farming System to Evaluate Productivity, Quality, and Sustainability of Agricultural Production. In *Flexible Automation and Intelligent Manufacturing: Establishing Bridges for More Sustainable Manufacturing Systems*; Silva, F.J.G., Ferreira, L.P., Sá, J.C., Pereira, M.T., Pinto, C.M.A., Eds.; Lecture Notes in Mechanical Engineering; Springer Nature Switzerland: Cham, Switzerland, 2024; pp. 938–945. [CrossRef]
77. Thomas, G. *Introduction to the Modbus Protocol*; Contemporary Control Systems, Inc.: Grove, IL, USA, 2008.
78. Valdez, J.; Becker, J. *Understanding the I2C Bus*; Texas Instruments Incorporated: Dallas, TX, USA, 2015.
79. Gabriel, K. Mycodo. 20 November 2023. Available online: https://github.com/kizniche/Mycodo (accessed on 21 November 2023).
80. Soni, D.; Makwana, A. A Survey on mqtt: A Protocol of Internet of Things (IoT). In Proceedings of the International Conference on Telecommunication, Power Analysis and Computing Techniques (ICTPACT-2017), Chennai, India, 6–8 April 2017; pp. 173–177.

sensors

*Article*

# Cloud-Based Machine Learning Methods for Parameter Prediction in Textile Manufacturing

**Ray-I Chang [1,*], Jia-Ying Lin [1] and Yu-Hsin Hung [2,*]**

[1] Department of Engineering Science and Ocean Engineering, National Taiwan University, No. 1, Sec. 4, Roosevelt Road, Taipei 10617, Taiwan; r05525122@ntu.edu.tw

[2] Department of Industrial Engineering and Management, National Yunlin University of Science and Technology, Yunlin 64002, Taiwan

* Correspondence: rayichang@ntu.edu.tw (R.-I.C.); hungyh@yuntech.edu.tw (Y.-H.H.)

**Abstract:** In traditional textile manufacturing, downstream manufacturers use raw materials, such as Nylon and cotton yarns, to produce textile products. The manufacturing process involves warping, sizing, beaming, weaving, and inspection. Staff members typically use a trial-and-error approach to adjust the appropriate production parameters in the manufacturing process, which can be time consuming and a waste of resources. To enhance the efficiency and effectiveness of textile manufacturing economically, this study proposes a query-based learning method in regression analytics using existing manufacturing data. Query-based learning allows the model training to evolve its decision-making process through dynamic interactions with its solution space. In this study, predefined target parameters of quality factors were first used to validate the training results and create new training patterns. These new patterns were then imported into the solution space of the training model. In predicting product quality, the results show that the proposed query-based regression algorithm has a mean squared error of 0.0153, which is better than those of the original regression-related methods (Avg. mean squared error = 0.020). The trained model was deployed as an application programing interface (API) for cloud-based analytics and an extensive auto-notification service.

**Keywords:** predictive maintenance; data communication; ensemble learning; process parameter; textile

**Citation:** Chang, R.-I.; Lin, J.-Y.; Hung, Y.-H. Cloud-Based Machine Learning Methods for Parameter Prediction in Textile Manufacturing. *Sensors* **2024**, *24*, 1304. https://doi.org/10.3390/s24041304

Academic Editors: Behnam Mobaraki and Jose Turmo

Received: 22 December 2023
Revised: 15 February 2024
Accepted: 16 February 2024
Published: 18 February 2024



## 1. Introduction

In the textile industry supply chain, the industrial chain begins with upstream processes that involve the processing of petrochemical and natural raw materials into fibers. Moving to the midstream, fabric production occurs through processes such as spinning, weaving, and dyeing. The downstream sector primarily revolves around garment manufacturing [1]. According to a recent report, as of 2022, the textile industry contributed a substantial USD 1.2 billion, accounting for 93.2% of the worldwide textile industry output [2]. The global textile industry maintains a high level of competitiveness, with increasing demands for product quality and flexible production capabilities [3]. Textile manufacturing involves several processes, raw materials, and equipment settings. In the midstream of textile manufacturing, the raw material must pass through several manufacturing steps, including warping, sizing, beaming, weaving, and inspection. All of these factors can influence the quality of the product (Table 1) [4–6]. The first three steps mentioned above constitute the preparatory phase. In each phase, technicians adjust the production parameters of different textile machines according to the raw yarn characteristics. For example, setting the warp beam winding tension of the warping machine is essential in the preparation phase. When the tension is low, the yarn winding force is loose. If the tension is too high, the yarn will be stretched thin, and can easily break in subsequent processes. In general, the setting of process parameters depends on the staff operation, which leads to a time-consuming, resource-wasting, trial-and-error process. Furthermore, new technicians cannot easily learn from senior technicians the appropriate production parameter settings for different types of raw yarn on textile machines.

**Table 1.** Description of the primary manufacturing process.

| Procedure | Action | Description |
|---|---|---|
| 1 | warping | The process of arranging and aligning yarns or threads in parallel to form the lengthwise foundation of a woven fabric [4,6]. |
| 2 | sizing | The process of applying a protective substance, known as sizing, to warp yarns before weaving [4–6]. |
| 3 | beaming | The process that follows the sizing process is part of the preparation of warp yarns for weaving [4,6]. |
| 4 | weaving | The process of interlacing two sets of yarns at right angles to create a fabric [6]. |
| 5 | inspection | The process of identifying and rectifying defects in the finished product early in the production stage. |

Note: Warping, sizing, and beaming are integrated in the preparation phase.

As digital transformation rapidly progresses [7–9], the manufacturing industry has evolving into a more cost-effective and value-enhancing business process by incorporating sensors, networks, and applications. Sensors are pivotal in transmitting substantial amounts of industrial data to the enterprise resource planning (ERP) system of the factory. Through self-learning, exploring valuable insights, and sharing with others, these data can significantly enhance the manufacturing process. The fundamental principle of digital transformation is value creation [10]. Thus, leveraging industrial data has proven to be an economical and efficient way to enhance business processes in the industry. This study aims to enhance the quality of the textile manufacturing process by using machine learning techniques to predict substandard products, thereby improving production efficiency and adding value to the existing digital infrastructure. This research not only uses machine learning with data but also enhances the computational performance through a query-based interaction approach. The midstream of the textile product manufacturing process serves as the use case. It employs manufacturing data and predictive models to forecast the quality of textile products using standard data analytics. The primary objectives of this study include intelligent production prediction and system design. This study highlights the top-performing predictive model, providing valuable insights to improve product quality. Additionally, the proposal introduces an online analytic module and a third-party application, facilitating the timely communication of analytics from machines to users. The contributions of this study can be summarized as follows:

- Investigated the prediction difference between the noninteractive and the interactive machine learning models.
- Developed an online analytical module for textile product manufacturing.
- Enhanced the value of the application by developing an office automation macro to enable automatic notifications.

The remainder of this study is organized as follows. Section 1 introduces the research background, research object, and contribution. Section 2 discusses related issues in the textile manufacturing industry. Section 3 outlines the data analytic procedure using the proposed approach and the system design for data communication. Section 4 details the experimental procedure, dataset used, and evaluation criteria. Section 5 presents the analysis results. Sections 6 and 7 discuss and conclude this paper by describing the significance of the study, its limitations, and suggestions for future work.

## 2. Related Work

### 2.1. Textile Production Management

Textile manufacturing processes such as warping and sizing are essential for the manufacturing site. Exact production information and procedure control can enhance manufacturing efficiency [4,11,12]. Taking weaving as an example, it encompasses a variety of processes such as winding, warping, sizing, and drawing. This intricate journey involves calculations, production motions, and procedures, all of which converge to set the stage for the weaving process to unfold [12]. Drean et al. (2022) provided practical calculations in weaving preparation to ensure consistent quality and prevent failure of events during subsequent processing [4]. Bathrinath et al. used the fuzzy analytical hierarchy process to analyze problems in the yarn winding process and designed a corresponding strategy to improve productivity [11]. Recently, predictive maintenance has been applied in textile production management, including machines [13,14] and scheduling maintenance [15]. The current generation of smart factories integrates sensors and computer algorithms into the entire production line. These factories use IoT devices [16,17] to collect real-time data from physical and cyber spaces, leverage cloud computing [18] to handle big data [19], and employ artificial intelligence for statistical analysis, decision making, and planning. As a result, these smart factories can intelligently produce highly customized products. A new trend is the use of CPS [20,21], where software and physical components are deeply intertwined and can interact intelligently with each other. In the textile industry, the development of CPS began with the use of data analytics and sensor technology [22,23]. Through heterogeneous network connections, all entities, including physical equipment and software systems in smart factories, are connected to a common communication interface for data exchange and remote control. All raw data and value-added information are integrated into the ERP systems. On the basis of these emerging technologies, a CPS can recognize the states of physical entities in an industrial chain. Furthermore, it can generate cyber twins for these physical entities, replacing physical tests with more cost-effective, faster software simulations for prediction and management before implementing any planned or unexpected change. Ślusarczyk et al. reported that Industry 4.0 has constructively enhanced the efficiency of both production and services within the textile industry [24]. The constructive collaboration of CPS, interoperability, smart city integration, and innovative product intelligence collectively yield positive outcomes for both production and services [24,25]. The processing power of internet communication technology can be significantly increased, particularly when machine learning techniques are used to analyze existing data.

### 2.2. Data Analytics in the Textile Manufacturing Process

In recent years, the machine learning approach has been employed in the manufacturing process. Eltayib et al. used linear regression models to predict fabric tear strength in the warp and weft directions [26]. They revealed that the tear strength of the fabric in the weft was influenced by the tensile strength of the yarn and the yarn count, whereas the tear strength in the warp direction was impacted by the linear density of the fabric and the yarn count [26]. Lu et al. used an artificial neural network and multiple linear regression models to predict the tensile strength of individual wool fibers [27]. The coefficients of determination, obtained from their back propagation neural network and stepwise regression, highlighted a strong correlation between the measured and predicted strengths of wool, with an error value within an acceptable range [27]. A multilinear regression model and a geometrical method were employed to forecast sewing thread consumption for overedge stitches [28]. These studies demonstrated the effectiveness of these models in extracting valuable insights from data within the textile industry [26–29]. Hoque et al. applied the least absolute shrinkage selector operator (LASSO) to predict the bursting strength of single jersey cotton plain knitted fabrics [30], whereas Rabaca et al. used logit ridge regression and LASSO to predict business failure [31]. In a study by Gorgül et al. [32], a kernel ridge regression (KRR) model produced satisfactory anomaly detection results. Gorgül et al.

recommended the use of KRR for modeling and anomaly detection, specifically in the context of temperature control in textile dyeing processes [32]. Their primary objective was to quickly identify issues with temperature controls, address failures in dyeing machines, and improve the efficiency of dyeing processes [32]. Ridge regression was used to predict the demand for textile products [33]. Although the algorithm was applied to textile manufacturing issues and demonstrated strong computational efficiency, its training mechanism is unidirectional and lacks interaction with the solution space. This limitation could lead to the emergence of local optimum problems. In contrast, recent advancements in reinforcement learning incorporate mechanisms for interaction with the environment. In this study, we introduce query-based learning. This approach enhances the predictive performance by fostering interaction with the solution space, validating with additional training instances, and exploring different solution spaces.

## 3. Materials and Methods

This study introduces a comprehensive workflow based on data from a case study. Edge data were transmitted to the ERP for data analytics and implementation notification. Machine learning approaches have proven to be highly accurate in certain instances. In this study, we used a regression-related method and our proposed method to predict issues in the textile manufacturing process. We processed the dataset according to the standard data workflow, the details of which are detailed in the experiment section. The process of knowledge discovery in databases involves data collection and preprocessing, modeling, model evaluation, and deployment (Figure 1). This approach can offer deeper insights into processes and allows exploration of the value of the imported data. The data are used to their fullest potential; both the original data and the results of the data analytics are transmitted to the API. Before a failure event occurs, the user receives an email notification and performs preventive maintenance.

### 3.1. Materials

In addition to developing functional fibers, the use case was dedicated to developing interweaving technology and the use of equipment to blend natural and chemical fibers to weave functional blended fiber textile products. Accordingly, the aim of this study was to improve the quality of products using machine learning to analyze production parameter data. Figure 2 shows the textile production line in the use case. The production process can be divided into three major phases: preparation, weaving, and inspection. In each phase, technicians adjust the production parameters of different textile machines according to the characteristics of the raw yarn, and the sensors transmit parameter data to the ERP system database.

In this study, the data were obtained from the ERP system, and the data format was a .csv type file. The obtained data were analyzed for early production diagnosis in textile manufacturing. Tables 2–4 describe the dataset and the specific prediction target. The data were obtained from the textile manufacturing process, encompassing its crucial phases, such as warping, sizing, beaming, and weaving. The original data had to be preprocessed. The different preprocessed datasets transformed into sixteen essential production attributes and thirteen raw material attributes were imported into the prediction model (Tables 3 and 4).

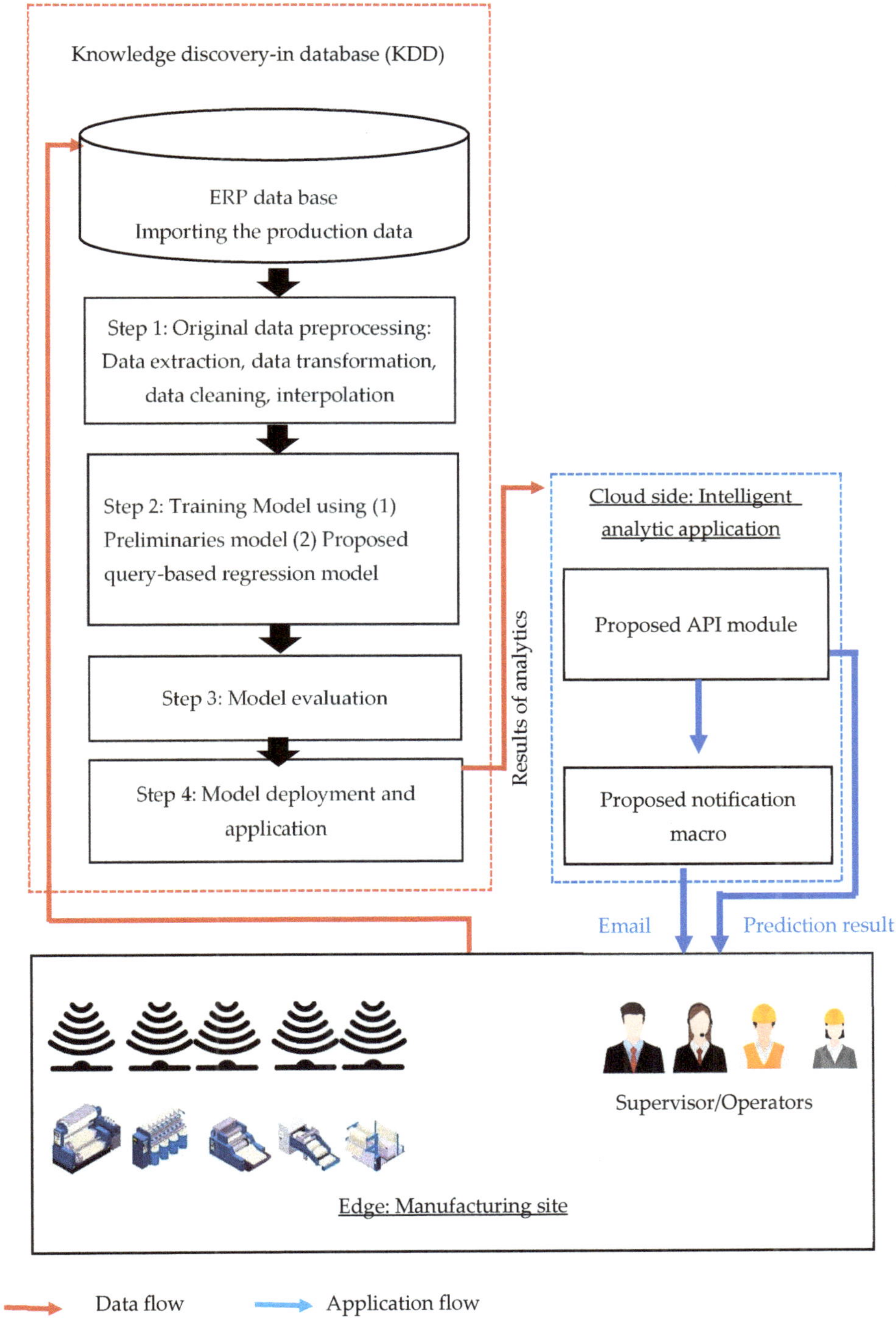


**Figure 1.** Flowchart of the data analytics of this study.

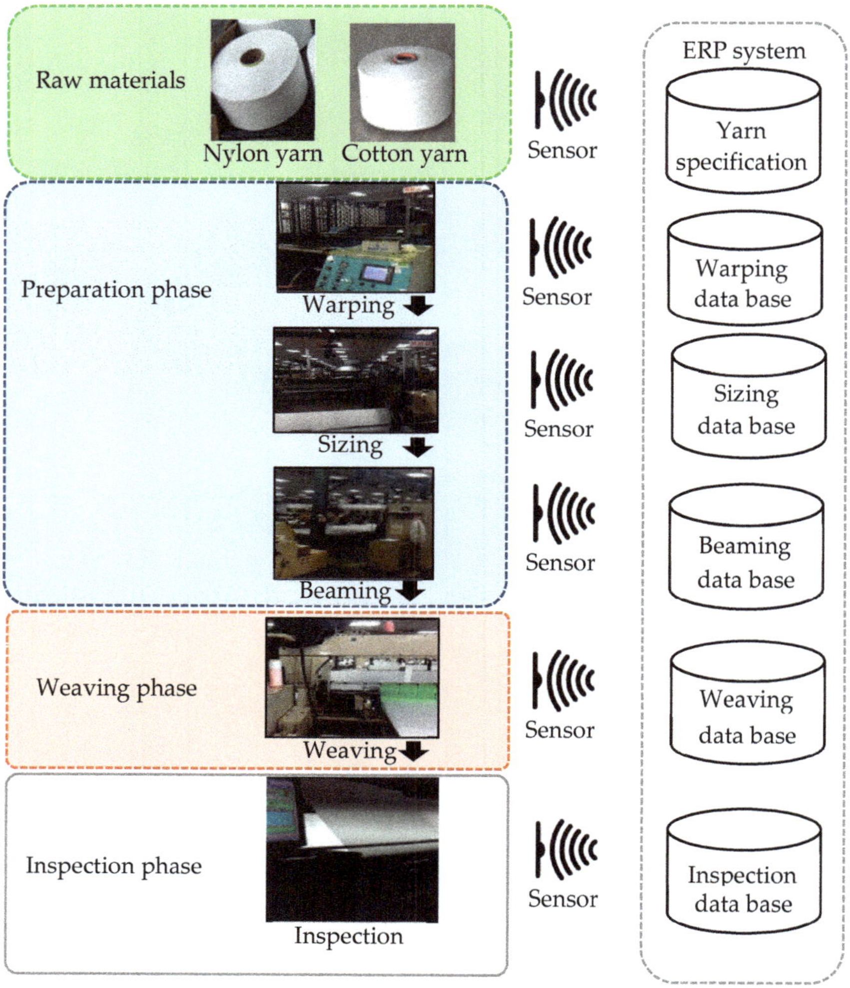


**Figure 2.** Data collection from textile manufacturing.

**Table 2.** Manufacturing dataset.

| Filename(.csv) | Columns | Raw | Description |
|---|---|---|---|
| warpop | 37 | 26,103 | production parameters in the warping process |
| sizeop | 46 | 15,950 | production parameters in the sizing process |
| beamop | 31 | 37,089 | production parameters in the beaming process |
| weaveop | 25 | 161,821 | production parameters in the weaving process |
| Inspection | 12 | 481,610 | fabric inspection results |

**Table 3.** Raw material parameters.

| Raw Material (13 Attributes) | Name | Description |
|---|---|---|
| Yarn Specification (13 attributes) | TOTALLENGTH | Actual length of warp [1] |
| | THEORYLENGTH | Theoretical total length warp [1] |
| | WARPTOTAL | Number of warps [1] |
| | YARNSPECDENIM | Denier [2] number of yarn specifications |
| | YARNSPECFIBERBASE | Fiber number of the yarn specifications |
| | DENIM | Theoretical denier [2] number of yarn specifications |
| | FIBERBASE | Fiber number of the yarn specifications |
| | UNITWEIGHT | Weight per unit |
| | GRANULARITY | Granularity of the yarn |
| | WARPLENGTH | Length of the warp |
| | WARPSTRIP | Length of beaming |
| | WARPLENGHT | Length of warping |
| | SIZINGLENGTH | Length of sizing |

Notes: [1] Warp refers to the set of yarns or threads that run lengthwise in a woven fabric. [2] Denier is a unit of measurement used in the textile industry to express the linear density of fibers or yarns.

**Table 4.** Equipment setting parameters.

| Process (16 Attributes) | Name | Description |
|---|---|---|
| Warping (5 attributes) | WARPSPEED | The speed of warping |
| | WARPPRES | The tension of the Warper's Beam |
| | SSTENSION | The tension of monofilament |
| | WARPTENSION | The tension of warping |
| | HYDRATENSION | The tension of hydraulic warping. |
| Sizing (6 attributes) | SIZINGSPEED | The speed of sizing. |
| | SIZINGBPRES | The pressure of sizing |
| | SIZINGATENSION | The tension of sizing (roll out) |
| | SIZINGBTENSION | The tension of sizing (winding) |
| | CONSISTENCY | The density of forming polymeric material |
| | DENSITY | The density of sizing |
| Beaming (4 attributes) | BEAMSPEED | The speed of beaming |
| | BEAMATENSION | The tension of roll-out |
| | BEAMBTENSION | The tension of winding |
| | BEAMTENSION | The tension of beaming |
| Weaving (1 attribute) | WEAVEBTENSION | The tension of weaving |
| Inspection (1 attribute) | QUALITYRATE | The value with range 0 to 1<br>1 = high product quality |

### 3.2. Data Preprocessing

The company under study operates service factories worldwide, with textile machines in each factory using sensors to gather production data. This sensor-collected data, along with the production line information from each factory, are uploaded to the company's ERP system for centralized management. The dataset used in this study was collected from the ERP system of the factory under investigation. The data source comprised five datasheets: warping, sizing, beaming, weaving, and inspection. These datasets contained 37, 46, 31, 25, and 12 columns, with 26,103, 15,950, 37,089, 161,821, and 481,610 records, respectively. The original data contained missing values and complicated content, and the weaving data were inherently complex [34]. Unclean or incorrect data can severely impact subsequent model training and prediction. Therefore, it was crucial to clean the original data [35] and identify their attributes using the extract–transform–load (ETL) method. ETL is a data preprocessing approach in which the original data are corrected, cleaned, and formatted. In this study, we extracted the data based on recommendations from senior experts in the production line of the factory under study. The main and frequently used attributes were extracted from the five datasets to form a new dataset. The extracted data were cleaned by removing incorrect, incomplete, malformed, or redundant data from the datasets. The cleaned data were then transformed into a machine-readable format. The preprocessing steps used in this study are as follows:

(1). Missing value process: The original dataset contained missing values and redundant columns (e.g., remark column). In this study, we filtered out data with missing values and removed redundant columns. The weaving machine could interweave a maximum of four yarns, resulting in many empty columns (variables) in the integrated dataset. For instance, when the weaving machine interweaves two yarns, the columns for the third and fourth production lines in the preparation phase are left empty.

(2). Feature interpolation: Blended fabrics are woven with different warp yarns, and each yarn material is assigned the same job order number. As a result, there are multiple rows of data with the same job order number, each representing a different set of operational parameters for these yarn materials during the preparation phase (warping, sizing, beaming). In this study, given a maximum of four different warp yarns, the original raw data may have contained empty values. Circular interpolation seamlessly interpolates cyclic numerical values. The main objective of circular interpolation is to achieve smooth, continuous interpolation within a numerical set, ensuring a steady trend in a constantly evolving environment. In this study, we filled the empty columns using circular interpolation, adhering to the practical production line schedule rule, as shown in Table 5. We also used circular interpolation to fill in the other empty values.

(3). Data transformation: The original data values were a mixture of descriptions and number values. Therefore, we used regular expressions (regex) to split the aggregated attributes of the production data [e.g., "0010/25N wine red"]) into individual attributes. For instance, the yarn specification attribute was the aggregate of three independent attributes: denier, fiber base, and material.

(4). Outlier Removal: The original data required the outlier removal process, and this study filtered out data beyond three standard deviations in each column.

(5). Normalization: We used min–max scaling to adjust the numerical range of each column between 0 and 1.

**Table 5.** Filling rules for different degrees of interweaving.

| Degree of Interweaving | Integrated Dataset | Fill Rules and Final Dataset |
|---|---|---|
| 1 | {A, 0, 0, 0} | [A, A, A, A] |
| 2 | {A, B, 0, 0} | [A, B, A, B] |
| 3 | {A, B, C, 0} | [A, B, C, A] |
| 4 | {A, B, C, D} | [A, B, C, D] |

Note: A, B, C, and D are the IDs of the production line.

### 3.3. Preliminary Models

In this study, we employed various regression algorithms to predict the production quality. Linear regression served as the foundation method for predicting the target variables. The LASSO addresses overfitting by eliminating redundant features, whereas ridge regression removes unnecessary features. Elastic net regression combines both strategies for enhanced prediction. Each regression algorithm has its own advantages and disadvantages, and we explored the differences in the prediction results of these regression methods in this case study.

#### 3.3.1. Linear Regression

Linear regression is a productive linear approach for modeling the relationships between independent and dependent variables [36]. As a data-driven technique, linear regression identifies a linear equation to predict the positions of future data points. In the context of two-dimensional data, the regression equation appears as a line. The difference between the existing data points and the line of the equation is the error function. The core of regression machine learning algorithms is to minimize this error function by determining the optimal equation. A common error calculation method is the least squares approach, which strategically minimizes the sum of squared errors.

The conceptual underpinning of this algorithm is illustrated in Figure 3 within a two-dimensional framework. Linear regression is used to derive a linear fitting equation. The error is the measured distance between the line and the data points. The primary goal is to minimize this error and identify an equation that closely aligns with the emerging data trend. This resultant equation forms the basis of the linear regression model. When new data points are introduced, the independent variable is input into the model, producing the dependent variable (the predicted value). During training, linear regression algorithms often face issues of underfitting and overfitting. Underfitting occurs when the equation does not adequately minimize the errors with the existing data points (Figure 3a). Conversely, overfitting occurs when the equation overly conforms to the existing data points, failing to capture the true data trend and resulting in substantial errors when new data inputs are introduced (Figure 3c). Figure 3b illustrates the ideal scenario. Ridge regression and LASSO were introduced to address these issues. Both methods operate by introducing an appropriate penalty term to shrink the model's feature parameters, a technique known as regularization, to minimize the error between the predicted and actual values.

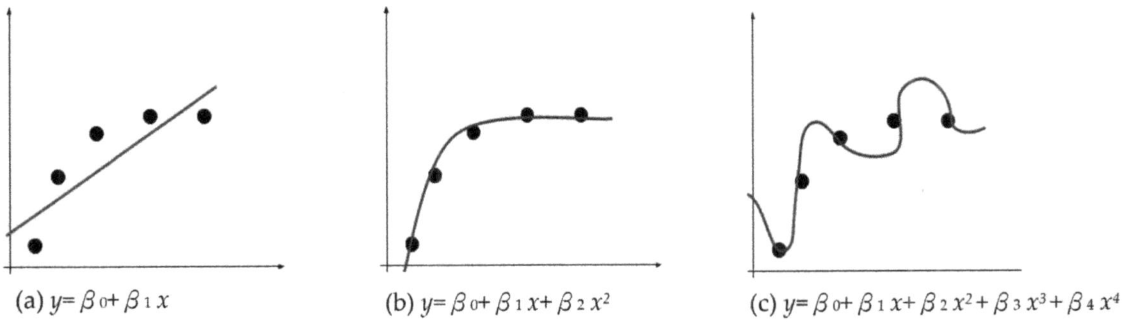


**Figure 3.** Linear regression's three (3) main phenomena. Note: The mathematical relationship between two variables $x$ and y, where $\beta_0$, $\beta_1$, $\beta_2$, and $\beta_3$ are coefficients.

#### 3.3.2. LASSO Regression

LASSO is a regression analysis algorithm [37,38] that employs regularization and variable selection to enhance the prediction accuracy [37]. This is a refined version of linear regression designed to counter potential overfitting in multiple linear regressions. The L1 penalty model inherent in the LASSO regression encourages specific feature coefficients to approach zero, introducing a sense of sparsity within the feature set. LASSO regression strategically uses the L1 penalty for feature selection, effectively addressing overfitting concerns in multiple linear regressions. LASSO is more effective for models with fewer features, resulting in streamlined and less complex structures.

#### 3.3.3. Ridge Regression

Ridge regression estimates the correlations between the predictor and observation variables [39]. Ridge and LASSO differ significantly; the key distinction lies in how they allocate penalties to feature coefficients. Ridge regression introduces a penalty term in the least squares sum, categorizing it under the category of L2 penalty models, whereas LASSO regression employs an L1 penalty term.

#### 3.3.4. Elastic Net Regression

Elastic net regression integrates LASSO and ridge regression to address issues such as multicollinearity and overfitting in linear regression models. This versatile technique in statistics and machine learning offers a dynamic solution to common regression challenges. LASSO enforces sparsity by pushing some coefficients to zero, whereas ridge regression penalizes large coefficients using a squared term. In contrast, the elastic net introduces a unique penalty term that encompasses both L1 (LASSO) and L2 (ridge) regularization. The strength of elastic net regression is its nuanced loss function, which combines the sum of squared errors with penalties from both L1 and L2 regularization. This balanced integration of penalties enables elastic net regression to balance feature selection (sparsity) and coefficient shrinkage. This is particularly useful in scenarios with many correlated features, where a nuanced level of feature selection is crucial. In summary, elastic net regression is a versatile and sophisticated approach that combines the strengths of LASSO and ridge regression to provide a comprehensive solution for regularization in linear regression models [40]. Van (2023) found that elastic net regression exhibits a strong ability to predict nitrogen concentration [41].

### 3.4. *Proposed Query-Based Regression Model*

Exiting machine learning algorithms have demonstrated good computation performances in real case studies. However, some of these algorithms are one-direction procedures from beginning to end [36,37,39,41], and the training models lack bi-direction interaction with the environment. In recent years, generative methods and interactive reinforcement methods have been applied to existing machine learning models to improve their computational performances [42–44]. This study proposes query-based learning, which takes advantage of the generative and interactive reinforcement methods by creating new learning instances and bi-direction interactions in the solution space. Query-based learning [45–49] is an active learning technique that effectively enhances the model performance by incorporating key data points. Figure 4 illustrates the concept of query-based learning using a simple binary classification example. The initial training dataset contains data points A and B from two different categories. A classification boundary R1 is obtained, and the actual classification boundary is R, as shown in Figure 4a. Query-based learning can select additional data points near the current boundary (e.g., data points C and D) to establish a better classification boundary, as shown in Figure 4b. By querying the oracle, the query-based learning identifies the categories of C and D and can then train a new classification boundary R2, as shown in Figure 4c. Query-based learning employs a learning protocol to dictate the method of information accumulation.

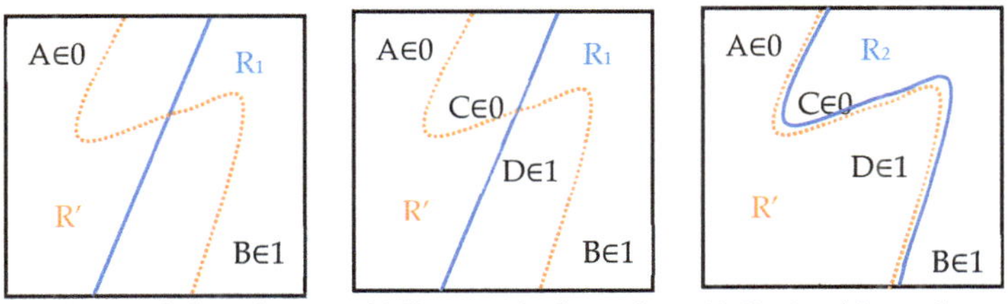


(**a**) Original classification (**b**) Tuning the boundary (**c**) Optimal boundary

**Figure 4.** Query-based learning concept. Note: (**a**) displays the original classification. R1 is the predicted classification boundary, and R′ is the actual boundary, with A and B representing the data points. As illustrated in (**b**), query-based learning can select additional data points near the current boundary (e.g., data points C and D) to establish a better classification boundary. Finally, as shown in (**c**), the new predicted classification boundary R2 can be obtained.

Query-based learning provides a deduction procedure for the training model to correct the computing path and improve the solution-exploring performance of each machine learning model during training. Under the learning protocol of query-based learning, the training model does not passively recognize correlations between data. Instead, it actively learns from a controlled environment by observing and recording the computation results (e.g., evaluation criteria). The query-based approach provides the learning protocol for the machine learning model. Input data can take the form of examples that illustrate the concept to be learned or oracles that indicate whether the data exemplifies the concept. Thus, the training model has the flexibility to use not only existing samples, but also additional samples generated by the oracle to enhance the training model. With the data point of the query set as Q, the oracle responds with Res(Q). The pair (Q, Res(Q)) refers to the queried sample (Figure 5). The sample query method is an incremental approach that dynamically adjusts the sample size extracted from each class [50]. We assume that the learner has the autonomy to inquire about the training samples based on a specific rule, rather than relying on random selection; training samples from decision boundaries yield the best training results. We set point Q such that Res(Q) = 0.5. In the proposed method, we first examine untrained samples to determine whether they are classified correctly. Because the output also indicates the probability of making a correct prediction for the samples, we can easily store these correctly classified samples in a priority line (max-heap). The stored points with the correct predictions are then selected as additional training samples. The learning process is considered complete when either the number of iterations exceeds the given threshold N or the obtained RMSE is below the given threshold RMSE. Figure 5 illustrates the framework of the query-based regression. The step-by-step process of the proposed algorithm is as follows; learners are classified into two types: strong learners (instances with low prediction error) and weaker learners (instances with high prediction error):

- Step 1. Initialize the parameters within the regression model using a random configuration. Let the iteration threshold be N and set the error threshold as the RMSE.
- Step 2. Configure the dataset as $D = \{Res_i \in Sample^n\}$, where n is the number of selected attributes. The data point of the query is set to Q, and the oracle responds with Res(Q). Collect the partial training sample set $DD \subset D$ through stratified random sampling.
- Step 3. Train the regression model employing the training sample set DD. If the error E is below the RMSE or the iteration number exceeds N, exit.
- Step 4. The untrained sample set $(D - DD)$ must be analyzed.
- Step 5. According to the two learning strategies, by adding samples to the DD, the training sample set consists of either (a) samples of the most correct prediction, which is learning from the strong learners [47,49], or (b) samples of the least correct prediction,

for which the weak learners are used to explore the fuzzy or unexplored solution spaces [48].

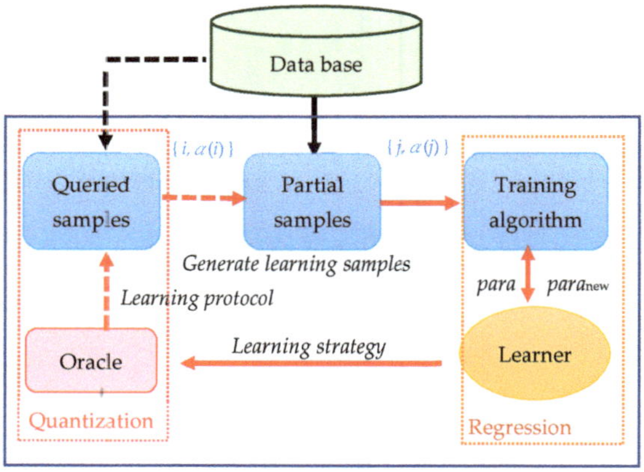


**Figure 5.** Procedure for the proposed query-based method in regression.

The original regression training is one-directional, import data training with an algorithm and a penalty model. Figure 5 demonstrates the interactive ability of query-based learning and the training model being dynamically adjusted according to each iteration outcome. The proposed method can improve the existing model using a self-learning mechanism.

### *3.5. Model Deployment and Application*

Further use of data is crucial for data-driven digital transformation at manufacturing sites. With the rapid development of the IoT and related applications, various types of third-party resources are widely adopted in different cocreation value applications. We deployed the trained model as an API and used a Django-based API server for other applications to access the model using the RESTful format. We developed an email macro to enhance the efficiency of production management. Upon receiving an anomalous product quality prediction, the user can automatically send a reminder email to their staff. Algorithm 1 shows the procedure of the proposed approach. Temporary cursors and counters were used to implement the algorithm. On the basis of the system analysis and design theory, we prebuilt the system context diagram, as shown in Figure 6.

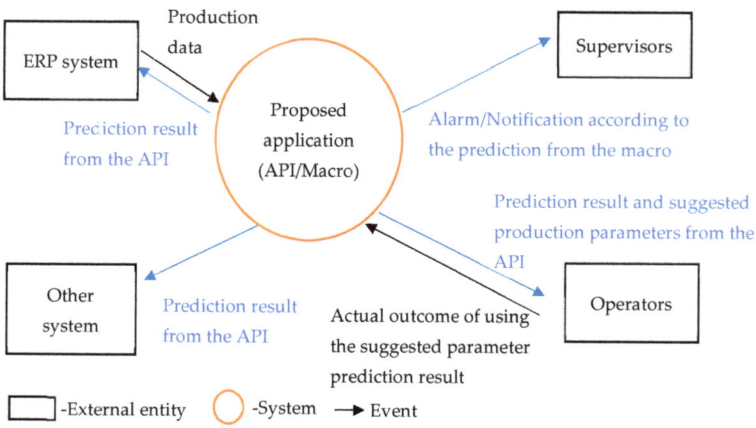


**Figure 6.** Diagram of the system context for the final stage.

**Algorithm 1** Anomaly prediction notification

**Description:** The API exports the result as a worksheet. The cells of the worksheet can be used in more applications, such as predictive maintenance. While users receive the predicted result from the cells of the worksheet and reach the threshold value, the proposed macro can automatically send notifications to other staff.

**Procedure** AlarmNotify (date_cursor, attriValue, attriThres)

```
Initialize:
Let emailAddrQueue be a queue of email address items
Let date_cursor be a temporary cursor of the currently used resource
Let sysDate be the current date time of running the algorithm
Let attriValue be the value of the target attribute
Let attriThres be the threshold value of the target attribute
sysDate = the current system date
   if (date_cursor = null) then
      dat_cursor ← sysDate
   endif
Begin:
   foreach Item of emailAddrQueue do
      if (attriValue ≥ attriThres) then
         date_cursor ← sysDate
         form the notification content
         execute the automatic notification
      endif
   endfor
End Procedure
```

## 4. Experiment

The dataset source for the experiment was the ERP system used in the proposed use case. The dataset involved textile manufacturing processes such as warping, sizing, and beaming. Our aim was to forecast the quality of the product in the textile manufacturing process. In this study, we used the quality of the finished product to tune the solution space of the model. Figure 7 shows the experiment flow, which consisted of two phases: the data analytic phase and the modeling phase. First, the dataset needed to be preprocessed and modeled. This study compared the proposed method with the preliminary methods. Predictive modeling was implemented using various regression techniques, including linear, LASSO, ridge, and elastic net regression. The preprocessed data were imported into the proposed query-based model. Finally, the trained model was deployed as an analytic API that could be integrated into the cloud system for further use, such as for early notification in cases with anomalous predictive outcomes.

### 4.1. Evaluation Criteria

We used the mean squared error (MSE) to evaluate the prediction performance. The evaluation criteria are described below. The MSE (Equation (1)) is a pivotal tool for error estimation; it gauges the disparity between predicted and observed values. In this study, The MSE is a loss function that offers a robust means of estimating and evaluating prediction performance:

$$\mathrm{MSE} = \frac{1}{S}\sum_{i=1}^{S}(\mathrm{var}_i - \mathrm{var'}_i)^2 \qquad (1)$$

where a vector of m predictions is generated from a sample of s data points for all variables, var is the observed value of the predicted variable, and var′ is the predicted value.

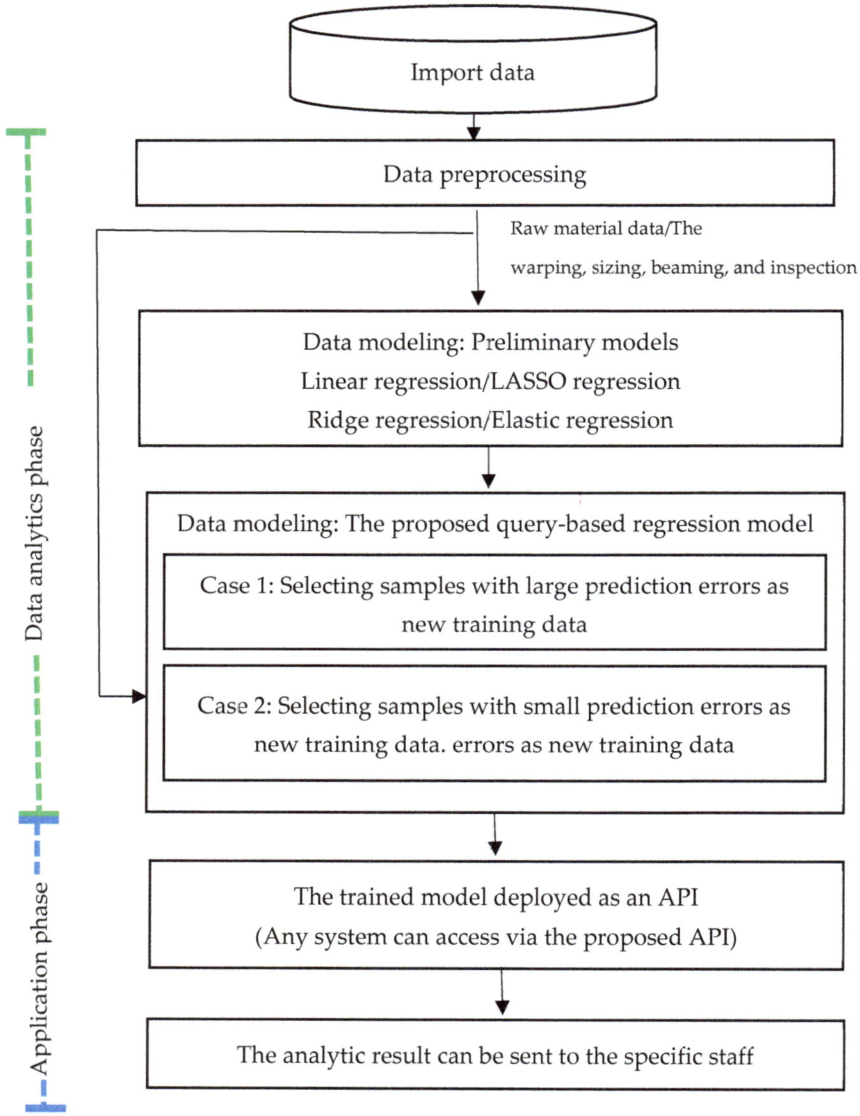


**Figure 7.** Procedure of the experiment.

We used the improvement rate, which compares the original model and the proposed method (Equation (2)), as a criterion to evaluate the performances of the algorithms:

$$\text{improvement rate} = \frac{\text{MSE}_{proposed} - \text{MSE}_{original}}{MSE_{original}} \quad (2)$$

*4.2. Validation*

K-fold cross-validation methodology is commonly used to compare and select predictive models. This involves iteratively resampling a dataset and recursively dividing the original set into multiple subgroups. This study employed 10-fold cross-validation to assess the trained model while mitigating potential biases stemming from data partitioning.

The overarching concept of 10-fold cross-validation is the recursive partitioning of a dataset, culminating in the computation of average values across these partitions. The procedure validation procedure is as follows [51].

(1). Data Division: The initial dataset is segmented into 10 subsets or folds, each maintaining an approximately uniform size.
(2). Model Training and Assessment: The model undergoes 10 training cycles, where each training session employs nine folds for training and reserves one fold for validation. During the initial iteration, the model is trained on Folds 2 to 10 and evaluated on Fold 1. Subsequently, in the following iterations, training encompasses Folds 1 and 3 to 10, with evaluation on Fold 2, and so forth. This process repeats until each fold serves as a validation set precisely once.
(3). Calculation of Performance Metrics: Evaluation criteria are computed for each iteration, generating 10 distinct performance scores.
(4). Compilation of Overall Performance Metrics: The 10 performance metrics are commonly averaged, yielding a singular performance score that encapsulates the model's comprehensive performance. This average performance score provides a more resilient estimate of the model's anticipated performance on unseen data.

## 5. Result

This section shows the predictive results of the regression-related approach and the proposed query-based approach in our case study; 60% of the preprocessed dataset was used as the training dataset, and 40% of the preprocessed dataset was used as the testing dataset. In this study, 10-fold cross-validation was used to evaluate the trained model to avoid possible bias from data segmentation. A total of 28 training attributes comprised yarn specification, warping, sizing, beaming, and weaving, and one predicted target was the quality rate of the product. In the model setting, the LASSO regression was trained with L1 regularization, and the alpha parameter value was set to 1. The ridge was trained with L2 regulation, and the alpha parameter value was set to 1. The elastic net regression model was trained with L1 regularization with alpha = 1.0. The MSE was the estimated error between the real quality variable and the predicted variable. The average MSE value of the original regression (Linear/LASSO/Ridge/Elastic Net) was 0.02, and that of the proposed method was 0.0153. Compared with unused query-based learning, the regression model that used query-based learning had an improvement rate of 22%. This result is beneficial for predictive maintenance in manufacturing (Table 6).

**Table 6.** Prediction results for the production quality.

| Model | MSE |
|---|---|
| Linear Regression | 0.0211 |
| LASSO Regression | 0.0191 |
| Ridge Regression | 0.0193 |
| Elastic Net Regression | 0.0190 |
| Query-Based Regression | 0.0153 |

Query-based learning is an active learning technique that effectively enhances the overall performance of a model by retraining it with the addition of new data. We compared various learning strategies using the following procedure: from 5898 data points, 60% were selected as the training data, and the remaining 40% served as the testing data. New data suitable for learning were identified from the neighboring points of the testing data. Two learning strategies were employed to generate new data in this experiment: selecting samples with large prediction errors as new training data and selecting samples with small prediction errors as new training data. Two learning strategies and a nonuse strategy (randomly generating new training data) were used in the experiment. The ability of the three methods to strengthen model prediction in query-based learning was compared. The results indicate that providing learning guidance by retraining using poorly performing samples yielded the best outcomes (Figure 8).

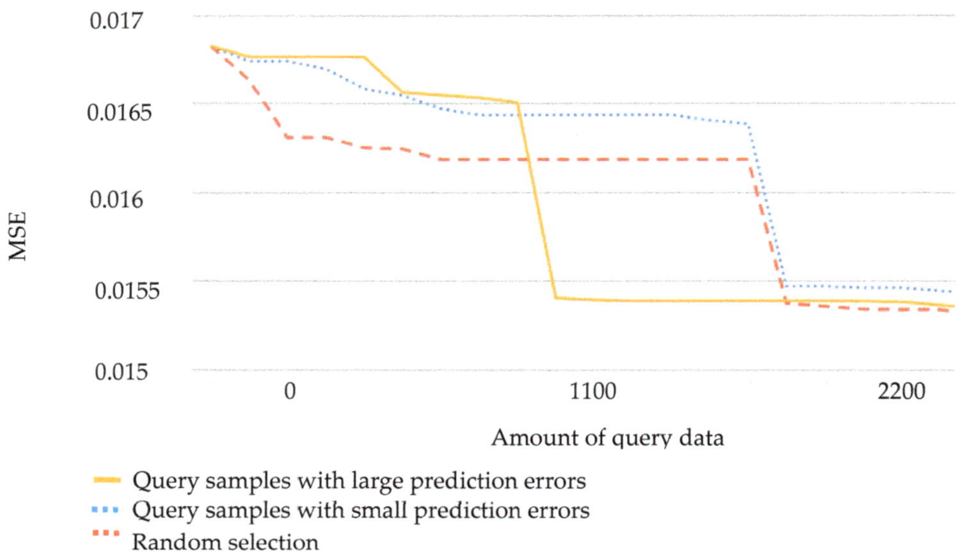


**Figure 8.** MSE of query-based regression in production quality prediction.

The trained model was deployed as a RESTful API that can be used in edge- or cloud-based analytics for the prediction of process parameters. The process of an API call starts when it is transmitted from the client side to the API endpoint. This endpoint, which is the final destination, is mainly found in web applications and servers. Consider, for example, a mobile client that initiates an API call. This call travels toward the specified API endpoint, which is often represented by the server. Upon reaching the server side, an API call undergoes a sequence of actions: it is processed, a request is executed, and a response is dispatched back to the client side. This dynamic interaction constitutes the essence of API calls and allows communication between the client side and the API endpoint. In this study, submission was detected by the server and automatically imported into the trained model based on the API message (Figure 9). Next, the server sends the predicted results to the client as a response.

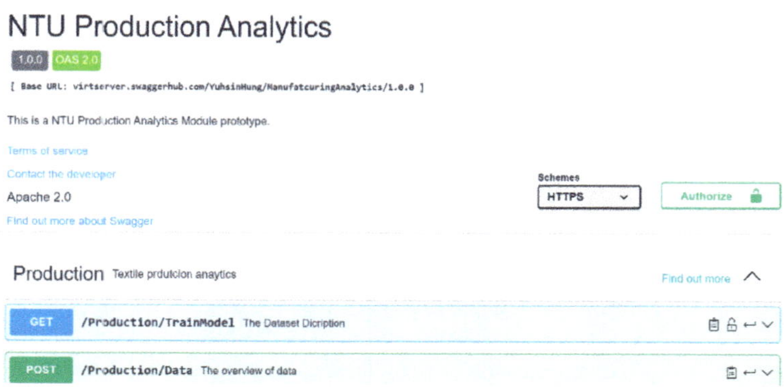


**Figure 9.** The trained model was deployed as an API for further use.

Numerous ERP systems offer features that enable users to export data seamlessly to Excel, thus opening avenues for in-depth analysis, customized reporting, and data manipulation. This is particularly beneficial for individuals who find Excel more intuitive for specific tasks. In contrast, most ERP systems facilitate data importation from Excel files. This functionality is valuable when users need to perform bulk updates or input data using the familiar, user-friendly interface of Excel for data preparation. In this study, we used Visual Basic for Applications, which is an office automation tool that automatically executes Excel's features using macros. The analytic results are then exported as a worksheet for further use. After receiving the predicted anomalies from the worksheet, a user can automatically send a notification to other staff. Figure 10 shows how emails are correctly sent to the user using the proposed anomaly notification macro.

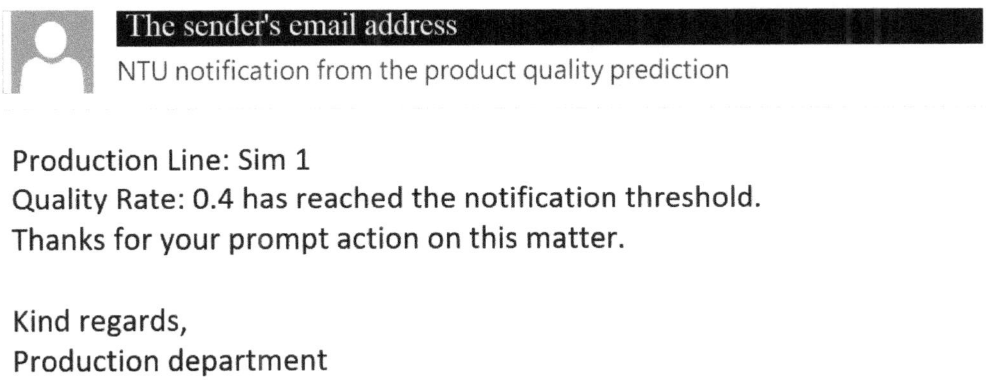


**Figure 10.** Result of early anomaly notification.

The main finding of this study is that the proposed method can predict product quality with promising advantages. Linear regression showed a better prediction performance than the other algorithms for product quality prediction. According to previous studies, query-based learning exhibits an excellent computing performance because of its advantages in generating new learning samples and training the model with different parameters [42–50]. This study obtained similar results. The proposed query-based regression model can search a larger solution space than the original preliminary function [42–50]. Thus, it exhibited a good prediction ability in the case study. Finally, the trained model can be widely applied for further use through the API and notification macro. The seamless flow of transmission data streams, which facilitates communication from edge-side equipment to cloud-side networks, is a pivotal catalyst for the continuous progress of the IoT. This interconnected data exchange not only forms the backbone of the IoT infrastructure, but also acts as a driving force behind its ongoing evolution [49]. This study fulfills the communication requirement between edge-side equipment and cloud-side applications. The efficacy and efficiency of the proposed API and macro fundamentally contribute to the dynamic landscape of the IoT, propelling advancements in data analytics and real-time decision making. Future studies will be conducted to analyze different data sources from other textile manufacturing processes, computerized machines, and manufacturing execution systems. This information will help determine the correlation between product quality and different manufacturing properties. Furthermore, the developed approach will be implemented and validated in an industrial case study and smart maintenance case studies to demonstrate its efficiency and capacity to support maintenance engineers and product inspectors.

## 6. Discussion

One key transformation for textile factories is their move to fine textile products, which usually means high quality, high customization, and delivery in a short time. However, the trial-and-error method for determining proper production parameters is time consuming and resource wasting. The proposed method has promising advantages in predicting product quality in textile manufacturing. Previous studies have stated that query-based learning leads to excellent computing performances because it has the advantage of tuning models to evolve dynamically [45–49]. The current study obtained similar results. The proposed query-based regression model searches a larger solution space than the original preliminary function [47,48], thus exhibiting a good prediction performance in our case study. The trained model can be further used through the API and notification macro. In this study, the seamless flow of transmission data streams facilitates communication from edge-side equipment to cloud-side networks. This interconnected data exchange constitutes the backbone of the IoT infrastructure and serves as a driving force behind its ongoing evolution [16–18,52].

### 6.1. Limitations of the Study

This study focused on investigating the influence of using the proposed query-based learning method with regression-related methods. In addition, the sizes and attributes of the industrial datasets limit this study. However, each industry has its own unique facilities, devices, products, and manufacturing processes. The study case was a midstream textile product manufacturer. The more manufacturing scenarios considered in a production line, the more data acquired in a factory.

### 6.2. Future Study

In the cloud-based environment, safeguarding sensitive manufacturing data in the realm of data security poses numerous challenges for organizations. The primary concerns revolve around the potential for data breaches arising from weak authentication and compromised credentials. Additional threats emerge from inadequate access controls and poorly configured permissions, which may result in unauthorized access to critical data while the user uses the API to access data analytics. Therefore, the proposed API uses authorized login account validation. In the future, API access will use the single-sign-in method to reduce the risk of unauthorized access, and the data value can be encrypted for data privacy protection. In addition, the uploaded data overview and confirmation can address incorrect data injection attacks, which may influence the prediction results.

## 7. Conclusions

For complex textile manufacturing, the production parameter setting in the manufacturing process is based on the operator's experience or trial-and-error testing. Human error may influence the quality of the product. To improve production efficiency, this study proposes a case study that demonstrates how to learn from existing data and integrate them as an application service into existing cloud infrastructure. This study has two findings: data analytics and system design. From the viewpoint of data analytics, we used query-based learning to reinforce model training. The proposed query-based method exhibits better prediction than the original model in high-quality production manufacturing because the proposed method adjusts the solution space based on dynamic integration with the solution space. From the viewpoint of system design, the trained module was deployed as an API for online analytics and further use. To add value to the existing environment, this study proposed an API that can be integrated with office automation. Thus, we developed an auto-notification macro for the user to obtain notifications when a predicted anomaly event occurs. Finally, the case results show that the proposed approach can preventively assist in product quality maintenance using data analytics and cloud technologies.

**Author Contributions:** Conceptualization, R.-I.C.; Methodology, R.-I.C., J.-Y.L. and Y.-H.H.; Software, R.-I.C., J.-Y.L. and Y.-H.H.; Validation, J.-Y.L.; Formal analysis, J.-Y.L.; Investigation, R.-I.C., J.-Y.L. and Y.-H.H.; Resources, R.-I.C.; Data curation, R.-I.C.; Writing—original draft, R.-I.C. and Y.-H.H.; Writing—review & editing, Y.-H.H.; Supervision, R.-I.C.; Project administration, R.-I.C. and Y.-H.H.; Funding acquisition, R.-I.C. and Y.-H.H. All authors have read and agreed to the published version of the manuscript.

**Funding:** This research was funded by the National Science and Technology Council, grant number NSTC 111-2622-8-002-029, 111-2221-E-224-033-MY2, and MOE "Teaching Practice Research" Subsidies Program, grant number PBM1110139.

**Institutional Review Board Statement:** Not applicable.

**Informed Consent Statement:** Not applicable.

**Data Availability Statement:** No new data were created or analyzed in this study. Data sharing is not applicable to this article.

**Acknowledgments:** The authors gratefully thank the financial support from the National Science and Technology Council, Taiwan (NSTC 111-2622-8-002-029, 111-2221-E-224-033-MY2).

**Conflicts of Interest:** The authors declare no conflicts of interest.

## References

1. Kritchanchai, D.; Wasusri, T. Implementing supply chain management in Thailand textile industry. *Int. J. Inf. Syst. Logist. Manag.* **2007**, *2*, 107–116.
2. Taiwan Textile Federation. *The Overview of Taiwan's Textile Industry in 2022*; Taiwan Textile Federation: Taipei City, Taiwan, 2023; pp. 1–3.
3. Khamrakulova, Z. Improving product quality by improving the working body of the spinning machine. In Proceedings of the International Conference on Developments in Education, Sciences and Humanities, Livorno, Italy, 15 June 2022; pp. 21–24.
4. Drean, J.Y.; Decrette, M. Weaving Preparation. In *Advanced Weaving Technology*; Springer International Publishing: Cham, Switzerland, 2022; pp. 3–80.
5. Goswami, B.C.; Anandjiwala, R.D.; Hall, D. *Textile Sizing*; CRC Press: Boca Raton, FL, USA, 2004.
6. Shaker, K.; Umair, M.; Ashraf, W.; Nawab, Y. Fabric manufacturing. *Phys. Sci. Rev.* **2016**, *1*, 20160024.
7. Kraus, S.; Jones, P.; Kailer, N.; Weinmann, A.; Chaparro-Banegas, N.; Roig-Tierno, N. Digital Transformation: An Overview of the Current State of the Art of Research. *SAGE Open* **2021**, *11*, 21582440211047576. [CrossRef]
8. Vial, G. Understanding digital transformation: A review and a research agenda. *J. Strateg. Inf. Syst.* **2021**, *28*, 118–144. [CrossRef]
9. Schwertner, K. Digital transformation of business. *Trakia J. Sci.* **2017**, *15*, 388–393. [CrossRef]
10. Stegmann, P.; Nagel, S.; Ströbel, T. The digital transformation of value co-creation: A scoping review towards an agenda for sport marketing research. *Eur. Sport Manag. Q.* **2023**, *23*, 1221–1248. [CrossRef]
11. Bathrinath, S.; Dhanasekar, M.; Koppiahraj, K.; Priyanka, R. A Fuzzy AHP Perspective on Improving Yarn Winding Productivity in Textile Industry. In Proceedings of the International Conference on Information and Communication Technology for Intelligent Systems, Ahmedabad, India, 27–28 April 2023; Springer Nature: Singapore, 2023; pp. 555–562.
12. Amjad, A.I.; Regar, M.L. Fabric preparatory. In *Textile Calculation*; Woodhead Publishing: Cambridge, UK, 2023; pp. 171–195.
13. Şuteu, M.D.; Baban, C.F.; Baban, M.; Dragomir, G.; Toth, K.E. *Predictive Maintenance of the Automated Sewing Machines in Textile Industry*; Universitatea din Oradea: Oradea, Romania, 2019.
14. Baban, C.F.; Baban, M.; Suteu, M.D. Using a fuzzy logic approach for the predictive maintenance of textile machines. *J. Intell. Fuzzy Syst.* **2016**, *30*, 999–1006. [CrossRef]
15. Baban, M.; Baban, C.F.; Suteu, M.D. Maintenance Decision-Making Support for Textile Machines: A Knowledge-Based Approach Using Fuzzy Logic and Vibration Monitoring. *IEEE Access* **2019**, *7*, 83504–83514. [CrossRef]
16. Wang, S.; Wan, J.; Li, D.; Zhang, C. Implementing Smart Factory of Industrie 4.0: An Outlook. *Int. J. Distrib. Sens. Netw.* **2016**, *12*, 3159805. [CrossRef]
17. Xu, L.D.; He, W.; Li, S. Internet of Things in Industries: A Survey. *IEEE Trans. Ind. Inform.* **2014**, *10*, 2233–2243. [CrossRef]
18. Aydin, G.; Hallac, I.R.; Karakus, B. Architecture and Implementation of a Scalable Sensor Data Storage and Analysis System Using Cloud Computing and Big Data Technologies. *J. Sens.* **2015**, *2015*, 834217. [CrossRef]
19. Agarwal, R.; Dhar, V. Editorial—Big Data, Data Science, and Analytics: The Opportunity and Challenge for IS Research. *Inf. Syst. Res.* **2014**, *25*, 443–448. [CrossRef]
20. Bullón Pérez, J.; González Arrieta, A.; HernándezEncinas, A.; Queiruga-Dios, A. Industrial Cyber-Physical Systems in Textile Engineering. In Proceedings of the International Joint Conference SOCO'16-CISIS'16-ICEUTE'16. SOCO 2016, ICEUTE 2016, CISIS 2016. Advances in Intelligent Systems and Computing, San Sebastián, Spain, 19–21 October 2016; Graña, M., López-Guede, J., Etxaniz, O., Herrero, Á., Quintián, H., Corchado, E., Eds.; Springer: Cham, Switzerland, 2017; Volume 527.
21. Davidson, S. Cyber-Physical System Design with Sensor Networking Technologies. *IEEE Des. Test* **2017**, *34*, 105–107. [CrossRef]

22. Ding, H.; Li, C. Cyber-Physical System and Its Application in Textile and Chemical Fiber Enterprises. *Open J. Soc. Sci.* **2017**, *5*, 352–360. [CrossRef]
23. Saggiomo, M.; Kemper, M.; Gloy, Y.; Gries, T. Weaving machine as cyber-physical production system: Multi-objective self-optimization of the weaving process. In Proceedings of the 2016 IEEE International Conference on Industrial Technology (ICIT), Taipei, Taiwan, 14–17 March 2016. [CrossRef]
24. Ślusarczyk, B.; Haseeb, M.; Hussain, H.I. Fourth industrial revolution: A way forward to attain better performance in the textile industry. *Eng. Manag. Prod. Serv.* **2019**, *11*, 52–69. [CrossRef]
25. Pivoto, D.G.S.; de Almeida, L.F.F.; Da Rosa Righi, R.; Rodrigues, J.J.P.C.; Lugli, A.B.; Alberti, A.M. Cyber-physical systems architectures for industrial internet of things applications in Industry 4.0: A literature review. *J. Manuf. Syst.* **2021**, *58*, 176–192. [CrossRef]
26. Eltayib, H.E.; Ali, A.H.M.; Ishag, I.A. The Prediction of Tear Strength of plain weave fabric Using Linear Regression Models. *Int. J. Adv. Eng. Res. Sci.* **2016**, *3*, 151–154. [CrossRef]
27. Lu, D.; Yu, W. Predicting the tensile strength of single wool fibers using artificial neural network and multiple linear regression models based on acoustic emission. *Text. Res. J.* **2021**, *91*, 533–542. [CrossRef]
28. Sarah, M.; Adolphe, D.C.; Boubaker, J. Prediction of Sewing Thread Consumption for Over-Edge Stitches Class 500 Using Geometrical and Multi-Linear Regression Models. *Autex Res. J.* **2021**, *21*, 150–162. [CrossRef]
29. Tang, X.; Kong, D.; Yan, X. Multiple regression analysis of a woven fabric sound absorber. *Text. Res. J.* **2019**, *89*, 855–866. [CrossRef]
30. Hoque, M.M.U.; Ahmed, T.; Shams, T.; Islam, I. Predicting bursting strength of single jersey 100% cotton plain knitted fabrics using different machine learning models. *World J. Adv. Res. Rev.* **2022**, *16*, 283–293. [CrossRef]
31. Rabaca, V.; Pereira, J.M.; Basto, M. Logit Ridge and Lasso in predicting business failure. *Glob. J. Account. Econ. Res.* **2023**, *4*, 33–46.
32. Gorgül, A.Ü.; Çom, M.; Sultanoğlu, S. Kernel Ridge Regression Based Modelling and Anomaly Detection for Temperature Control in Textile Dyeing Processes. In Proceedings of the27th International Conference on System Theory, Control and Computing (ICSTCC), Timisoara, Romania, 11–13 October 2023; pp. 392–397.
33. Medina, H.; Peña, M.; Siguenza-Guzman, L.; Guamán, R. Demand Forecasting for Textile Products Using Machine Learning Methods. In Proceedings of the International Conference on Applied Technologies, Quito, Ecuador, 27–29 October 2021; Springer International Publishing: Cham, Switzerland, 2021; pp. 301–315.
34. Kovačević, S.; Schwarz, I. Weaving Complex Patterns—From Weaving Looms to Weaving Machines. In *Cutting Edge Research in Technologies*; InTech: Rijeka, Croatia, 2015. [CrossRef]
35. Greenwood-Nimmo, M.; Shields, K. An Introduction to Data Cleaning Using Internet Search Data. *Aust. Econ. Rev.* **2017**, *50*, 363–372. [CrossRef]
36. Stanton, J.M. Galton, Pearson, and the Peas: A Brief History of Linear Regression for Statistics Instructors. *J. Stat. Educ.* **2001**, *9*, 11910537. [CrossRef]
37. Tibshirani, R. Regression shrinkage and selection via the lasso. *J. R. Stat. Soc. Ser. B Stat. Methodol.* **1996**, *58*, 267–288. [CrossRef]
38. Meier, L.; Van De Geer, S.; Bühlmann, P. The group lasso for logistic regression. *J. R. Stat. Soc. Ser. B Stat. Methodol.* **2008**, *70*, 53–71. [CrossRef]
39. Vigneau, E.; Devaux, M.F.; Qannari, E.M.; Robert, P. Principal component regression, ridge regression and ridge principal component regression in spectroscopy calibration. *J. Chemom.* **1997**, *11*, 239–249. [CrossRef]
40. Balamurugan, P. Large-Scale Elastic Net Regularized Linear Classification SVMs and Logistic Regression. In Proceedings of the 2013 IEEE 13th International Conference on Data Mining, Dallas, TX, USA, 7–10 December 2013. [CrossRef]
41. Van, D.M. Predicting Nitrogen Concentrations Using Machine Learning Techniques. Ph.D. Thesis, Tilburg University, Tilburg, The Netherlands, 2021.
42. Ohno, H. Auto-encoder-based generative models for data augmentation on regression problems. *Soft Comput.* **2020**, *24*, 7999–8009. [CrossRef]
43. Ayoub, A.; Jia, Z.; Szepesvari, C.; Wang, M.; Yang, L. Model-based reinforcement learning with value-targeted regression. In Proceedings of the International Conference on Machine Learning, Virtual, 13–18 July 2020; pp. 463–474.
44. He, J.; Zhao, H.; Zhou, D.; Gu, Q. Nearly minimax optimal reinforcement learning for linear Markov decision processes. In Proceedings of the International Conference on Machine Learning, Honolulu, HI, USA, 23–29 July 2023; pp. 12790–12822.
45. Chang, R.I. Disease Diagnosis Using Query-Based Neural Networks. In Proceedings of the Advances in Neural Networks—ISNN 2005 Lecture Notes in Computer Science, Chongqing, China, 30 May–1 June 2005; pp. 767–773. [CrossRef]
46. Chang, R.-I.; Hsu, H.-M.; Lin, S.-Y.; Chang, C.-C.; Ho, J.-M. Query-Based Learning for Dynamic Particle Swarm Optimization. *IEEE Access* **2017**, *5*, 7648–7658. [CrossRef]
47. Chang, R.; Huang, C.; Lai, L.; Lee, C. Query-Based Machine Learning Model for Data Analysis of Infrasonic Signals in Wireless Sensor Networks. In Proceedings of the 2nd International Conference on Digital Signal Processing—ICDSP 2018, Tokyo, Japan, 25–27 February 2018. [CrossRef]
48. Chang, R.-I.; Lin, S.-Y.; Hung, Y. Particle swarm optimization with query-based learning for multi-objective power contract problem. *Expert Syst. Appl.* **2012**, *39*, 3116–3126. [CrossRef]
49. Lai, L.; Chang, R.; Kouh, J. Mining Data by Query-Based Error-Propagation. In Proceedings of the Lecture Notes in Computer Science Advances in Natural Computation 2005, Changsha, China, 27–29 August 2005; pp. 1224–1233. [CrossRef]
50. Ratsaby, J. Incremental learning with sample queries. *IEEE Trans. Pattern Anal. Mach. Intell.* **1998**, *20*, 883–888. [CrossRef]

51. Jung, Y. Multiple predicting *K*-fold cross-validation for model selection. *J. Nonparametric Stat.* **2018**, *30*, 197–215. [CrossRef]
52. Rehman, M.H.U.; Ahmed, E.; Yaqoob, I.; Hashem, I.A.T.; Imran, M.; Ahmad, S. Big Data Analytics in Industrial IoT Using a Concentric Computing Model. *IEEE Commun. Mag.* **2018**, *56*, 37–43. [CrossRef]

*sensors*

MDPI

*Article*

# Sybil Attacks Detection and Traceability Mechanism Based on Beacon Packets in Connected Automobile Vehicles

**Yaling Zhu [1], Jia Zeng [1], Fangchen Weng [1], Dan Han [1], Yiyu Yang [1,2], Xiaoqi Li [1] and Yuqing Zhang [1,2,*]**

1 The School of Cyberspace Security, Hainan University, Haikou 570208, China; zhuyl@nipc.org.cn (Y.Z.); zengj@nipc.org.cn (J.Z.); hand@nipc.org.cn (D.H.); yangyy@nipc.org.cn (Y.Y.); csxqli@gmail.com (X.L.)
2 The National Computer Intrusion Protection Center, University of Chinese Academy of Sciences, Beijing 101408, China
* Correspondence: zhangyq@ucas.ac.cn

**Abstract:** Connected Automobile Vehicles (CAVs) enable cooperative driving and traffic management by sharing traffic information between them and other vehicles and infrastructures. However malicious vehicles create Sybil vehicles by forging multiple identities and sharing false location information with CAVs, misleading their decisions and behaviors. The existing work on defending against Sybil attacks has almost exclusively focused on detecting Sybil vehicles, ignoring the traceability of malicious vehicles. As a result, they cannot fundamentally alleviate Sybil attacks. In this work we focus on tracking the attack source of malicious vehicles by using a novel detection mechanism that relies on vehicle broadcast beacon packets. Firstly, the roadside units (RSUs) randomly instruct vehicles to perform customized key broadcasting and listening within communication range. This allows the vehicle to prove its physical presence by broadcasting. Then, RSU analyzes the beacon packets listened to by the vehicle and constructs a neighbor graph between the vehicles based on the customized particular fields in the beacon packets. Finally, the vehicle's credibility is determined by calculating the edge success probability of vehicles in the neighbor graph, ultimately achieving the detection of Sybil vehicles and tracing malicious vehicles. The experimental results demonstrate that our scheme achieves the real-time detection and tracking of Sybil vehicles, with precision and recall rates of 98.53% and 95.93%, respectively, solving the challenge of existing detection schemes failing to combat Sybil attacks from the root.

**Keywords:** CAVs; Sybil attacks; traceability; attacker; security

**Citation:** Zhu, Y.; Zeng, J.; Weng, F.; Han, D.; Yang, Y.; Li, X.; Zhang, Y. Sybil Attacks Detection and Traceability Mechanism Based on Beacon Packets in Connected Automobile Vehicles. *Sensors* **2024**, *24*, 2153. https://doi.org/10.3390/s24072153

Academic Editors: Behnam Mobaraki and Jose Turmo

Received: 6 February 2024
Revised: 17 March 2024
Accepted: 26 March 2024
Published: 27 March 2024



## 1. Introduction

Traffic congestion and accidents are common problems faced in metropolitan areas In the United States, according to the statistics of the National Highway Traffic Safety Administration (NHTSA) [1], billions of traffic waiting times cause the unnecessary consumption of more than 3.1 billion gallons of fuel each year. On the other hand, about 35,000 people are killed, and nearly 4 million people are injured due to traffic accidents, and the average annual economic loss is more than USD 836 billion. Therefore, governments and researchers actively seek solutions, such as more intelligent roads and traffic signals With the expansion and extension of Internet applications and the support of the new generation of information technology represented by 5G, CAV technology can effectively solve the above problems. According to a report by Allied Market Research, the global self-driving car market is estimated to be USD 54.23 billion in 2019 and is expected to reach USD 556.67 billion by 2026 [2].

However, as CAVs grow, the cybersecurity risks they face are becoming more pronounced. The Upstream 2022 report shows that over 900 CAV cybersecurity incidents occurred in 2021 alone [3]. This grim reality has drawn the close attention of many researchers and prompted them to explore in depth the security threats faced by CAVs to propose effective prevention and response strategies [4–11]. Among the many threats,

Sybil attacks [12] are particularly challenging. During information sharing and cooperative driving by CAVs, attackers forge multiple identities or location information and launch this relatively low-cost attack by broadcasting false data. In this attack mode, a malicious vehicle can control or influence a large number of normal nodes by using only a small number of nodes, posing a serious hazard to the CAV system. Therefore, it is considered one of the top threats in Telematics. In order to deal with the potential threat of Sybil attacks in CAV networks, we must conduct more in-depth research on attack detection and countermeasure strategies. Benadla et al. have provided a brief description of the impact of Sybil attacks on vehicular networks and a detailed categorization of Sybil attack detection methods in VANETs [13]. These studies provide valuable references but still need to be further explored in depth in order to establish a more complete and effective defense mechanism.

During information sharing and cooperative driving between CAVs, an attacker may create and broadcast false data and thus launch a Sybil attack. In such attacks, attackers aim to control or influence a large number of normal nodes using only a small number of nodes by forging multiple identities or location information. In the CAV networks, a malicious vehicle is a physical vehicle that can obtain multiple legitimate identities illegally, and it is fully capable of launching Sybil attacks by forging vehicle location information and simulating the operating characteristics of normal vehicles. As a result, the road condition monitoring and decision-making of the CAV systems will be easily confused and misled, causing DOS attacks [14] on the CAV systems, even leading to traffic accidents, casualties, and property losses. As shown in Figure 1, before the Sybil attack is launched, CAVs know that the current traffic is relatively smooth through information sharing among them. As depicted in Figure 2, malicious vehicles broadcast traffic packets with false information to interfere with the CAV driving status. This leads some vehicles to misinterpret traffic conditions as being more congested, resulting in reduced speed or lane changes, thus causing inefficiency in the entire traffic fleet. Once a normal vehicle trusts a Sybil vehicle, a malicious vehicle can successfully mislead a normal vehicle. Therefore, the detection of Sybil attacks in CAVs is necessary.

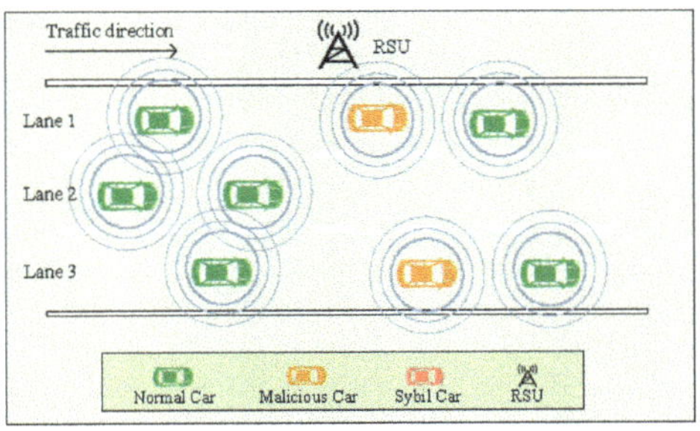


**Figure 1.** Before the Sybil attacks are launched.

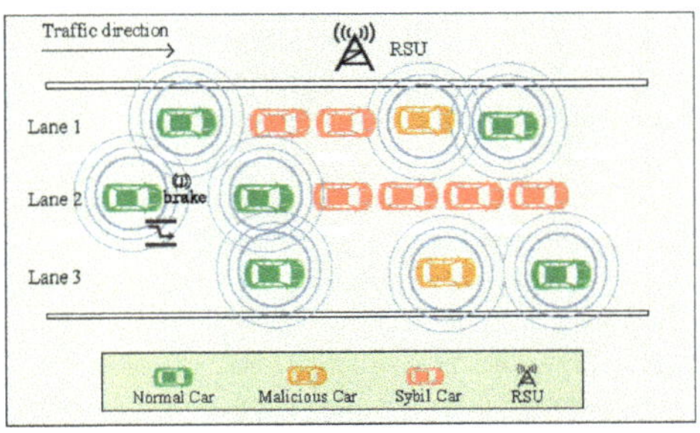


**Figure 2.** After the Sybil attacks are launched.

Once the normal vehicles trust the Sybil vehicles, malicious vehicles can successfully mislead normal vehicles. Therefore, the detection of Sybil attacks in CAVs is necessary. In addition, the collusion [15] and separation behaviors between malicious and Sybil vehicles make malicious vehicles exhibit similar behavioral characteristics as normal vehicles. This is the main reason why many mitigation solutions against Sybil attacks can only detect Sybil vehicles but cannot accurately track malicious vehicles [16–21]. Notably, malicious vehicles are the source of Sybil attacks. Yang et al. organized Sybil attacks regarding the traceability of malicious vehicles [22], pointing out that existing schemes have apparent drawbacks for the traceability of malicious vehicles. This allows malicious vehicles to continue to exist in the vehicular network and continue Sybil attacks. Zhang et al. [23] also suggested that accurately tracking malicious vehicles is one of the pressing challenges in solving Sybil attacks. Currently, most of the schemes rely on RSSI values to accomplish the detection. Yuan proposed an edge computing-based Sybil detection scheme [24], where the vehicle sends a control packet to two nearby edge nodes. The nodes use the Jake model to calculate the RSSI, transmit it to each other, and determine the range of normal RSSI ratios through multiple rounds of computation to detect malicious vehicles. However, RSSI values are susceptible to real-world scenario factors such as traffic, attacker density, etc. Krishnan et al. proposed a collaborative strategy to detect malicious vehicles [25]. Since malicious vehicles need to maintain more Sybil vehicles, they are more likely to have the longest list of nearest vehicles. However, this scheme mainly targets Sybil attacks in the presence of a single malicious vehicle, and its detection rate decreases once there is a conspiracy between malicious vehicles. Rakhi et al. proposed a Sybil detection method based on LCSS similarity computation and RSSI time-series variation point detection [26]. Without power control, malicious vehicles are detected by finding the similarity of RSSI nodes. However, when the distance between vehicles is relatively small, RSSI sequences received from normal and malicious nodes show high similarity, and it is difficult to identify malicious nodes from normal nodes. To address the single detection factor limitation, Chen et al. proposed a multi-scale data fusion detection framework for the Sybil attack [27]. By acquiring BSM, map data, and sensor data, the detection of malicious vehicles is accomplished using machine learning classification models. However, the limitation is that the framework can only be used in the exact location where the Sybil attack occurred. Secondly, the detectors laid on the roads are costly. Finally, since the scheme incorporates machine learning, when the attack samples in the training dataset with low attack density are much smaller than normal samples, it will prevent the model from learning the attack behavior sufficiently, resulting in lower accuracy and recall. While attempting to address the problem of malicious vehicle tracing, these works have limitations, as they mainly rely on RSSI values or machine learning that requires extensive training. In contrast, our research is based on a vehicle broadcast beacon packet detection mechanism that is not susceptible to traffic and attack density and only requires a few training samples. We can only stop Sybil attacks from launching Sybil attacks at their source by accurately identifying malicious vehicles.

We observe that a key feature of Sybil attacks is that Sybil vehicles are essentially beacon data packets forged by malicious vehicles. These fake vehicles lack the broadcasting and listening capabilities of real ones and must rely on malicious vehicles to mimic the behavior of normal ones. Given this, we propose a beacon packet-based detection method to trace malicious vehicles based on trusted RSUs combined with information interactions performed between CAVs. Firstly, the roadside unit (RSU) enables vehicles within the detection range to communicate with each other in a point-to-multipoint (PMP) manner and exchange a specific field in the beacon data packet to prove their physical existence. We refer to this specific field in the beacon packet as *Key*. If the exchange of *Key* is successful, it is considered that both communication parties are physical nodes, and we build an edge for them in the adjacency matrix. Note that since Sybil vehicles do not have actual broadcasting and listening capabilities, this will result in the inability to establish edges between Sybil vehicles and normal vehicles. Therefore, we can detect Sybil vehicles simply by traversing the neighbor relationships in the adjacency matrix. At this point, the RSU will assign the Sybil vehicle a *Key* that identifies its identity, and the Sybil vehicle will share the *Key* with the malicious vehicles. In order to maintain the existence of Sybil vehicles, malicious vehicles will then help broadcast the *Key* on behalf of Sybil vehicles, making normal vehicles believe that Sybil vehicles also have the broadcasting capability, mistakenly. Therefore, we can trace the malicious vehicles based on the *Key* sharing behavior between the Sybil vehicles and the malicious vehicles. The traceability of malicious vehicles can curb the occurrence of Sybil attacks from the source.

After classification using the Sybil vehicle detection algorithm, we perform a fine-grained classification of "original Sybs" and "original Hons". First, "original Sybs" are categorized at a fine-grained level based on the probability of failure to communicate successfully between Sybil vehicles and normal vehicles. Second, Original Hons are categorized at a fine-grained level based on the key-sharing behavior between Sybil vehicles and malicious vehicles. The fine-grained classification process accurately identifies malicious vehicles while improving the detection rate of Sybil vehicles. Our detection mechanism can detect Sybil vehicles and trace malicious vehicles in real time and efficiently under different Sybil attack densities, malicious vehicle densities, and vehicle density attacks.

The main contributions of this work are as follows.

- We propose a beacon packet-based scheme to trace malicious vehicles. The F1 of the malicious vehicle reaches 96.38%, which helps to resist the Sybil attack from its source.
- The detection rate of Sybil vehicles is improved while tracing the malicious vehicles, and the F1 for Sybil vehicles reaches 98.11%, which is 2% higher than in the literature [17].
- Neighborhood graphs are formed instantaneously and independently without reference to the vehicles' historical trust values, reducing the privacy risk associated with historical data.

The full text of this paper is organized as follows: Section 2 introduces the progress of related research work; Section 3 elaborates on our threat models; Section 4 introduces our tracking mechanism for malicious vehicles; Section 5 is our experimental part; Section 6 compares the performance of our proposed method with existing methods; and Section 7 is the summary and outlook of the full text.

## 2. Related Work

Much research has recently been conducted on Sybil attacks at home and abroad. Nonetheless, many schemes have focused only on detecting Sybil vehicles faked by malicious vehicles. Existing detection schemes for detecting Sybil vehicles can be divided into two categories: direct identity detection mechanisms and indirect identity detection mechanisms.

### 2.1. Direct Identity Detection Mechanisms

Direct identity detection mechanisms mainly achieve identity detection through the authentication of vehicle certificates or keys [28]. Santhosh [29] proposed a hybrid cryptographic management mechanism to detect Sybil vehicles. In this scheme, the base station generates a public key for the vehicle that wants to enter the network to communicate and uses the node's identity to encrypt the public key. Nevertheless, this scheme does not consider the security threat to the private key during the generation process. In order to solve this problem, Cheng [30] proposed an RSU authentication scheme based on elliptic curve cryptography, which effectively reduces the security threats of generating pseudonyms and private keys. Regardless, there are overhead and delay problems when RSU authenticates a large number of vehicle beacon packets. Although the direct identity detection schemes [31–33] are the most direct way to verify the identity of illegal vehicles in real time, they cannot meet the needs of rapid authentication when vehicles are gathered in a short time or move at high speeds. In addition, since malicious vehicles have legitimate identities, schemes through direct identity detection fail to detect malicious vehicles.

### 2.2. Indirect Identity Detection Mechanisms

Compared with the overhead of direct identity detection mechanisms, indirect identity detection mechanisms are more lightweight, such as the detection methods based on vehicle cooperation [18,21,34–36]. Panchal [37] proposed a method of separating Sybil attacks using adjacent information in VANETs. In this scheme, RSU discovers its neighbor vehicles through the vehicle ID and then evaluates the trust degree of neighbor vehicles to detect Sybil attacks. However, this scheme cannot avoid the situation of avoiding detection due to collusion between Sybil vehicles. It is also impossible to avoid the credibility evaluation of the vehicle itself. If it is a malicious vehicle, the credibility of its neighbor vehicle list is unconvincing. In order to solve that reliability problem of the vehicle itself, Huo [38] designed and implemented an identity authentication mechanism based on vehicle history information. The vehicle generates the historical message hash value *HD1* from the historical message sent by itself and then hashes the message to be sent and the historical information into *HD2* to send to the agent. The agent detects the Sybil vehicle by comparing *HD2* with *HD1*. However, the historical messages in this scheme have too much influence on future decision-making, and it is impossible to avoid the long-term latency of malicious vehicles. To solve the problem of relying on historical messages, Verchok [17] proposed a scheme for detecting Sybil nodes to verify the existence of each other through local peer-to-peer communication. In this scheme, the server randomly instructs any pair of nodes within the communication threshold to perform a customized beacon packet broadcast and listen. The beacon packet contains a *Key* that identifies the vehicle's identity, and the success or failure edge is established by whether the *Key* is exchanged successfully. Finally, a neighbor graph is formed. Due to the unequal information between Sybil nodes and normal nodes, the formed neighbor graph will make Sybil nodes extreme, and then Sybil nodes can be detected.

The existing solutions mainly focus on detecting Sybil vehicles because malicious vehicles in Sybil attacks typically exhibit normal behavior, which increases the difficulty of detecting malicious vehicles. Angappan [39] proposed a Sybil attack detection scheme that combines RSSI and neighbor information, detects Sybil attacks by comparing the similarity of RSSI, and regards the vehicle with the most neighbor entries as a malicious vehicle. Yet, multiple vehicles may have the same RSSI value only if these vehicles belong to the same vehicle, which cannot defend against collusion in Sybil attacks. In addition, since power affects RSSI, this scheme can only detect malicious vehicles without power control but cannot detect malicious vehicles with transmission power control [40]. Zhang [41] proposed a detection method based on Basic Safety Message (BSM) packets in 2023, which mainly detects the detected vehicle by receiving the receiver (*Recv*) of the BSM message broadcast by the detected vehicle. The *Recv* calculates the transmission distance ($T_d$) of the BSM message according to the sending time and receiving time information of the BSM message. Next, the distance ($D_{N_i}$) between the surrounding neighbor vehicles ($N_i$) and *Recv*, respectively, is calculated. Finally, we take the $N_i$ with the minimum error between $D_{N_i}$ and $T_d$ as *Recv* predicts the BSM package broadcast source. Finally, an integral strategy is implemented for all receivers ($Recv_i$) of the BSM message, and the neighbor vehicle $N_i$ with the highest integral value is selected as the final predicted broadcast source of the BSM message broadcast by the detected vehicle. Sybil vehicles and malicious vehicles are detected by comparing whether the predicted broadcast source is consistent with the broadcast source marked in the BSM message. Regardless, the defect of this scheme is that the proposed spatio-temporal model is prone to errors when it is affected by vehicle densities.

In summary, existing Sybil attack detection schemes mainly focus on detecting Sybil vehicles. However, only by tracing malicious vehicles can Sybil attacks be resisted from their source. Compared with existing solutions, our proposed traceability mechanism considers collusive behavior and separation behavior between malicious vehicles. It does not need to analyze many vehicle trajectories, nor is it affected by vehicle densities to complete the detection of Sybil vehicles and trace back to malicious vehicles, resisting attacks from the source of Sybil attacks.

## 3. Threat Model and Attack Categories

First, we describe the assumptions on which this study is based. Next, our threat model is presented. Finally, the attack types of Sybil attacks are explored.

### 3.1. Assumptions

Our research is based on the following assumptions:

- All actual vehicles are equipped with a high-precision Global Positioning System.
- The RSU has the power to indicate the vehicle's broadcast or listening status.
- Keys listened to by malicious vehicles can be shared between malicious vehicles, creating opportunities for other malicious vehicles in a listening state to maintain Sybil vehicles [17].
- The success rate of communication between the malicious vehicle and the Sybil vehicle is set to 0.6 [17].

### 3.2. Threat Model

In order to attack successfully, malicious vehicles usually use illegal means to obtain legitimate identities. While the literature [20] mentions that malicious attacks by vehicles with illegal identities can be prevented by Public Key Infrastructure (PKI), vehicles with legal identities cannot be prevented from broadcasting false information. Our application scenario is that all vehicles have legitimate identities.

We classify the vehicles into three categories: normal vehicles, malicious vehicles and Sybil vehicles. The properties of the vehicle are shown in Table 1.

**Table 1.** Vehicle property table.

| | Physical Location | Legal Identity | Broadcast and Listen |
|---|---|---|---|
| Normal Vehicles | ✓ | ✓ | ✓ |
| Malicious Vehicles | ✓ | ✓ | ✓ |
| Sybil Vehicles | × | ✓ | × |

We define a vehicle with a physical location as an actual vehicle. In our research, normal vehicles only broadcast authentic and credible beacon packets related to themselves. Although malicious vehicles broadcast authentic and credible beacon packets related to themselves, in order to control the driving status of normal vehicles, they often also broadcast some fake beacon packets in the VANETs. These beacon packets are Sybil vehicles.

In this research, we do not need to consider the authentication of illegal vehicle identity when the vehicle joins the VANETs, and the direct identity detection schemes have already solved this problem. Instead, our detection mechanism focuses on detecting the Sybil attacks launched by the attacker by sending false location messages after entering the VANETs. This kind of internal attack often causes more damage to the VANETs.

The purpose of malicious vehicles launching Sybil attacks is to obtain network resources that are disproportionate to normal vehicles by forging Sybil vehicles. Furthermore, it is to use a small number of vehicles to control and influence as many normal vehicles in the VANETs as possible, and finally decide on the control systems. Figure 3 is an example of an attacker model, where vehicles *N1*, *N2*, and *N3* are normal vehicles, and *M1* is a malicious vehicle.

As shown in Figure 3, *M1* forges a beacon packet at time *t1* and broadcasts it to *N1*, *N2*, and *N3*, where the pseudonym (*S1*) in the beacon packet is a valid pseudonym obtained by *M1* through illegal means. *Pos* is the forged position information of *M1*. When *N2* receives the beacon packet, it thinks that the vehicle *S1* is also in the same lane. In order to avoid collision, *N2* takes a deceleration or emergency braking operation. After analyzing the beacon packets, *N1* and *N3* find that the road conditions they are on are not affected, and they continue to drive normally. Finally, *M1* successfully interferes with the driving state of *N2* by generating a Sybil vehicle (*S1*) by forging its location.

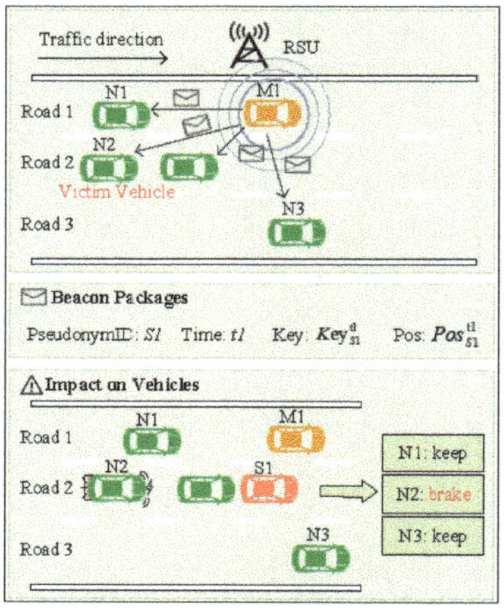


**Figure 3.** The principles and impactions of Sybil attacks.

### 3.3. *Attack Categories*

We consider the collusion and separation behaviors of malicious vehicles as different attack categories.

- Collusion behaviors: Malicious vehicles can share information through particular internal communication.
    - (a) Malicious vehicles help Sybil vehicles broadcast. For example, when a Sybil vehicle is instructed to broadcast by the RSU, in order to reduce the possibility of the Sybil vehicle being exposed, the nearby malicious vehicle in the listening state will use the identity of the Sybil vehicle instead of the Sybil vehicle to perform the *Key* broadcast so that normal vehicles mistakenly believe that the Sybil vehicle also has normal communication ability.
    - (b) Malicious vehicles help Sybil vehicles listen. For example, when the RSU instructs a Sybil vehicle to listen, the nearby malicious vehicles will share the listened lists with the Sybil vehicle, making the RSU mistakenly believe that the Sybil vehicle also can listen.
- Separation behaviors: We consider the separation behavior between malicious vehicles and Sybil vehicles and set the communication success rate between malicious vehicles and Sybil vehicles to 0.6 [17], which can reduce the clustering between malicious vehicles and Sybil vehicles, which is more in line with the selfish behavior of malicious vehicles in Sybil attacks.

## 4. Beacon Packet-Based Traceability Mechanism

A beacon packet is a type of data packet used in vehicle-to-vehicle communication technology. We have customized the format of the beacon packet message as {*PseudonymID*, *Time*, *Key*, *Pos*}, where *PseudonymID* is the pseudonym ID of the vehicle broadcaster; *Time* is the sending time of the beacon packet; *Key* is a particular field allocated by RSU to the vehicle in the broadcast state, where we believe that the *Key* is only known by the vehicle itself and the RSU and cannot be forged; and *Pos* is the location information when the vehicle broadcaster broadcasts the beacon packet, which can be forged by malicious vehicles.

In this study, we mainly complete the detection of Sybil attacks based on the location of the vehicles and further trace the malicious vehicle on this basis. In the first step, we apply the Sybil detection algorithm proposed in the literature [17] to the CAVs and achieve good results. Through a large number of experiments, we find that the detection algorithm is not sensitive to the detection of malicious vehicles. It means that malicious vehicles can continue to launch Sybil attacks at the right time, and the security risks to VANETs still exist. Our research aims to complete the traceability of malicious vehicles and to resist Sybil attacks from the source to improve vehicle safety and traffic efficiencies.

### 4.1. Execution of Communication

In order to allow each vehicle to have sufficient opportunities to communicate with each other, RSU adopts a dichotomy method to assign the broadcast or listen state to the vehicle (the broadcast state is assigned a *Key* that can identify the vehicle's identity, and the listen state is assigned "listen"). First, the RSU divides the vehicles into approximately equal broadcast and listen groups. In the first round, half of the vehicles are set to broadcast and the other half to listen. In the second round, the vehicle status is reversed. After every two rounds, RSU divides the subgroups divided in the previous round and performs state distribution so that each vehicle can have a sufficient opportunity to communicate with other vehicles within the time complexity of log ($N$). Among them, $N$ is the number of detected vehicles.

We use an example to explain the process of RSU assigning a broadcast or listen status to vehicles using the dichotomy method. As shown in Figure 4, the vehicles to be detected are $A$, $B$, $C$, and $D$. In the first round, the vehicles to be detected are dichotomized. The subgroup composed of $A$ and $B$ is set as the broadcast state, the assigned particular fields are $K1$ and $K2$, respectively, and the subgroup composed of $C$ and $D$ is set as the listening state. In the second round, the vehicle status is reversed. The subgroup composed of $A$ and $B$ is in the listening state, the subgroup composed of $C$ and $D$ is in the broadcasting state, and the assigned particular fields are $K3$ and $K4$, respectively. In the third round, the subgroup composed of $A$ and $B$ is dichotomized again, and subgroup $A$ is set to be in the broadcast state. The particular field assigned is $K5$, and the subgroup $B$ is in the listening state. In the fourth round, the vehicle's status is reversed, subgroup $A$ is in the listening state, subgroup $B$ is in the broadcasting state, and the assigned particular field is $K7$. Similarly, we perform the dichotomy method on the subgroups formed by $C$ and $D$.

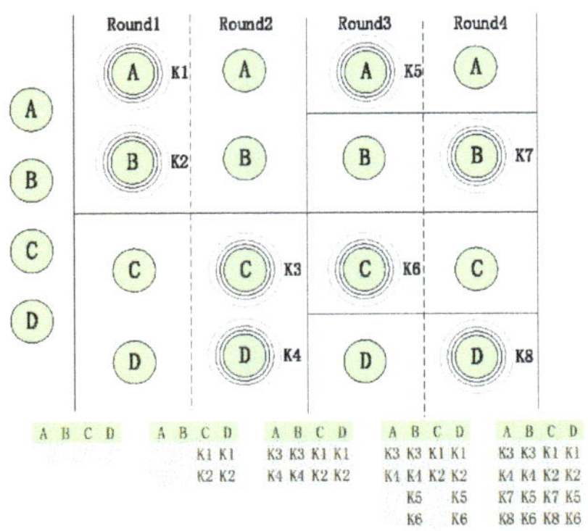


**Figure 4.** Broadcasting or listening state allocation diagram (each round is executed for $T$ time; $T$ is the minimum time interval).

### 4.2. Construction of the Neighborhood Graph

After the communication execution in Section 4.1 is completed, all the vehicles to be detected report to the RSU the *Key* lists they have listened to, and the RSU builds neighbor edges for vehicles according to the *Key* lists and finally completes the neighbor graph construction. The neighbor graph comprises all the vehicles ($N$) to be detected and the edges $E$ formed between the vehicles, that is, $G = (N, E)$, where $\mathrm{N} = \{N_1, N_2 \ldots, N_n\}$,

$E$ is the directed edges formed between two vehicles, and $E_{(i,j)}$ represents the directed edges from $N_i$ to vehicle $N_j$. If $N_j$ listens to the *key* of $N_i$, it means that $N_i$ can form a successful edge to $N_j$, that is, $E_{(i,j)} = S$; otherwise, a failed edge is formed, which is recorded as $E_{(i,j)} = F$.

We use an example to describe the process of RSU constructing a neighborhood graph for a vehicle as shown in Figure 5. Assume that *N1* and *N2* are normal vehicles, *M* is a malicious vehicle, and *S* is a Sybil vehicle generated by *M*. With *S* not maintained by *M*, its interactions with *N1* and *N2* will fail. The failure from *N2* to *M* is due to the communication failure caused by environmental factors. The success from *S* to *M* is the communication success of the malicious vehicle with a probability of 0.6 to reduce the possibility of Sybil vehicle exposure. The failure from *M* to *S* is a separation behavior between the malicious vehicle and the Sybil vehicle, and the malicious vehicle fails to communicate with the Sybil vehicle to reduce the cluster.

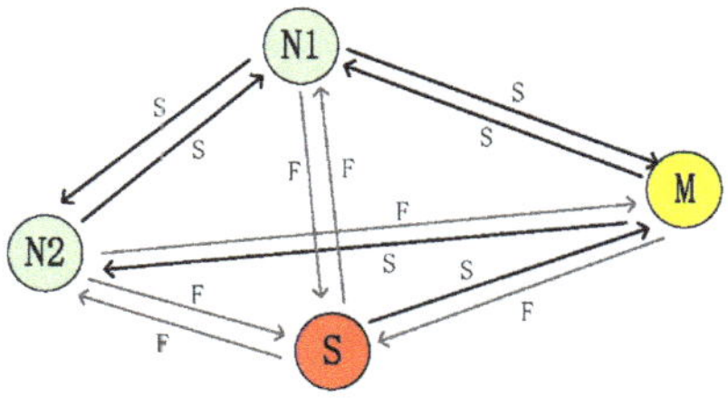


**Figure 5.** Neighbor graph.

### 4.3. Construction of Probabilistic Neighborhood Graph

With Section 4.2, we can now construct a neighbor graph for vehicles. Next, we have to transform the neighbor graph into a probabilistic neighbor graph, which in turn computes the vehicle's confidence level.

#### 4.3.1. Distances and Probabilities Relationships

The traceability mechanism in this article is based on broadcast communication between vehicles, involving signal attenuation during wireless communication [1]. Newport [42] proposed that the probability of beacon reception does indeed decay with the distance between the transmitters and receivers. To explore the relationship between vehicle distances and communication success, we experiment with two cars, *A* and *B*. Vehicle *A* broadcasts beacon packets at a frequency of 0.05 s in place. In contrast, vehicle *B* moves at 10 m/s in the opposite direction, allowing us to observe how well vehicle *B* received these packets. We consider signal fading caused by paths and obstacles and analyze the packet loss rate every 50 m. As shown in Figure 6, we find the relationship between the vehicle distances and the communication success probabilities.

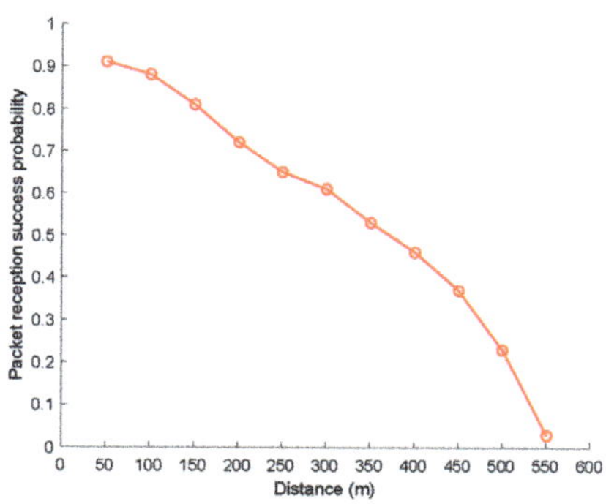


**Figure 6.** Distances and Communication Success Probabilities model.

According to the distances between vehicles, interpolating the distance intervals for the proposed Distances and Communication Success Probabilities model is the probability of success of the edges between vehicles.

4.3.2. Probabilistic Neighborhood Graph

Consider that when a vehicle broadcasts, it forms an outgoing edge for itself, and when it listens, it forms an incoming edge. In each round, the number of outgoing edges formed as broadcasters is much smaller than that of incoming edges formed as listeners. Therefore, we analyze the input edge set with more data than the output edge set to reduce the errors.

According to the Distances and Communication Success Probabilities model mentioned in Section 4.3.1, we consider the probability of edge success as a function of the distances between vehicles. Assuming that the vehicle ($N_j$) listens to the probability of the vehicle ($N_i$) broadcasting the beacon packet expressed as $P_{(i,j)}$, the calculation is shown in Equation (1):

$$P_{(i,j)} = Prob[dist(E_{(i,j)})] \tag{1}$$

where $E_{(i,j)}$ represents the incoming edge formed from $N_i$ to $N_j$, dist() is a function to calculate the distance between $N_i$ and $N_j$, and Prob[] is the probability of successful communication based on the distance between vehicles combined with the Distances and Communication Success Probabilities model in Section 4.3.1. Therefore, the calculation of the success or failure probability of $E_{(i,j)}$ is shown in Equation (2):

$$\mathrm{Edge}\left(E_{(i,j)}\right) = \begin{cases} P_{(i,j)}, & E_{(i,j)} = S \\ 1 - P_{(i,j)}, & E_{(i,j)} = F \end{cases} \tag{2}$$

The total probability product of a vehicle's incoming edge set during the entire communication process is shown in Equation (3):

$$P_{Total} = \prod_{i \neq j}^{N} Edge(E_{(i,j)}) \tag{3}$$

For the convenience of calculation, the logarithm of $P_{total}$ is calculated and recorded as $P_{Pval}$ as shown in Equation (4):

$$Pval = \ln\left(P_{Total}\right) = \sum_{i \neq j}^{N} \ln\left(Edge(E_{i,j})\right) \tag{4}$$

We refer to Pval as the vehicle's credibility after calculating the credibility of each vehicle by Equation (4). The neighbor graph in Section 4.2 is transformed into a probabilistic neighbor graph as shown in Figure 7.

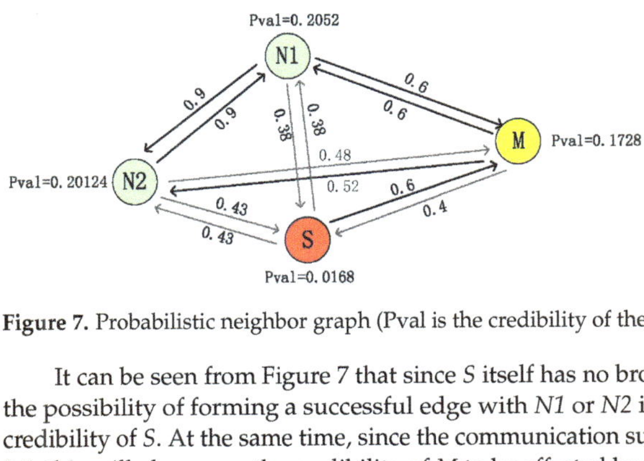


**Figure 7.** Probabilistic neighbor graph (Pval is the credibility of the vehicle).

It can be seen from Figure 7 that since $S$ itself has no broadcast and listen capabilities, the possibility of forming a successful edge with $N1$ or $N2$ is tiny, which leads to the low credibility of $S$. At the same time, since the communication success rate between $M$ and $S$ is 0.6, this will also cause the credibility of $M$ to be affected by $S$.

### 4.4. Traceability Mechanism of Malicious Vehicles

After we apply the Sybil vehicle detection algorithm in [17] to our traceability mechanism, we can divide the vehicles into two categories: the original predicted Sybil vehicle group ("original Sybs") and the original predicted normal vehicle group ("original Hons"). However, after many experiments, we find that there are wrongly predicted vehicles (such as malicious vehicles or normal vehicles) in "original Sybs"; similarly, there are also wrongly predicted vehicles (such as malicious vehicles and Sybil vehicles) in "original Hons". In addition, we also find that the algorithm has a poor detection rate for malicious vehicles, which means that malicious vehicles may relaunch Sybil attacks at the right time. Therefore, it is impossible to eradicate Sybil attacks from the root, which is also our focus: to detect "original Sybs" and "original Hons" further to detect malicious vehicles and improve the precision of Sybil vehicles.

Our traceability mechanism is divided into two stages. The first stage subdivides the "original Sybs", and the second stage subdivides the "original Hons". The traceability mechanism for malicious vehicles is shown in Figure 8:

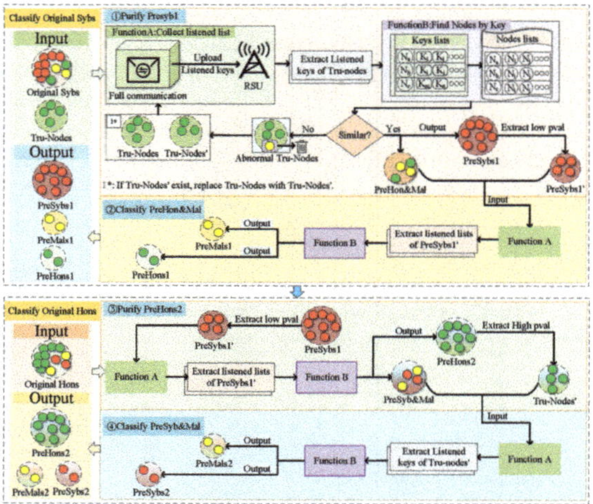


**Figure 8.** Malicious vehicle traceability mechanism diagram.

#### 4.4.1. Subcategory "Original Sybs"

Vehicles misclassified in "original Sybs" may be malicious or normal vehicles. Malicious or normal vehicles have normal broadcast and listening capabilities. For Sybil vehicles, without the maintenance of malicious vehicles, they will not have normal broadcast and listening capabilities. Therefore, our strategy is first to screen out vehicles with normal broadcast and listen capability from "original Sybs" as Predictive Normal and Malicious Vehicle Groups (*PreHon&Mal*). The remaining vehicles are classified as Predictive Sybil Vehicle Group 1 (*PreSybs1*). Secondly, *PreHon&Mal* is further subdivided into Predictive Malicious Vehicle Group 1 (*PreMals1*) and Predictive Normal Vehicle Group 1 (*PreHons1*).

Because normal vehicles cannot communicate with Sybil vehicles but they can communicate with malicious vehicles and normal vehicles, we need to select a certain number of normal vehicles to screen out Sybil vehicles. According to the *Pval* mentioned in Section 4.3.2, the larger the *Pval*, the more likely the vehicle is a normal vehicle. We use the parameter $\alpha_{Tru\text{-}Nodes} = 0.1$ to select vehicles with higher *Pval* values, record them as *Tru-Nodes*, and collocate them as listening states. Since setting all "original Sybs" to the broadcast state can easily raise suspicion of malicious vehicles, we use the dichotomy method mentioned in Section 4.1 to allocate the communication state of "original Sybs". Without the malicious vehicles' help, Sybil vehicles do not have broadcast capability, and *Tru-Nodes* cannot listen to the key broadcast by Sybil vehicles. After sufficient communication between the "original

Sybs" and *Tru-Nodes* groups, RSU extracts and analyzes the list of listened-to keys uploaded by *Tru-Nodes* to determine the list of vehicles listened to by each trusted vehicle.

Considering that there is a communication failure probability of 0.4 between malicious vehicles and Sybil vehicles, this may lead to the *Pval* value of such malicious vehicles not being affected by Sybil vehicles and then appearing in the selected *Tru-Nodes*. When *Tru-Nodes* fully communicate with the "original Sybs", there is still a communication success probability of 0.6 between malicious vehicles and Sybil vehicles, which leads to the malicious vehicles being able to listen to the majority of Sybil vehicles in the "original Sybs". In addition, malicious vehicles in the listening state may also help Sybil broadcast, allowing normal vehicles to listen to Sybil vehicles, which may cause this part of normal vehicles to be misclassified as suspiciously malicious. Therefore, we use the parameter $\beta_{similar}$ = 0.9 to ensure that all extracted *Tru-Nodes* are normal vehicles as much as possible. If the number of "original Sybs" listened to by a vehicle in the *Tru-Nodes* is greater than Len(Original Sybs)*$\beta_{similar}$, the vehicle is considered a suspicious mal. The remaining vehicles after these vehicles are eliminated constitute a $SuplsToppval$ and reassign the communication status of the $SuplsToppval$. Otherwise, *Tru-Nodes* are considered trustworthy vehicles. At this point, the vehicles listened to by *Tru-Nodes* are classified as *PreHon&Mal*, and the remaining vehicles are *PreSybs1*. Further classifications of misclassified vehicles can help improve the accuracy of Sybil vehicles. The process of using *Tru-Nodes* to screen *PreHon&Mal* with broadcast capability is shown in Algorithm 1.

**Algorithm 1** Screening vehicles that can broadcast.

```
DETECTCANBDCASTNODES(nodes, IdToPval)
    CanBdcastNodes ← []
    n ← len(IdToPval) * 0.1
    For i = 0 → n − 1
        TopPvalId ← IdToPval[i]
    EndFor
    NodeSum ← nodes + TopPvalId
    Rounds ← 2 * log_2(len(NodeSum))
    For i ← Rounds
        For j ← TopPvalId
            If CommPlan[j, rnd] ≠ "listen"
                broadkeys.APPEND(ConnSim[j, rnd])
            Else
                ListenedKeys.APPEND(ConnSim[j, rnd])
            EndIf
        EndFor
    EndFor
    ListenedKeys.REMOVE(broadkeys)
    For i ← ListenedKeys
        CanBdcastNodes.APPEND(KeyToId[i])
    EndFor
    CleanNodes ← nodes.REMOVE(CanBdcastNodes)
    Return CanBdcastNodes, CleanNodes
```

Next, we need to classify *PreHon&Mal* further. Considering the malicious vehicles and Sybil vehicles have a success rate of 0.6, while the probability of communication between normal vehicles and Sybil vehicles is 0, we further classify *PreHon&Mal* through Sybil vehicles.

The environment easily influences vehicle communication. Despite using the dichotomy method in Section 4.1 to ensure effective communication, a few normal or malicious vehicles may fail to connect with *Tru-Nodes* in each round. These vehicles might be wrongly categorized in *PreSybs1*. To reduce the impact of these vehicles on subsequent detection, we use the parameter $\gamma_{syb}$ = 0.1 to select the vehicle with a lower *Pval* value in *PreSybs1* and record them as *PreSybs1′*. We use *PreSybs1′* to assist in the classification of *PreHon&Mal*.

We first assign the communication status to *PreSybs1′* through the dichotomy method in Section 4.1 and broadcast status to *PreHon&Mal*. Then, after sufficient communication between *PreSybs1′* and *PreHon&Mal*, the RSU extracts the listened lists of *PreSybs1′*. Because there are Sybil vehicles in *PreSybs1′* in the broadcast state, this means that they have *Keys*. Therefore, we need to remove the *keys* in this section from the listened lists and use the remaining *keys* to search for the corresponding vehicles, categorizing the listened vehicles as *PreMals1*, and the remaining vehicles as *PreHons1*, which achieves fine-grained classifications of "original Sybs".

#### 4.4.2. Subcategory "Original Hons"

We analyze that misclassified vehicles in "original Hons" may be malicious or Sybil vehicles in Section 4.1. Their characteristic is that there is a possibility of successful communication with Sybil vehicles, while the probability of successful communication between normal vehicles and Sybil vehicles is 0. Therefore, our strategy is to first screen out the vehicles in "original Hons" that can interact with Sybil vehicles and set them as predicted Sybil vehicles and malicious vehicle group (*PreSyb&Mal*), and the remaining vehicles are classified as predicted normal vehicle group2 (*PreHons2*). Secondly, *PreSyb&Mal* is subdivided into predicted malicious vehicle group2 (*PreMals2*) and Sybil vehicle group2 (*PreSybs2*).

Because Sybil vehicles can communicate with malicious or Sybil vehicles but cannot communicate with normal vehicles, we use *PreSybs1′* in Section 4.1 to assist in classification. We still allocate the status of *PreSybs1′* through the dichotomy method and set "original Hons" to the broadcast status. After sufficient communication between the "original Hons" and *PreSybs1′*, we collect and analyze the *keys* listened by *PreSybs1′*. We remove the *keys* assigned to *PreSybs1′* from the listened lists and use the remaining *keys* to search for vehicles. These vehicles are classified as *PreSyb&Mal*, and the remaining vehicles are classified as *PreHons2*. The process of screening out *PreSyb&Mal* by *PreSybs1′* is shown in Algorithm 2:

**Algorithm 2** Screening vehicles that can be listened to.

```
DETECTCANBELISDNODES(nodes, PreCleanSybs)
  CanBeLisdNodes ← []
  NodeSum ← nodes + PreCleanSybs
  Rounds ← 2 * log2(len(NodeSum))
  For i ← Rounds
    For j ← PreCleanSybs
      If CommPlan[j, rnd] ≠ "listen"
        broadkeys.APPEND(ConnSim[j, rnd])
      Else
        ListenedKeys.APPEND(ConnSim[j, rnd])
      EndIf
    EndFor
  EndFor
  ListenedKeys.REMOVE(broadkeys)
  For i ← ListenedKeys
    CanBeLisdNodes.APPEND(KeyToId[i])
  EndFor
  CleanNodes ← nodes.REMOVE(CanBeLisdNodes)
  Return CanBeLisdNodes, CleanNodes
```

Next, we will further classify *PreSyb&Mal*. Considering the possibility of successful communication between malicious vehicles and normal vehicles, while the probability of communication between Sybil vehicles and normal vehicles is 0, our goal is to extract vehicles as close as possible to Hon to assist in verification. Therefore, after updating the *Pval* value of *PreHons2*, we extract the vehicles with the top 0.1 *Pval* values and mark them as *Tru-Nodes*″ (based on the experience of extracting *Tru-Nodes* in Section 4.4.1). And set we *Tru-Nodes*″ to the listening state and *PreSyb&Mal* to the broadcasting state. After sufficient communication between these two groups, we extract the listening list of *Tru-Nodes*″, and classify the monitored vehicles as *PreMals2* and the remaining vehicles as *PreSybs2*, thereby completing the fine-grained classification of "Original Hons".

According to Sections 4.4.1 and 4.4.2, our detection results for Sybil vehicles are composed of *PreSybs1* and *PreSybs2*, the detection results for malicious vehicles are composed of *PreMals1* and *PreMals2*, and the detection results for normal vehicles are composed of *PreHons1* and *PreHons2*. Our goal is to trace malicious vehicles and improve the precision of Sybil vehicles.

#### 4.4.3. Update Of Pval Value

In order to reduce the impact of Sybil vehicles or malicious vehicles on the *Pval* of normal vehicles, when there is a suspicious vehicle, we ignore the edge formed by the suspicious vehicle and use Equation (4) to recalculate the vehicle's credibility. In this way, reducing the impact of suspicious vehicles on the *Pval* of normal vehicles can also make the *Pval* of suspicious vehicles increasingly extreme. The algorithm for updating *Pval* is shown in Algorithm 3.

**Algorithm 3** Updating the Pvals.

RECALCUPVAL(**Suspinodes, nodes**)
 $IdtoPval \leftarrow []$
 **If** $nodes \notin Suspinodes$
  $idtoPval \leftarrow nodePval$(idtoEdges(nodes),Suspinodes)
 **EndIf**
 **Return** $idtoPval$

## 5. Experiments

We conduct simulation experiments to verify the effectiveness of our proposed scheme. We evaluate the proposed scheme under different Sybil attack densities, malicious vehicle densities, and vehicle density attack scenarios.

In designing our detection mechanism, we realize the importance of scalability to ensure the long-term effectiveness of the system and to adapt to future changes. Sybil Vehicle Detection and Malicious Vehicle Traceability adopt a modular design ideology, allowing each functional module to operate independently and flexibly expand and combine to meet the needs of CAV systems of varying sizes and complexities. Regarding Sybil vehicle detection, when a vehicle executes the broadcast of a beacon packet, the module analyzes the key list reported by the vehicle to identify potential Sybil vehicles accurately. Regarding tracing malicious vehicles, the detected Sybil vehicles are assigned states and traced based on their collusive behavior. Our scheme also has good openness. If a better Sybil vehicle detection algorithm appears, we can easily replace the existing detection module without making large-scale changes to the system. The information table of our experimental equipment is shown in Table 2.

**Table 2.** Experimental equipment information.

| Device | Detail |
|---|---|
| CPU | Intel® Core™ i7-10750H CPU @ 2.60GHz |
| OS | Debian GNU/Linux 11 (bullseye) |
| OS Type | 64-bit |
| Memory | 3.8 GB |

### 5.1. Simulation Design

Sybil attacks have a greater impact on areas with high traffic densities, such as being more likely to cause vehicle collisions. Therefore, we analyze urban areas with higher traffic densities. Veins [43] is an open-source framework for vehicle network simulation. We use the Veins simulator to simulate the running state of the car, which is based on the road traffic simulator SUMO and the event network simulator OMNeT++. As shown in Figure 9, we select a route map for some areas of Haidian Island through OpenStreetMap and use SUMO traffic scenarios to describe vehicle trajectories. Our simulation parameters are shown in Table 3.

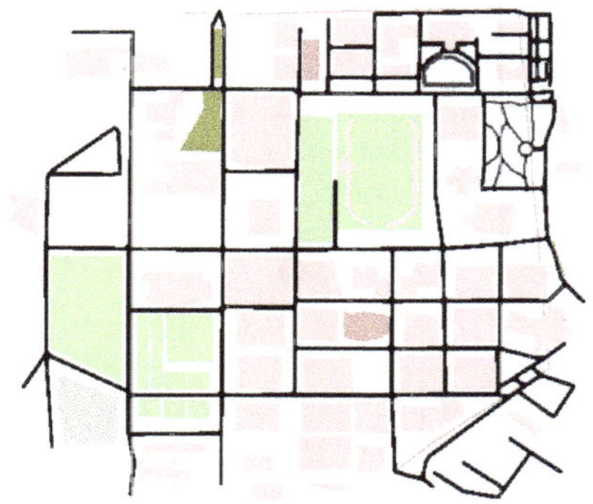

**Figure 9.** Haidian Island area.

**Table 3.** Simulation parameter settings.

| Parameter | Value |
|---|---|
| Simulation Time | 100 s |
| Attack Probability | 5%, 10%, 20%, 30%, 40% |
| Simulation Area | 400 m × 400 m |
| Obstacle Shadowing | Simple Path Loss Model |
| MAC Implementation | IEEE 802.11p |
| Minimum Receive Power | −110 dBm |
| Carrier Frequency | 5.9 GHz |
| Noise Floor | −98 dBm |
| Antenna Height | 1.895 m |
| Simulator | Veins |
| Path loss index | 2 |
| Obstacle loss index | 0.4 |

We randomly select a circle with a center position and radius $R$ as our detection range. Vehicles in the range are to be detected. According to the Distances and Communication Success Probabilities model in Section 4.3.2, to ensure that the vehicle is within a high probability of communication success, we set $R$ to 200 m. This allows vehicles to prove their physical presence by broadcasting. When a Sybil vehicle is instructed to broadcast, a malicious vehicle in a listening state replaces the Sybil vehicle for broadcasting. When a Sybil vehicle is instructed to listen, the malicious vehicle in an actual listening state shares the listened *key* with the Sybil vehicle.

### 5.2. Evaluation Metrics

We evaluate the performance of our detection mechanism using three indicators: precision, recall, and F1-score (F1). Precision is the ratio of correctly predicted positive samples to the total number of predicted positive samples, recall is the ratio of predicted positive samples among actually positive samples, and the F1-score is the harmonic mean of precision and recall rates. The precision and recall calculation formulas for Sybil vehicles are represented by Equations (5) and (6), respectively:

$$P_s = \frac{TP_s}{TP_s + FP_s} \tag{5}$$

$$R_s = \frac{TP_s}{TP_s + FN_s} \tag{6}$$

The precision and recall calculation formulas for malicious vehicles are expressed as Equations (7) and (8), respectively:

$$P_m = \frac{TP_m}{TP_m + FP_m} \tag{7}$$

$$R_m = \frac{TP_m}{TP_m + FN_m} \tag{8}$$

The F1 calculation formulas for Sybil vehicles and malicious vehicles are represented by Equations (9) and (10), respectively:

$$F1_s = \frac{2 \times P_s \times R_s}{P_s + R_s} \tag{9}$$

$$F1_m = \frac{2 \times P_m \times R_m}{P_m + R_m} \tag{10}$$

The overall precision calculation formula for Sybil vehicles and malicious vehicles is as follows (11):

$$P_{s\&m} = \frac{TP_s + TP_{\mathrm{m}}}{TP_s + FP_s + TP_{\mathrm{m}} + FP_{\mathrm{m}}} \tag{11}$$

The overall recall calculation formula for Sybil vehicles and malicious vehicles is Equation (12):

$$R_{s\&m} = \frac{TP_s + TP_{\mathrm{m}}}{TP_s + FN_s + TP_{\mathrm{m}} + FN_{\mathrm{m}}} \tag{12}$$

The overall F1 calculation formula for Sybil vehicles and malicious vehicles is Equation (13):

$$F1_{s\&m} = \frac{2 \times P_{s\&m} \times R_{s\&m}}{P_{s\&m} + R_{s\&m}} \tag{13}$$

*Note*: The Hunting scheme [17] does not classify malicious vehicles. We select malicious vehicles from "original Sybs" as the predicted malicious vehicles in the Hunting scheme.

## 6. Performance Evaluation

In order to reduce the situation where normal vehicles cannot generally run due to detection errors, our goal is to increase the recall as much as possible while ensuring precision. Our comparative schemes are the Hunting scheme [17] and the Eliminate scheme [41]. The Hunting scheme [17] is chosen due to its graph-based detection mechanism. This mechanism avoids the need for extensive data training like machine learning-based schemes. It offers efficient detection and a high detection rate for Sybil vehicles. Our work is based on the Hunting scheme [17] to trace malicious vehicles and has improved the detection rate of

Sybil vehicles compared to this scheme. On the other hand, the Eliminate scheme [41] is chosen because although both this scheme and our proposed traceability mechanism are based on beacon packet analysis, this scheme utilizes the unique signal source of beacon packets to trace malicious vehicles. In contrast, our traceability mechanism traces malicious vehicles through the possibility of successful communication between Sybil vehicles and malicious vehicles. We mainly evaluate the solution's performance from Sybil vehicle proportions, malicious vehicle proportions, and vehicle densities.

### *6.1. Increasing the Proportion of Sybil Vehicles*

We use different attack densities of Sybil vehicles to verify the effectiveness of the schemes. The normal and malicious vehicles ratio is 5:1, with 25 and 5 vehicles, respectively. The Sybil vehicle ratios are 5%, 10%, 20%, 30%, and 40% for performance verifications.

As the proportion of Sybil vehicles increases, malicious vehicles cannot continuously maintain a large number of Sybil vehicles, making them more susceptible to exposure. Therefore, all three schemes can stably detect Sybil vehicles.

As shown in Figure 10, on the detection rate of Sybil vehicles, the graphs $P_s$, $R_s$, $F1_s$ reflect the precision rate, recall rate, and F1 value of the three schemes for Sybil vehicle detection under different Sybil vehicle attack densities, respectively. Among them, the Hunting [17] scheme has a stable overall performance in the three metrics. This is because as the number of Sybil vehicles increases, the malicious vehicles cannot maintain the gradually increasing number of Sybil vehicles, resulting in a decrease in the likelihood of them successfully establishing connections with normal vehicles. This makes the trustworthiness of Sybil vehicles behave more extremely in the neighbor graph. As a result, there is a slight upward trend in the three metrics of the scheme for different Sybil vehicle attack densities. It also reveals that the Hunting scheme can maintain a relatively stable performance when dealing with different densities of Sybil vehicles. The point strategy of the Eliminate scheme [41] depends on the designation of neighboring vehicles. As the density of Sybil vehicles increases, the probability that neighboring vehicles will be mixed with Sybil vehicles increases. However, since a Sybil vehicle is just a beacon packet, it needs to depend on the maintenance of malicious vehicles to impact the decision of the points strategy. Therefore, this scheme performs more consistently in the three metrics for detecting Sybil vehicles. Our Trace scheme is based on the Hunting scheme for the fine-grained classification of predicted Sybil vehicles. Even when the density of Sybil vehicles is low, our scheme maintains a high detection rate compared to the other two, suggesting that it is more sensitive to stealthy attacks. For $F1_s$, our scheme also improves by 0.9% and 3% over the Hunting and Eliminate schemes, respectively, further proving its superiority in Sybil vehicle detection.

On the detection rate of malicious vehicles, Figures $P_m$, $R_m$, $F1_m$ then reflect the precision rate, recall rate, and F1 value of the three schemes for malicious vehicle detection, respectively. The Hunting [17] scheme takes into account both the selfish behavior of malicious vehicles, prioritizing their broadcasting tasks to avoid exposure, and the clustering tendency between malicious and Sybil vehicles, resulting in reduced interaction between them. However, with the increase in Sybil vehicles, malicious vehicles still need to maintain more Sybil vehicles, so the detection rate of malicious vehicles in this scheme is on the rise. Nevertheless, the selfish behavior of malicious vehicles leads to a poor overall detection rate. Furthermore, because the Hunting scheme has a good detection rate for Sybil vehicles, the $P_s$ in Figure 10 is close to 1. Nevertheless, the screening rate of malicious vehicles from the "original Sybs" is very low, which leads to approaching 0. We use max(x,1) to treat the denominator non-0, so the calculated $P_m$ approaches 0 and $P_{s\&m}$ approaches $P_s$.

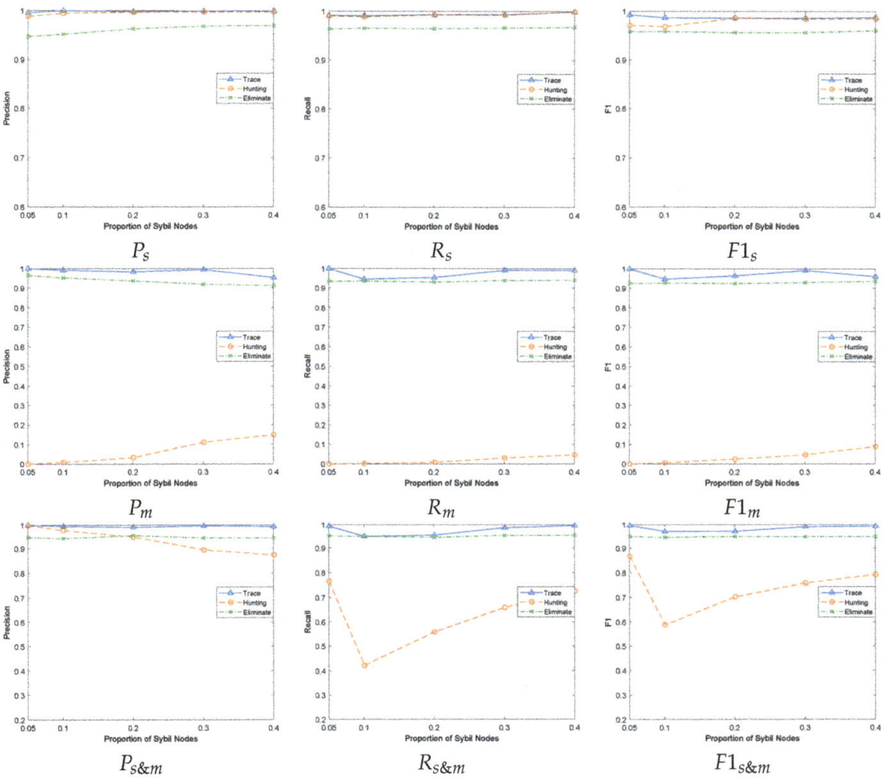


**Figure 10.** Performance comparisons under different Sybil attack densities.

In the Eliminate [41] scheme, which utilizes suspicious Sybil vehicles to locate the source of packets that are further sent as malicious vehicles, the detection rate of malicious vehicles remains stable, as the overall detection rate of Sybil vehicles remains stable. However, after detecting Sybil vehicles, our scheme leverages the conspiracy between Sybil vehicles and malicious vehicles to trace the malicious vehicles. This is despite the malicious vehicle randomly reducing its interactions with the Sybil vehicle while maintaining the Sybil vehicle to reduce the risk of being detected. The communication state and number of rounds we set in Sections 4.4.1 and 4.4.2 provide ample opportunities for their interactions. For $F1_m$, our traceability mechanism is improved by 93.9% and 4.3% compared to the Hunting [17] and Eliminate [41] schemes, respectively. The effectiveness of using the conspiracy behavior between Sybil and malicious vehicles to trace malicious vehicles is demonstrated.

Figures $P_{s\&m}$, $R_{s\&m}$, and $F1_{s\&m}$ comprehensively present the three schemes' precision, recall, and F1 values regarding the overall detection of Sybil vehicles and malicious vehicles. Our scheme performs well in both Sybil vehicle detection and malicious vehicle tracing. This result is the uniqueness of our use of fine-grained classification. Regarding detecting Sybil vehicles, the Trace scheme and Hunting scheme perform better than the Eliminate scheme. This is mainly due to their use of constructed graphs, which are more accurate than distance-dependent detection. The communication relationship between nodes is constructed as a graph, which enables more accurate identification of Sybil vehicles. However, distance-based detection is more susceptible to interference from various factors, such as the environment and communication delays, which can decrease precision.

### 6.2. Increasing the Proportion of Malicious Vehicles

Due to the fact that Sybil vehicles are only beacon packets, they need to be maintained by a malicious vehicle in order to exhibit the characteristics of a normal vehicle. Therefore, we want to verify whether an increase in the proportion of malicious vehicles will affect the detection rate of Sybil and malicious vehicles. Therefore, we adopt a ratio of 5:1 for normal vehicles and Sybil vehicles, with 25 and 5 vehicles, respectively, and 5%, 10%, 20%, 30%, and 40% for malicious vehicles, respectively.

As shown in Figure 11, the graphs $P_s$, $R_s$, and $F1_s$ reflect the precision rate, recall rate, and F1 value of the three schemes for Sybil vehicle detection under different malicious vehicle attack densities, respectively. As the proportion of malicious vehicles increases, the detection rate of the Hunting scheme on Sybil vehicles shows a significant decreasing trend. The reason is that the increase in the density of malicious vehicles enables more malicious vehicles to help Sybil vehicles broadcast, which makes Sybil vehicles maintain a high level of confidence in the neighbor graph. As a result, the Hunting scheme increases the likelihood of misclassifying this number of Sybil vehicles into normal vehicles, leading to a decrease in the detection rate of Sybil vehicles. In the Eliminate [41] scheme, as the proportion of malicious vehicles increases, the voting probability of malicious vehicles participating in the integral strategy will fluctuate, resulting in a slight fluctuation in the detection rate of Sybil vehicles in this scheme. Although our scheme has a fluctuating trend, overall, it maintains a high detection rate. That is because as the proportion of malicious vehicles increases, it increases the probability of malicious vehicles maintaining Sybil vehicles continuously, making a small number of Sybil vehicles behave less extremely. Our strategy for less extreme Sybil vehicles is to extract trusted groups with higher *Pval* in through Section 4.4.2 instead of directly using and then further classify *PreSyb&Mal*. The advantage of this approach is that it can reduce the impact of Sybil vehicles successfully disguised by malicious vehicles, as their *Pval* is difficult to exceed the *Pval* of normal vehicles. However, we also know the limitations, where malicious vehicles may appear in trusted groups when they do not maintain Sybil vehicles. In the future, we will consider implementing additional authentication mechanisms, such as private key authentication, on the extracted trusted groups to ensure they contain only normal vehicles. For $F1_s$, our traceability mechanism has improved by 2.3% and 3.47% compared to the Hunting [17] and Eliminate [41] schemes, respectively.

In terms of the detection rate of malicious vehicles, as the proportion of malicious vehicles increases, more malicious vehicles in the Hunting [17] scheme can maintain Sybil vehicles, increasing their interaction. Therefore, there is an overall upward trend. However, due to the selfish behavior of malicious vehicles, this scheme only affects the credibility of malicious vehicles through Sybil vehicles, which is far from achieving a high detection rate for malicious vehicles. For $F1_m$, our traceability mechanism is increased by 91.87% and 3.52% compared to the Hunting [17] and Eliminate [41] schemes, respectively. All three scenarios show a decreasing trend when the percentage of malicious vehicles increases. This is mainly due to the disadvantage of relying on detecting neighboring vehicles, which are prone to vote manipulation by malicious vehicles.

### 6.3. Increasing the Vehicles Density

Since our proposed traceability mechanism is based on vehicle location information, we would like to verify whether the detection rate of Sybil vehicles and malicious vehicles is affected when the vehicle density increases. Therefore, we adopt ratios of 5:1 between normal and malicious vehicles, with Sybil vehicles accounting for 20% of both.

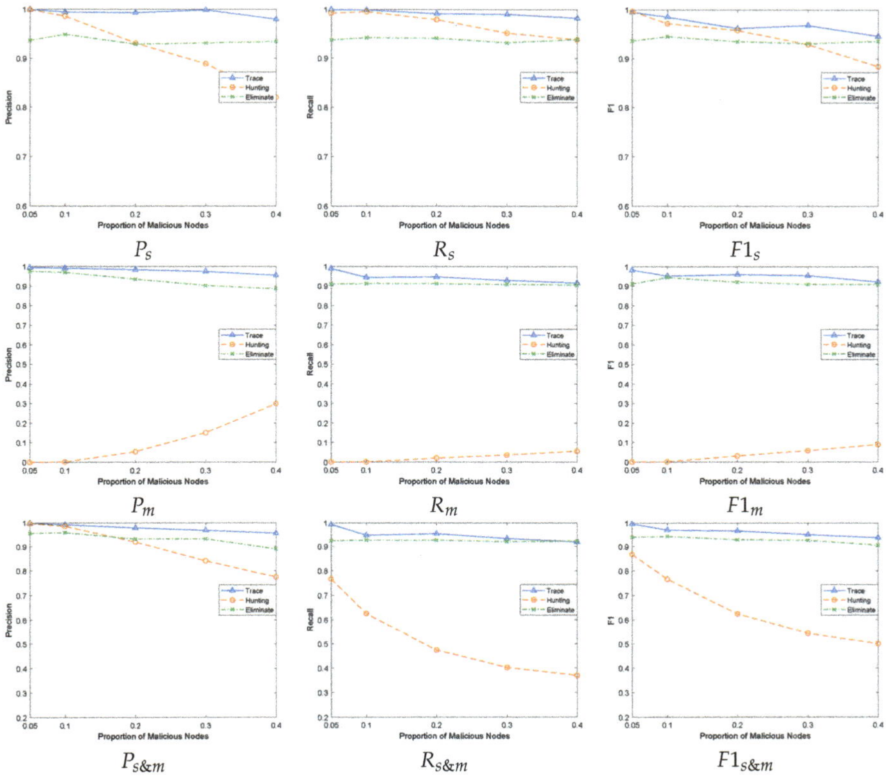


**Figure 11.** Performance comparisons under different attack densities of malicious vehicles.

As shown in Figure 12, graphs $P_s$, $R_s$, and $F1_s$ reflect the precision rate, recall rate, and F1 value of the three schemes for Sybil vehicle detection, respectively. Figures $P_m$, $R_m$, and $F1_m$ reflect the precision rate, recall rate and F1 value of the three schemes for malicious vehicle detection, respectively. And Figures $P_{s\&m}$, $R_{s\&m}$, and $F1_{s\&m}$ combine the precision, recall, and F1 values of the three schemes for Sybil vehicles and malicious vehicles in general. Among them, the Eliminate [41] scheme is most affected by vehicle density and shows a decreasing trend overall. This is mainly due to the fact that the spatio-temporal model adopted by the Eliminate scheme is easily affected by signal transmission. When the vehicle density increases, the complexity and interference of signal transmission also increase, leading to a decrease in the accuracy of the spatio-temporal model. To compensate for this shortcoming, the authors propose an integration strategy that attempts to reduce the error through the information of neighboring vehicles. However, this integration strategy is more sensitive to vehicle density and is prone to false alarms in the case of high vehicle density. When detecting Sybil vehicles, if the theoretical distance between the Sybil vehicles and the vehicles to be detected has the minimum error compared to the spatio-temporal model distance, the Sybil vehicle will be mistaken for the source of the beacon packet and classified into normal vehicles. In this case, Sybil vehicles and malicious vehicles cannot be detected. In addition, when there is a suspicious Sybil vehicle to search for the source of the beacon packet, if the theoretical distance between the normal vehicle and the vehicle to be detected has the minimum error compared to the spatio-temporal model distance, the normal vehicle will be mistaken for the source of the Sybil vehicle and classified as a malicious vehicle. Our traceability mechanism focuses on proving the physical existence between vehicles through broadcasting *keys*. The greater the density of vehicles, the lesser the possibility that successful communication between the vehicles is affected by the distance, and the more advantageous our mechanism. Our traceability mechanism is increased by 0.4% and 3.22%, respectively, compared to the Hunting [17] and Eliminate [41] schemes in $F1_s$, and by 95.39% and 3.29% respectively compared to the Hunting [17] and Eliminate [41] schemes in $F1_m$.

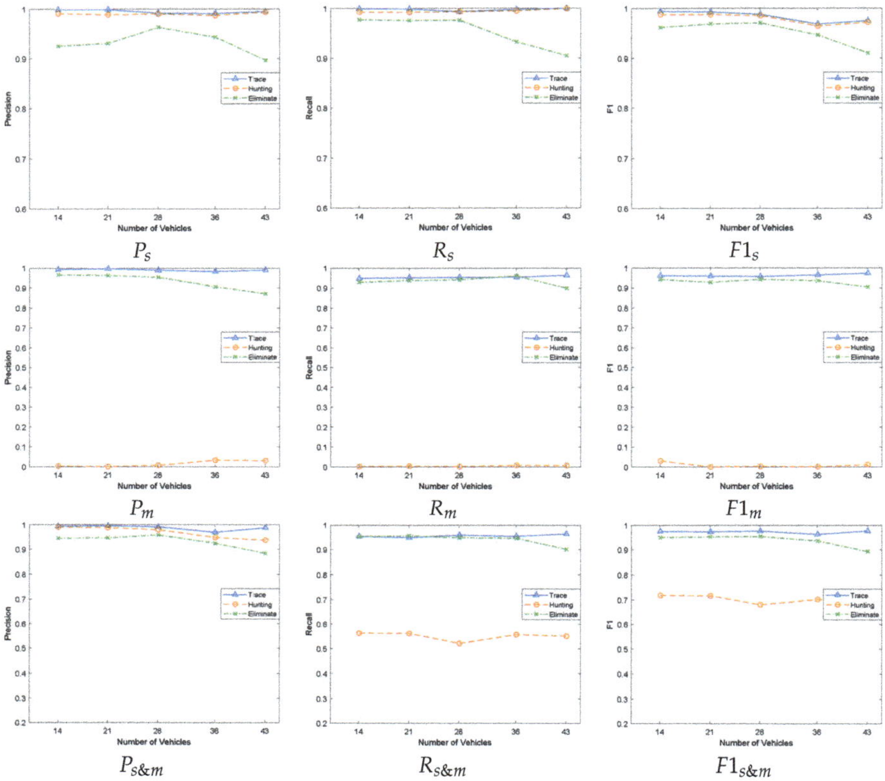


**Figure 12.** Performance comparisons under different vehicle densities.

Due to our traceability mechanism's focus on tracing malicious vehicles in Sybil attacks, we compare the detection rates of three schemes for malicious vehicles using a table as shown in Table 4. Under the three attack densities of Sybil vehicle proportions, malicious vehicle proportions, and vehicle densities, the arithmetic mean of $P_m$, $R_m$, and $F1_m$ is calculated respectively and expressed with $AvgP_m$, $AvgR_m$, and $AvgF1_m$.

**Table 4.** Comparisons of three schemes for detecting malicious vehicles.

| | $AvgP_m$ | $AvgR_m$ | $AvgF1_m$ |
|---|---|---|---|
| Trace (ours) | **0.9853** | **0.9593** | **0.9638** |
| Hunting [17] | 0.0591 | 0.015 | 0.3637 |
| Eliminate [41] | 0.9345 | 0.927 | 0.9264 |

Table 4 shows that the precision of our proposed scheme for malicious vehicles is as high as 98.53%. This excellent performance is mainly attributed to the stability of our scheme, which is not easily affected by the fluctuation of vehicle density. Meanwhile, when the number of malicious vehicles increases, we effectively reduce the risk of being disguised as Sybil vehicles by malicious vehicles by extracting trusted groups, thus ensuring efficient detection of malicious vehicles. In contrast, the Hunting scheme has a lower detection rate for malicious vehicles and is still deficient in malicious vehicle tracing. The detection rate of the Eliminate scheme is slightly lower than that of the Trace scheme because the proposed spatio-temporal model is easily affected by environmental disturbances and vehicle density.

In order to evaluate the overall performance of the three schemes, we calculate the arithmetic mean values of $P_{s\&m}$, $R_{s\&m}$, and $F1_{s\&m}$ under the proposed three attack densities. Then, we express them in $AvgP_{s\&m}$, $AvgR_{s\&m}$, and $AvgF1_{s\&m}$. A comparison of the overall performance and detection time of the three schemes is shown in Table 5.

**Table 5.** Overall performance and detection time comparison of three schemes.

| | $AvgP_{s\&m}$ | $AvgR_{s\&m}$ | $AvgF1_{s\&m}$ | Time |
|---|---|---|---|---|
| Trace (ours) | **0.9879** | **0.9601** | **0.9725** | 1.35 s |
| Hunting [17] | 0.9371 | 0.5685 | 0.2578 | 0.86 s |
| Eliminate [41] | 0.9376 | 0.9402 | 0.9385 | 34.59 s |

The comprehensive analysis of the experimental results shows that our proposed traceability mechanism exhibits good classification performance results in detecting Sybil vehicles and malicious vehicles under three attack densities, and our scheme is not easily affected by vehicle density. The average precision of Sybil and malicious vehicles is as high as 98.79%, which is about 5% higher than the two comparative schemes. Our scheme also exceeds 96% in terms of overall average recall and F1, which proves that our traceability mechanism is feasible for detecting Sybil vehicles and tracing malicious vehicles. In terms of running time, our scheme takes only 1.35 s to achieve more than 95% of the F1 value, which can satisfy the demand for real-time detection. In contrast, the Eliminate scheme is time-consuming because it needs first to analyze the packets broadcasted by all neighboring vehicles, calculate a broadcaster with the smallest distance error as the source of suspicious packets, and then make a collective decision to detect Sybil vehicles through the integral strategy. This complex processing flow poses a considerable challenge to the real-time nature of Sybil attack detection. Our scheme shows obvious advantages in detection rate, accuracy, and running time and provides a practical solution for Sybil vehicle detection and malicious vehicle tracing.

### 6.4. Threshold Analysis

We conduct an in-depth experimental analysis of the introduced thresholds, aiming to find the optimal threshold setting through data validation to improve the precision of malicious vehicle tracing further. Figure 13a demonstrates the change in the extraction probability of honest vehicles for different values of $\alpha_{Tru\text{-}Nodes}$. Figure 13b reflects the change in the extraction probability of honest vehicles for different values of $\beta_{similar}$. Figure 13c depicts explicitly the effect of different values of $\gamma_{syb}$ on the extraction probability of Sybil vehicles.

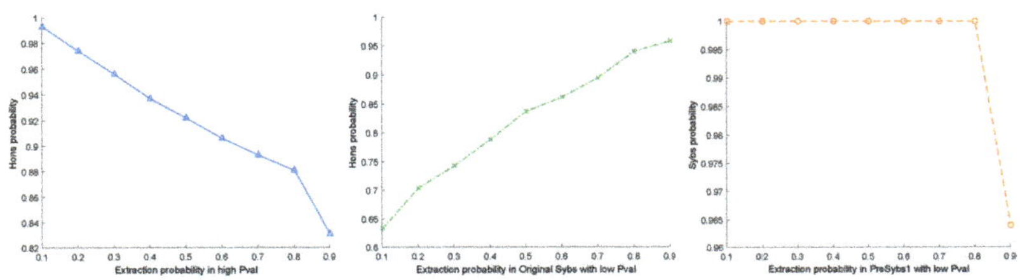


(a) (b) (c)

**Figure 13.** Threshold analysis. (**a**) $\alpha_{Tru\text{-}Nodes}$. (**b**) $\beta_{similar}$. (**c**) $\gamma_{syb}$.

#### 6.4.1. $\alpha_{Tru\text{-}Nodes}$

According to the *Pval* value mentioned in Section 4.3.2, the higher the *Pval* value, the greater the likelihood that the vehicle is normal. The $\alpha_{Tru\text{-}Nodes}$ value affects the accuracy of *Tru-Nodes*, which is crucial for the algorithm's performance. Therefore, this section conducts an experimental analysis on the threshold of $\alpha_{Tru\text{-}Nodes}$ at different levels. We take the value of each probability at intervals of 0.1 and conduct 100 independent repeated experiments. The results are shown in Figure 13a. It can be seen that with the increase in the $\alpha_{Tru\text{-}Nodes}$ value, the probability that the vehicle is normal is lower. This is because malicious vehicles will randomly maintain Sybil vehicles, and the *Pval* value of such malicious vehicles will be affected by Sybil vehicles, resulting in a lower ranking. However, when $\alpha_{Tru\text{-}Nodes}$ is less than 0.1, an effective number of vehicles will not be selected. To ensure the stability of the algorithm, the $\alpha_{Tru\text{-}Nodes}$ value starts from 0.1. When $\alpha_{Tru\text{-}Nodes}$ value = 0.1, the probability of Hon extraction is the best. So, our $\alpha_{Tru\text{-}Nodes}$ value is set to 0.1.

#### 6.4.2. $\beta_{similar}$

In the algorithm, a parameter $\beta_{similar}$ is designed to ensure that the extracted Tru-Nodes are all normal vehicles as much as possible, and the remaining vehicles after elimination by $\beta_{similar}$ constitute the $\text{Supls}_{\text{Toppval}}$. In order to evaluate the effect of leaving as many normal vehicles as possible in the $\text{Supls}_{\text{Toppval}}$ as well as the malicious vehicles culling effect, *ProA* and *ProB* are introduced, respectively. The formulas are as in (14):

$$\begin{cases} ProA = \dfrac{Supls_{\text{Toppval}} \cap Tru_{\text{Hons}}}{Tru\text{-}Nodes} \\ \\ ProB = \dfrac{Supls_{\text{Toppval}} \cap Tru_{\text{Hons}}}{Supls_{\text{Toppval}}} \end{cases} \tag{14}$$

where $\text{Tru}_{\text{Hons}}$ are actual hon vehicles. Ultimately, the credibility of *Tru-Nodes* is as in Equation (15):

$$Cred_{\text{Tru-Nodes}} = \frac{ProA + ProB}{2} \tag{15}$$

As shown in Figure 13b, the higher the $\beta_{similar}$, the higher the confidence of *Tru-Nodes*. This is because by eliminating suspicious malicious vehicles in *Tru-Nodes*, the probability that malicious vehicles can maintain the normal broadcast of Sybil vehicles is lower. Thus, the probability that normal vehicles can listen to Sybil vehicles is negligible. That is, with the lower probability that normal vehicles are affected by malicious vehicles, an upward trend is shown. Hence, our $\beta_{similar}$ is set to 0.9.

#### 6.4.3. $\gamma_{syb}$

According to the *Pval* value mentioned in Section 4.3.2, it can be seen that the smaller the *Pval* value, the higher the probability that the vehicle is a Sybil vehicle, and the $\gamma_{syb}$ value designed in the algorithm affects the accuracy of *PreSybs1'*. We choose to take values for each probability at intervals of 0.1 and perform 100 independent repetitions of the experiment. We want to extract vehicles more likely to be Sybil vehicles, so the $\gamma_{syb}$ value is set to 0.1. However, a slight decrease in the $\gamma_{syb}$ value after 0.8 can be seen from Figure 13c, which means that the parameter is not very sensitive to the algorithm. Consequently, this parameter's influence on the algorithm is not critical.

## 7. Conclusions and Future Work

Existing detection mechanisms only detect Sybil vehicles and cannot trace malicious vehicles. We propose a scheme to trace malicious vehicles based on vehicle broadcast beacon packets by analyzing the differences between Sybil vehicles, malicious vehicles, and normal vehicles. The experimental results show that under three attack densities, namely, increasing the Sybil proportion, malicious vehicle proportion, and vehicle density, the traceability mechanism achieves an average checking accuracy and completeness of

98.53% and 95.93%, respectively. Our traceability mechanism performs better than the latest solution for tracking malicious vehicles, particularly showing a more stable detection rate under high vehicle density. This implies enhanced defense against Sybil attacks from initiators, which is crucial for improving the security and reliability of intelligent transportation systems.

The malicious vehicle-tracing mechanism proposed in this paper demonstrates superior accuracy and stability in experimental comparisons with other schemes. However, it also exhibits limitations and shortcomings.

- During detection, we discovered that neighbor-based collective witnessing reduces errors from a few vehicles but is vulnerable to manipulation by malicious vehicles. Future research will focus on a detection scheme independent of neighboring vehicles, necessitating significantly enhancing individual vehicles' detection capabilities. This implies stronger data processing and analysis for each vehicle to independently assess its surroundings, posing algorithm design and performance challenges. It necessitates deeper research and meticulous debugging to address diverse scenarios and attacks.
- The proposed mechanism depends on trusted RSUs for key management in vehicle broadcasting, making it vulnerable to single-point failure. Once the RSU is attacked or malfunctions, it is easy to cause the entire detection mechanism to malfunction. Decentralization is a promising approach to alleviating this problem. It distributes key distribution and management tasks across multiple nodes, minimizing the impact of a single point of failure.
- Current research focuses on simulations and lacks real-world validation. This hinders the mechanism's feasibility and reliability in practical settings. Future studies should incorporate real data and scenario testing to ensure the mechanism's effectiveness in real environments.

**Author Contributions:** Conceptualization, Y.Z. (Yaling Zhu), J.Z. and Y.Z. (Yuqing Zhang); Methodology, Y.Z. (Yaling Zhu); Software, Y.Z. (Yaling Zhu); Validation, Y.Z. (Yuqing Zhang); Formal analysis, J.Z., Y.Y. and X.L.; Investigation, Y.Z. (Yaling Zhu), J.Z., F.W. and D.H.; Resources, D.H.; Writing—original draft, Y.Z. (Yaling Zhu); Writing—review & editing, Y.Z. (Yaling Zhu), J.Z., F.W., D.H., Y.Y., X.L. and Y.Z. (Yuqing Zhang); Supervision, F.W. and Y.Y.; Project administration, X.L.; Funding acquisition, Y.Z. (Yuqing Zhang). All authors have read and agreed to the published version of the manuscript.

**Funding:** This work was supported by the Key Research and Development Science and Technology of Hainan Province (GHYF2022010), the Research Startup Foundation of Hainan University (RZ2100003340), the National Natural Science Foundation of China (U1836210).

**Institutional Review Board Statement:** Not applicable.

**Informed Consent Statement:** Not applicable.

**Data Availability Statement:** Data are contained within the article.

**Acknowledgments:** Zixi Jin is responsible for simulating the running state of vehicles with Veins. This work played a vital role in the smooth progress of the whole experiment. We are very grateful for her contribution.

**Conflicts of Interest:** The authors declare no conflicts of interest.

## Abbreviations

The following abbreviations are used in this manuscript:

| | |
|---|---|
| CAVs | Connected Automobile Vehicles |
| RSU | Roadside Unit |
| NHTSA | National Highway Traffic Safety Administration |
| PMP | Point-to-Multipoint |
| BSM | Basic Safety Message |
| PKI | Public Key Infrastructure |

## References


1. Ali, G.M.N.; Ayalew, B.; Vahidi, A.; Noor-A-Rahim, M. Analysis of reliabilities under different path loss models in urban/sub urban vehicular networks. In Proceedings of the 2019 IEEE 90th Vehicular Technology Conference (VTC2019-Fall), Honolulu, HI USA, 22–25 September 2019; pp. 1–6.
2. Sadaf, M.; Iqbal, Z.; Javed, A.R.; Saba, I.; Krichen, M.; Majeed, S.; Raza, A. Connected and automated vehicles: Infrastructure applications, security, critical challenges, and future aspects. *Technologies* **2023**, *11*, 117. [CrossRef]
3. Bari, B.S.; Yelamarthi, K.; Ghafoor, S. Intrusion detection in vehicle controller area network (can) bus using machine learning A comparative performance study. *Sensors* **2023**, *23*, 3610. [CrossRef] [PubMed]
4. Giannaros, A.; Karras, A.; Theodorakopoulos, L.; Karras, C.; Kranias, P.; Schizas, N.; Kalogeratos, G.; Tsolis, D. Autonomous vehicles: Sophisticated attacks, safety issues, challenges, open topics, blockchain, and future directions. *J. Cybersecur. Priv.* **2023** *3*, 493–543. [CrossRef]
5. Ahmad, J.; Zia, M.U.; Naqvi, I.H.; Chattha, J.N.; Butt, F.A.; Huang, T.; Xiang, W. Machine learning and blockchain technologies for cybersecurity in connected vehicles. *Wiley Interdiscip. Rev. Data Min. Knowl. Discov.* **2024**, *14*, e1515. [CrossRef]
6. Sheik, A.T.; Maple, C.; Epiphaniou, G.; Dianati, M. A Comprehensive Survey of Threats in Platooning—A Cloud-Assisted Connected and Autonomous Vehicle Application. *Information* **2023**, *15*, 14. [CrossRef]
7. Sheik, A.T.; Maple, C.; Epiphaniou, G.; Dianati, M. Securing Cloud-Assisted Connected and Autonomous Vehicles: An In-Depth Threat Analysis and Risk Assessment. *Sensors* **2023**, *24*, 241. [CrossRef] [PubMed]
8. Bendiab, G.; Hameurlaine, A.; Germanos, G.; Kolokotronis, N.; Shiaeles, S. Autonomous vehicles security: Challenges and solutions using blockchain and artificial intelligence. *IEEE Trans. Intell. Transp. Syst.* **2023**, *24*, 3614–3637. [CrossRef]
9. Jayanthi, D.B. A review of attacks and their countermeasures in autonomous vehicles. *IJCSPUB-Int. J. Curr. Sci. (IJCSPUB)* **2023** *13*, 287–294.
10. Gupta, S.; Maple, C.; Passerone, R. An investigation of cyber-attacks and security mechanisms for connected and autonomous vehicles. *IEEE Access* **2023**, *11*, 90641–90669. [CrossRef]
11. Gupta, S.; Maple, C. A Survey of Security Mechanisms for Edge Computing based Connected Autonomous Vehicles. *Authorea Preprints* **2023**. [CrossRef]
12. Douceur, J.R. The sybil attack. In Proceedings of the Peer-to-Peer Systems: First InternationalWorkshop, IPTPS 2002, Cambridge MA, USA, 7–8 March 2002; Revised Papers 1; Springer: Berlin/Heidelberg, Germany, 2002; pp. 251–260.
13. Benadla, S.; Merad-Boudia, O.R. The impact of sybil attacks on vehicular fog networks. In Proceedings of the 2021 International Conference on Recent Advances in Mathematics and Informatics (ICRAMI), Tebessa, Algeria, 21–22 September 2021; pp. 1–6.
14. Dey, M.R.; Patra, M.; Mishra, P. Efficient Detection and Localization of DoS Attacks in Heterogeneous Vehicular Networks. *IEEE Trans. Veh. Technol.* **2023**, *72*, 5597–5611. [CrossRef]
15. Tao, Y.; Javanmardi, E.; Lin, P.; Nakazato, J.; Jiang, Y.; Tsukada, M.; Esaki, H. Zero-Knowledge Proof of Traffic: A Deterministic and Privacy-Preserving Cross Verification Mechanism for Cooperative Perception Data. *IEEE Access* **2023**, *11*, 142846–142861 [CrossRef]
16. Zhong, Y.; Yang, H.; Li, Y.; Yang, B.; Li, X.; Yue, Q.; Hu, J.; Zhang, Y. Sybil Attack Detection in VANETs: An LSTM-Based BiGAN Approach. In Proceedings of the 2023 International Conference on Data Security and Privacy Protection (DSPP), Xi'an, China, 16–18 October 2023; pp. 113–120.
17. Verchok, N.; Orailoğlu, A. Hunting Sybils in Participatory Mobile Consensus-Based Networks. In Proceedings of the 15th ACM Asia Conference on Computer and Communications Security, Taipei, Taiwan, 5–9 October 2020; pp. 732–743.
18. Faisal, S.M.; Gupta, B.K.; Zaidi, T. A hybrid framework to prevent VANET from Sybil Attack. In Proceedings of the 2022 5th International Conference on Multimedia, Signal Processing and Communication Technologies (IMPACT), Aligarh, India, 26–27 November 2022; pp. 1–6.
19. Lim, K.; Islam, T.; Kim, H.; Joung, J. A Sybil attack detection scheme based on ADAS sensors for vehicular networks. In Proceedings of the 2020 IEEE 17th Annual Consumer Communications & Networking Conference (CCNC), Las Vegas, NV, USA, 10–13 January 2020; pp. 1–5.
20. Kamel, J.; Haidar, F.; Jemaa, I.B.; Kaiser, A.; Lonc, B.; Urien, P. A misbehavior authority system for sybil attack detection in c-its. In Proceedings of the 2019 IEEE 10th Annual Ubiquitous Computing, Electronics & Mobile Communication Conference (UEMCON), New York City, NY, USA, 10–12 October 2019; pp. 1117–1123.
21. Minu, R.; Nagarajan, G.; Munshi, A.; Venkatachalam, K.; Almukadi, W.; Abouhawwash, M. An Edge Based Attack Detection Model (EBAD) for Increasing the Trustworthiness in IoT Enabled Smart City Environment. *IEEE Access* **2022**, *10*, 89499–89508 [CrossRef]
22. Yang, H.; Zhong, Y.; Yang, B.; Yang, Y.; Xu, Z.; Wang, L.; Zhang, Y. An overview of sybil attack detection mechanisms in vfc. In Proceedings of the 2022 52nd Annual IEEE/IFIP International Conference on Dependable Systems and Networks Workshops (DSN-W), Baltimore, MD, USA, 27–30 June 2022; pp. 117–122.
23. Zhang, Z.; Lai, Y.; Chen, Y.; Wei, J.; Feng, Y. A Real-Time Detection Method for Sybil Attacks with High Traceability. Available online: https://ssrn.com/abstract=4511059 (accessed on 6 January 2024).
24. Yuan, M.; Lin, L.; Wu, Z.; Ye, X. A novel sybil attack detection scheme based on edge computing for mobile iot environment. *arXiv* **2019**, arXiv:1911.03129.

25. Krishnan, R.P.; Kumar, A.R.P. A collaborative strategy for detection and eviction of Sybil attacker and Sybil nodes in VANET. *Int. J. Commun. Syst.* **2021**, *34*, e4621.
26. Rakhi, S.; Shobha, K. LCSS Based Sybil Attack Detection and Avoidance in Clustered Vehicular Networks. *IEEE Access* **2023**, *11*, 75179–75190. [CrossRef]
27. Chen, Y.; Lai, Y.; Zhang, Z.; Li, H.; Wang, Y. MDFD: A multi-source data fusion detection framework for Sybil attack detection in VANETs. *Comput. Netw.* **2023**, *224*, 109608. [CrossRef]
28. Song, L.; Sun, G.; Yu, H.; Du, X.; Guizani, M. Fbia: A fog-based identity authentication scheme for privacy preservation in internet of vehicles. *IEEE Trans. Veh. Technol.* **2020**, *69*, 5403–5415. [CrossRef]
29. Santhosh, J.; Sankaran, S. Defending against sybil attacks in vehicular platoons. In Proceedings of the 2019 IEEE International Conference on Advanced Networks and Telecommunications Systems (ANTS), Goa, India, 16–19 December 2019; pp. 1–6.
30. Cheng, H.; Liu, Y. An improved RSU-based authentication scheme for VANET. *J. Internet Technol.* **2020**, *21*, 1137–1150.
31. Hicks, C.; Garcia, F.D. A vehicular DAA scheme for unlinkable ECDSA pseudonyms in V2X. In Proceedings of the 2020 IEEE European Symposium on Security and Privacy (EuroS&P), Genoa, Italy, 7–11 September 2020; pp. 460–473.
32. Yao, Y.; Chang, X.; Mišić, J.; Mišić, V.B.; Li, L. BLA: Blockchain-assisted lightweight anonymous authentication for distributed vehicular fog services. *IEEE Internet Things J.* **2019**, *6*, 3775–3784. [CrossRef]
33. Ali, I.; Li, F. An efficient conditional privacy-preserving authentication scheme for Vehicle-To-Infrastructure communication in VANETs. *Veh. Commun.* **2020**, *22*, 100228. [CrossRef]
34. Haddaji, A.; Ayed, S.; Fourati, L.C. Blockchain-based multi-levels trust mechanism against sybil attacks for vehicular networks. In Proceedings of the 2020 IEEE 14th International Conference on Big Data Science and Engineering (BigDataSE), Guangzhou, China, 29 December–1 January 2020; pp. 155–163.
35. Parham, M.; Pouyan, A.A. An effective privacy-aware Sybil attack detection scheme for secure communication in vehicular ad hoc network. *Wirel. Pers. Commun.* **2020**, *113*, 1149–1182. [CrossRef]
36. Gu, K.; Dong, X.; Jia, W. Malicious node detection scheme based on correlation of data and network topology in fog computing-based VANETs. *IEEE Trans. Cloud Comput.* **2020**, *10*, 1215–1232. [CrossRef]
37. Panchal, A.; Singh, D. Segregation of Sybil Attack using Neighbouring Information in VANET. *Int. Adv. Res. J. Sci. Eng. Technol. ISSN* **2017**, *4*, 172–180. [CrossRef]
38. Yujia, H.; Yongfeng, H.; Fu, C. Research on node authentication of MQTT protocol. In Proceedings of the 2020 IEEE 11th International Conference on Software Engineering and Service Science (ICSESS), Beijing, China, 16–18 October 2020; pp. 405–410.
39. Angappan, A.; Saravanabava, T.; Sakthivel, P.; Vishvaksenan, K. Novel Sybil attack detection using RSSI and neighbour information to ensure secure communication in WSN. *J. Ambient Intell. Humaniz. Comput.* **2021**, *12*, 6567–6578. [CrossRef]
40. Benadla, S.; Merad-Boudia, O.R.; Senouci, S.M.; Lehsaini, M. Detecting Sybil Attacks in Vehicular Fog Networks Using RSSI and Blockchain. *IEEE Trans. Netw. Serv. Manag.* **2022**, *19*, 3919–3935. [CrossRef]
41. Zhang, Z.; Lai, Y.; Chen, Y.; Wei, J.; Wang, Y. Detection method to eliminate Sybil attacks in Vehicular Ad-hoc Networks. *Ad Hoc Netw.* **2023**, *141*, 103092. [CrossRef]
42. Newport, C.; Kotz, D.; Yuan, Y.; Gray, R.S.; Liu, J.; Elliott, C. Experimental evaluation of wireless simulation assumptions. *Simulation* **2007**, *83*, 643–661. [CrossRef]
43. Sommer, C.; German, R.; Dressler, F. Bidirectionally coupled network and road traffic simulation for improved IVC analysis. *IEEE Trans. Mob. Comput.* **2010**, *10*, 3–15. [CrossRef]

sensors

MDPI

*Article*

# EPOPTIS: A Monitoring-as-a-Service Platform for Internet-of-Things Applications

**Petros Zervoudakis [1], Nikolaos Karamolegkos [1], Eleftheria Plevridi [1,2], Pavlos Charalampidis [1] and Alexandros Fragkiadakis [1,*]**

1 Institute of Computer Science, Foundation for Research and Technology-Hellas (FORTH), GR70013 Heraklion, Greece; zervoudak@ics.forth.gr (P.Z.); nkaram@ics.forth.gr (N.K.); eleftheria@ics.forth.gr (E.P.); pcharala@ics.forth.gr (P.C.)

2 Department of Computer Science, University of Crete, GR70013 Heraklion, Greece

* Correspondence: alfrag@ics.forth.gr

**Abstract:** The technology landscape has been dynamically reshaped by the rapid growth of the Internet of Things, introducing an era where everyday objects, equipped with smart sensors and connectivity, seamlessly interact to create intelligent ecosystems. IoT devices are highly heterogeneous in terms of software and hardware, and many of them are severely constrained. This heterogeneity and potentially constrained nature creates new challenges in terms of security, privacy, and data management. This work proposes a Monitoring-as-a-Service platform for both monitoring and management purposes, offering a comprehensive solution for collecting, storing, and processing monitoring data from heterogeneous IoT networks for the support of diverse IoT-based applications. To ensure a flexible and scalable solution, we leverage the FIWARE open-source framework, also incorporating blockchain and smart contract technologies to establish a robust integrity verification mechanism for aggregated monitoring and management data. Additionally, we apply automated workflows to filter and label the collected data systematically. Moreover, we provide thorough evaluation results in terms of CPU and RAM utilization and average service latency.

**Keywords:** internet of things; blockchain; smart contracts; data integrity; management; monitoring; FIWARE

**Citation:** Zervoudakis, P.; Karamolegkos, N.; Plevridi, E.; Charalampidis, P.; Fragkiadakis, A. EPOPTIS: A Monitoring-as-a-Service Platform for Internet-of-Things Applications. *Sensors* **2024**, *24*, 2208. https://doi.org/10.3390/s24072208

Academic Editors: Behnam Mobaraki and Jose Turmo

Received: 28 February 2024
Revised: 23 March 2024
Accepted: 26 March 2024
Published: 29 March 2024



## 1. Introduction

In recent years, the technology landscape has been dynamically reshaped by the rapid growth of the Internet of Things (IoT), introducing an era where everyday objects, equipped with smart sensors and connectivity, seamlessly interact to create intelligent ecosystems. The evolution of the IoT has revolutionized how technology is perceived and utilized and opened the door for innovative and intelligent applications. From smart cities [1] and buildings [2], precision agriculture [3], and advanced energy management [4] to data acquisition and monitoring systems for hydrogen generators [5] and the bioleaching industry [6], the scope of IoT has expanded, fostering a connected world that leverages data-driven insights for enhanced efficiency; it is estimated (https://transformainsights.com/news/iot-market-24-billion-usd15-trillion-revenue-2030 (accessed on 25 March 2024)) that the number of IoT devices that will be in operation by 2030 will reach 21.1 billion, with a revenue of USD 1.5 trillion. Moreover, as Industry 4.0 combines traditional industries with cutting-edge technologies enabling smart processes and product realization, the IoT is a major driving force for efficient and massive decentralized data collection, supporting a complex system of diverse systems and devices [7].

The fundamental building blocks of the interconnected IoT network that makes up the IoT are smart devices and sensors that are equipped with smart communication abilities, make gathering and sharing data easy, and create a network of remarkable complexity, in this way necessitating the implementation of (semi-)automatic mechanisms to manage their diverse components effectively. Moreover, such networks often function in distant and challenging environments, making sensors susceptible to failures and malfunctions. Consequently, autonomous monitoring and maintenance are crucial to mitigate the risk of disruptions and ensure adherence to Service-Level Agreements between IoT application providers and consumers. Monitoring can follow either a passive or active approach, encompassing the collection and logging of data from different network subsystems. Data generated by modern IoT systems are typically stored in cloud storage services (CSSs) for subsequent processing and analysis tasks, including anomaly detection [8], fault diagnosis [9], predictive maintenance [10], and more. Utilizing external CSSs helps to address the difficulties associated with local storage management; however, it also introduces an increased risk of data manipulation on remote servers. Unfortunately, such tampering can reduce the precision and reliability of subsequent processing or analysis, compromising decision-making effectiveness; thus, data integrity becomes crucial for the efficient cloud storage of the collected data. IoT devices are highly heterogeneous in terms of software and hardware (https://datatracker.ietf.org/doc/html/rfc7228 (accessed on 25 March 2024)), and many of them are severely constrained (processor, memory, storage). This heterogeneity and potentially constrained nature creates new challenges in terms of security, privacy, and data management. Moreover, there is an increase in users' privacy concerns [11,12] as personal identifiable information is collected by IoT devices (e.g., wearables) and stored in cloud locations users are not aware of.

There is a plethora of IoT platforms and protocols ([13]), with their scope falling in three main categories: (i) those used for pure data collection, (ii) those used for management purposes, and (iii) hybrid, used for data collection as well as for management. Regarding data management, the protocols used mainly split into two categories: (i) IoT data collection protocols and (ii) IoT network management protocols. IoT data collection refers to the protocols and the support mechanisms for collecting IoT sensory data, including everything from determining the type and format of the data to be collected and deciding how frequently data should be collected to the methods by which data are transmitted from devices to the network. Additionally, IoT data collection protocols also encompass the issue of commands for actuation purposes, such as controlling a valve based on the data collected from the sensors. Several protocols of this type have been developed, such as the COAP [14], LwM2M (https://omaspecworks.org/what-is-oma-specworks/iot/lightweight-m2m-lwm2m (accessed on 25 March 2024)), MQTT (https://mqtt.org (accessed on 25 March 2024)), etc. On the other hand, the IoT network management protocols oversee the status, configuration, and performance of individual IoT devices on the network, including tasks such as device provisioning, firmware updates, and error handling. They also focus on the overall health and performance of the network as a whole, including tasks such as traffic management, resource allocation, and security. Overall, effective data management protocols are essential for ensuring that IoT devices function properly and that IoT data are collected and processed in a secure and efficient manner. Popular protocols include SNMP [15], NETCONF [16], RESTCONF [17], CORECONF [18], etc.

This work proposes EPOPTIS (EPOPTIS is the Greek word for supervisor), a Monitoring-as-a-Service (MaaS) platform for both monitoring and management purposes, offering a comprehensive solution for collecting, storing, and processing monitoring data from heterogeneous IoT networks for the support of diverse IoT-based applications. To ensure a flexible and scalable solution, we leverage the FIWARE open-source framework (https://www.fiware.org (accessed on 25 March 2024)), also incorporating blockchain (BC) and smart contract (SC) technologies to establish a robust integrity verification mechanism for aggregated monitoring and management data. Additionally, we apply automated workflows to filter and label the collected data systematically. The existing literature on data integrity verification for IoT data stored in cloud storage predominantly relies on encryption techniques, often coupled with the trustworthiness of Third-Party Auditors (TPAs). In contrast, BC-based data integrity schemes offer a compelling alternative by eliminating the need for TPAs, thereby addressing the trust issue. However, they have to face the issues of significant storage overhead, especially when considering the direct storage of the raw data in a BC ledger. To address the challenges mentioned, our solution adopts a strategic approach that involves the handling of raw IoT data in distinct time-windows. For integrity verification purposes, we generate a set of verified tags that are stored in the BC ledger. This methodology is designed with the goal to minimize the cost associated with data storage. Our main contributions are summarized as follows: (i) we propose a Monitoring-as-a-Service platform for IoT applications based on FIWARE, (ii) we utilize BC and SC technologies for data integrity verification purposes, and (iii) we thoroughly evaluate the platform in terms of CPU and RAM utilization and average service latency.

The remainder of the paper is organized as follows. Section 2 provides a summary of related works. In Section 3, we describe the system requirements and logical architecture of the proposed platform. In Section 4, the implementation details of the platform are presented. In Section 5, we present and discuss the performance evaluation results, and finally, the conclusions and further work appear in Section 6.

## 2. Related Works

This section presents works related to the proposed platform contributions, which mainly consist of (i) generic platforms for the support of IoT applications that do not utilize BC and SC technologies and (ii) platforms that use BC technology to accomplish automatic data integrity verification.

The authors in [19] propose an IoT platform based on edge computing to support various types of devices located within cities, introducing a cloud-native environment to solve operation and maintenance management problems. The proposed platform is divided into three layers: (i) a Terminal Device Layer, (ii) an Edge Layer, and (iii) a Cloud Layer, performing tasks such as data collection, pre-processing and partial storage of the incoming data, etc. The authors also describe a Smooth Weight Round-Robin algorithm for load balancing. Nevertheless, no evaluation results or details on various fundamental components of the platform (i.e., identity management and access control) are provided. Article [20] proposes MEWiN, a platform for precision agriculture that is based on FIWARE, using components such as the Cosmos Generic Enabler, Orion Context Broker, etc.; however, the evaluation results presented are limited to the power consumption of the devices, battery voltage discharge, and specific types of data collection (e.g., water content values). A Cloud-IoT-based sensing service for health monitoring is presented in [21], which consists of various layers (data collection, data management, application service), supporting wearables and smart devices. In [22], the authors propose a FIWARE-based platform for remote patient monitoring, considering various users such as physicians, medical operators, and patients. The platform consists of three logic layers: (i) a Front-end Layer that includes all required components for the interaction with the medical and paramedical staff, (ii) an Elaboration Layer, which implements the functionalities for health data gathering and storage, and (iii) a Security Layer that provides access control and identity management; however, no performance evaluation results are provided. The authors in [23] propose a FIWARE-based platform for sensor data monitoring in seaports, employing various FIWARE components, such as the Orion Context Broker, Cosmos, etc. The collection of various types of data, such as the wave height, ship orientation, etc., is demonstrated but without any evaluation results to showcase the scalability of the platform. The authors in [24] describe a monitoring platform for energy management that consists of three layers (acquisition, transmission, management), using the RS-485 and MODBUS-RTU protocols for the data communication between the acquisition devices and the data center services. OpenHab [25] is a freeware IoT platform that implements an open-source solution to the Eclipse SmartHome framework, using Apache Karaf and Eclipse Equinox runtime [13]. SmartThings [26] is a proprietary home automation platform developed by Samsung that follows the producer/consumer paradigm, supporting sensors and actuators. There are various other commercial platforms, such as the Apple HomeKit [27], Amazon Web Services IoT [28], IBM Watson [29], etc., which require commercial agreements or subscriptions. All of the aforementioned platforms and services have two common characteristics: (i) their core operations are centralized and hence threats that emerge as single-point of failures are possible (also considering the increased number of cyber-attacks worldwide) (https://www.enisa.europa.eu/publications/enisa-threat-landscape-2023 (accessed on 25 March 2024)), and (ii) they are vertical implementations having a very narrow scope, servicing only specific applications (e.g., healthcare).

Several other contributions use BC technology to accomplish automatic data integrity verification. Article [30] presents a BC-based scheme where integrity verification is performed by SCs; however, there are several limitations: (i) this is a standalone service that has not been evaluated within a complete IoT platform, (ii) third parties need to use a BC-compliant client in order to communicate with the BC, (iii) fees (Gas) have to be paid for the services as the Ethereum (https://ethereum.org (accessed on 25 March 2024)) BC is used, and (iv) there is a significant overhead for data encryption/decryption as the (hashed and encrypted) data are stored in the public ledger. The authors in [31] propose a data integrity detection model based on BC, with a process divided into three parts: data generation, data storage to the BC, and data fetching from the BC. In this work, data are directly stored in the BC, which is not efficient, and the evaluation is performed in a simulated environment that is not realistic given the current advances in BC technology. The HyperLedger Fabric (https://www.hyperledger.org/projects/fabric (accessed on 25 March 2024)) BC (HLF) is used in [32] to provide data integrity and storage, also employing IPFS (https://ipfs.tech/distributedstoragesystem (accessed on 25 March 2024)) for distributed storage. This platform consists of four layers (physical, network, middle, application), and the authors have demonstrated its scalability in terms of throughput and transaction time and average latency; however, a limitation is that the integrity verification process is "hardcoded", merely based on BC and HLF, as no SCs are employed. The authors in [33] propose a BC-based data verification scheme that splits into three stages (setup, processing, verification), using bilinear mapping techniques; however, their solution has two weaknesses: (i) various cryptographic operations (e.g., digital signatures) are required by clients that collect data, so it may be infeasible for this scheme to support constrained IoT devices, and (ii) the BC technology used is not based on a real BC such as HLF, Ethereum, etc., and rather it has been simulated using the Python programming language, so its robustness and usage in real environments is questionable. In [34], the authors present an integrity verification scheme that uses three types of SCs for: (i) checking data owners' legitimacy (using the RSA algorithm), (ii) checking data repository's (cloud servers) reliability using Merkle hash tree, and (iii) preventing replay attacks using bilinear mapping. Here, a different approach (as compared to EPOPTIS) is taken as only the data owners can request their data, so support for third-party applications (e.g., within a smart city context) cannot be easily supported. The authors have evaluated SC execution in an Ethereum network, but it is not clear if the core functionalities of the scheme can execute on Ethereum. The authors in [35] present a BC-based data integrity verification scheme for smart home applications defining five basic components: smart devices, trusted third parties, cloud service providers, home gateways, and blockchains. They utilize the home gateway to aggregate all data information and formulate homomorphic verifiable tags for verification purposes. However, there are several limitations: (i) limited evaluation results are provided, (ii) as a private bitcoin network is used, it cannot support a high number of transactions due to the Proof-of-Work consensus algorithm used, (iii) no flexibility is provided as SCs are not used, and (iv) the scheme is not integrated or tested within a real platform. Finally in [36], the authors propose a platform for smart city applications that is based on two BC levels: (i) private BCs that store the IoT data provided by the various city organizations (e.g., water management, energy management, etc.) and (ii) a consortium BC that stores the IoT data provided by the private BCs. The drawbacks of this approach are that multiple consensus algorithms have to execute prior to persistent IoT data storage, thus complexity and latency increase, and IoT devices have to send their data through BC transactions that may not be feasible if the devices are constrained (in terms of memory and processing). The related contributions that do not utilize BC technology for data integrity verification are shown in Table 1, while those that do utilize it are summarized in Table 2, along with their limitations.

**Table 1.** Related contributions that do not utilize BC technology.

| Contribution | Scope | Commercial |
|---|---|---|
| City-level IoT [19] | Generic | No |
| MEWiN [20] | Agricultural water management | No |
| [21] | Health monitoring | No |
| [22] | Remote patient monitoring | No |
| SmartPort [23] | Seaport data monitoring | No |
| [24] | Energy management | No |
| Openhab [25] | Smart home applications | No |
| SmartThings [26] | Smart home applications | Yes |
| Apple HomeKit [27] | Smart home applications | Yes |
| Amazon Web Services IoT [28] | Generic | Yes |
| IBM Watson [29] | Generic | Yes |

**Table 2.** Related contributions that utilize BC technology.

| Contribution | Scope | Limitations |
|---|---|---|
| [30] | Generic | Standalone service not evaluated within a real platform; BC-compliant client required; fees have to be paid; significant overhead for encryption/decryption; whole data records stored in the public ledger |
| [31] | Generic | Data stored in BC; simulated evaluation system |
| [32] | Generic | Integrity verification process not supported by SCs |
| [33] | Generic | Resource-intensive cryptographic operations required for the clients; simulated BC used |
| [34] | Generic | Gas required for the SC execution; not clear which BC is used for the core functionalities; only data owners can retrieve data, thus there is no support for authorized third-party applications; this a standalone verification service not integrated within a real platform |
| [35] | Smart homes | Limited evaluation results provided; a private bitcoin network is used for BC that cannot support a high number of transactions; no flexibility is provided as SCs are not used; the scheme is not integrated or tested with a real platform |
| [36] | Smart cities | Increased latency due to multiple consensus algorithms' execution; IoT devices required to send BC transactions |

## 3. System Requirements and Logical Architecture

This section presents the system requirements and the logical architecture of the EPOPTIS platform.

### 3.1. System Requirements

Following the common convention, system requirements are categorized into two types: *functional* and *non-functional*. *Functional* requirements define specific functions and capabilities that the platform must possess to meet the needs of its users, while the *non-functional* ones address constraints and performance characteristics that the system must exhibit to ensure its overall quality and user experience.

#### 3.1.1. Functional Requirements

- **Heterogeneous device support:** The platform should consider the inherent heterogeneity of native IoT device hardware platforms and ensure seamless integration, offering flexible enrollment for various device flavors.
- **Device virtualization:** Physical devices and their resources should be appropriately virtualized, adhering to a standardized information model that allows easy querying though common syntax.
- **Ubiquitous access and transparent communication:** Devices should be accessible irrespective of constraints imposed by end-node network topology and configuration. In scenarios where devices do not support IPv6 or are not IPv6 routable, devices should be accessed though transparent gateways that provide protocol translation or Network Address Translation.
- **Multi-tenancy:** Multi-tenancy mechanisms should be supported, providing complete resource isolation of different tenants and preventing unauthorized access.
- **Data processing of large-scale monitoring data:** The platform should be able to process and analyze large amounts of monitoring data and provide useful insights based on user-defined metrics.
- **Visualization of monitoring data:** The platform should visualize monitoring data, data integrity verification, and event-based notifications in a comprehensible and user-friendly manner (e.g., graphs, labels, etc.).
- **Data integrity:** The platform should ensure the integrity and consistency of monitoring data throughout its life cycle by developing robust mechanisms to prevent unauthorized modifications. This enhances the reliability and trustworthiness of the monitoring data, maintaining its accuracy and validity when visualized.
- **Real-time alerting:** A robust event-based notification system that operates on predefined rules should be incorporated, promptly alerting users about the violations of these rules. The platform should be capable of generating real-time alerts, ensuring timely communication of critical events or sensory data deviations from predefined rules.

#### 3.1.2. Non-Functional Requirements

- **Security and privacy:** The platform should integrate robust authentication and authorization mechanisms for users and IoT devices. Control and application data communication should be encrypted and integrity-protected.
- **High availability:** The platform should provide high availability of monitoring services as well as an IoT data storage service, taking into consideration diverse network conditions.
- **Fault tolerance:** The platform should be resilient and able to recover from faults and failures on both cloud and IoT devices.
- **Scalability:** The platform should be capable of handling large volumes of data in terms of storage, retrieval, and processing capabilities.
- **Quality of service (QoS):** QoS should be maintained as high as possible in terms of interactions with cloud services (e.g., low latency of data retrieval, near-real-time notification mechanism) and IoT devices (e.g., low latency of IoT data updates).

### 3.2. *Logical Architecture*

Here, we present the logical architecture of the proposed MaaS platform, with its components shown in Figure 1.

The *service management* layer provides appropriate application programming interfaces (APIs) for the interaction of external applications/users with the services offered by the platform. It acts as the main endpoint and service orchestrator, handling incoming requests and discovering and orchestrating all the internal services.

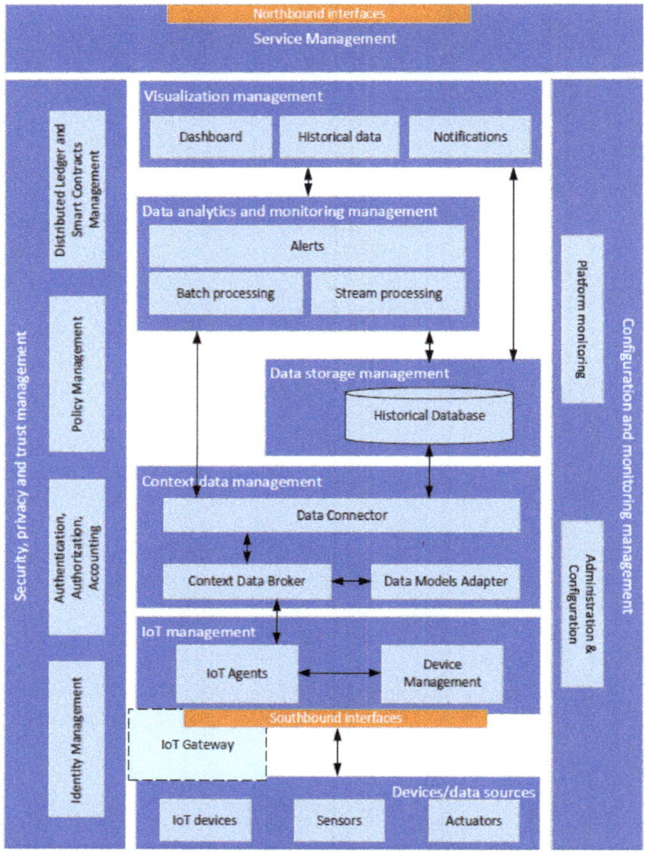


**Figure 1.** Logical architecture of the EPOPTIS platform.

The *visualization management* layer provides various mechanisms for visualizing context information pertaining to IoT networks managed by the platform. It encompasses a variety of functions, including the integration of IoT network status and measurements in a user-friendly dashboard. Additionally, it provides charts that showcase historical IoT sensory data and its integrity validity. Moreover, it presents notifications generated by the platform, ensuring that users are promptly informed about critical events and updates.

The *data analytics and monitoring management* layer provides a robust mechanism for processing the monitoring data collected from the IoT networks. The primary goal of this layer is to provide valuable insights into network performance and the sensory data collected. Additionally, it supports event-based notifications in response to abnormal situations or operation errors. Data processing can be performed either (near) real-time, as context data are collected, or on historical data persistently stored by the platform.

The *data storage management* layer provides essential functionalities for the secure, reliable, and persistent storage of context information originating from heterogeneous IoT networks.

The *context data management* layer is responsible for managing context information collected from IoT networks, offering a unified and standardized interface to access this information. Generally, contextual information is defined as the current state of all entities across the platform and it is represented in a structured manner using appropriate data models.

The *IoT management* layer is an entity responsible for bridging the communication gap between IoT devices and the *context data management* layer. Its primary goal is to ensure the compatibility and availability of these devices on the platform by implementing suitable protocol/data model translators. Additionally, this entity facilitates the management of heterogeneous IoT devices, encompassing functionalities such as registration, configuration, and monitoring.

The *devices/data sources* layer consists of IoT devices, including end devices equipped with sensors and/or actuators that enable the monitoring of their operational environment.

The *security, privacy, and trust management* layer is a cross-layer entity responsible for all operations, policies, and mechanisms implemented to ensure security, privacy, and trust in the platform's pillars (i.e., the cloud services of the platform and the IoT networks). This entity offers critical security and trust mechanisms for user authentication, authorization, access control, secure device bootstrapping, and data integrity verification. Security and system robustness is provided by utilizing the BC and SC technologies as (i) the data are stored in the immutable ledger (actually stored in replicas in multiple nodes), and hence data corruption is not feasible as it would require the corruption of the data in all possible locations and the re-calculation of the cryptographic hashes of all previous block stores in the ledger; (ii) the consensus algorithm executed by the BC peers guarantees that no malicious or faulty transactions can be validated within the BC network as long as the majority of the peers are properly functioning; (iii) SC states are protected through the immutable ledger and the consensus algorithm and their outputs are valid as long as the majority of the peers are honest; and (iv) the data producers (e.g., IoT devices) are authenticated and authorized by utilizing suitable authorization tokens and FIWARE components such as the Keyrock, while their data are encrypted in transit using transport layer security (TLS). Privacy preservation can become feasible through the use of ephemeral bearer authentication tokens that do not reveal user/device identities; moreover, only the hashed values of the collected data are stored in the ledger, and thus the actual data are protected in case a compromised BC peer is present.

The *configuration and monitoring management* layer is also a cross-layer entity responsible for managing and monitoring the configuration parameters of the platform and its components. This entity allows the definition of specific Key Performance Indicators to evaluate a platform's quality of services, reliability, and availability. By managing the configuration and continuously monitoring parameters, it ensures the system operates optimally and detects any deviations from normal conditions.

## 4. Implementation Details of the EPOPTIS Platform

In this section, we detail the system architecture of the proposed MaaS platform, initially, by defining the data model that describes the contextual information and then presenting its functional architecture and discussing the functional entities and their interactions.

### 4.1. Context Information Model

The *Context Information Model* (CIM) establishes a standardized and structured representation of contextual information, facilitating the capture, organization, and sharing of such information between IoT networks and the monitoring platform. Within the CIM, context information is represented using entities, attributes, and relationships. Entities can correspond to conceptual abstractions or physical objects, while attributes describe the properties or characteristics of the entity. The relationships in this model define the semantic associations and dependencies between different entities. A CIM representation is illustrated in Figure 2, appropriately adjusted to align with the NGSI (Next-Generation Service Interface) protocol (https://fiware.github.io/specifications/ngsiv2/stable (accessed on 25 March 2024)). This adjustment ensures that the platform's entities can efficiently communicate and integrate context information, fostering a harmonious and interconnected monitoring ecosystem. At the core of this data model, the *Device* entity acts as a fundamental abstraction, representing a generic device that encompasses essential attributes common to all devices. Depending on the nature of the physical devices, the *Device* entity can be specialized into two other distinct entities, each one tailored to specific characteristics. The *Sensing Device* entity represents specialization in sensor-related characteristics, facilitating the representation of data measurements. Additionally, further specialization for the *Sensing Device* entity provides a specific representation of air quality measurements, as defined in the *AirQSensor* entity. On the other hand, the *Gateway* entity specializes as a *Device*, representing all the characteristics required for a networking device. In this context, each *Device* entity can be associated with 0 to *n Notification* entities, indicating that a *Device* can have multiple *Notifications* related to it. The association between *Devices* and *Notifications* enables the effective monitoring and alerting within the platform, allowing users to receive timely and relevant information about potential issues or critical events associated with specific *Devices*.

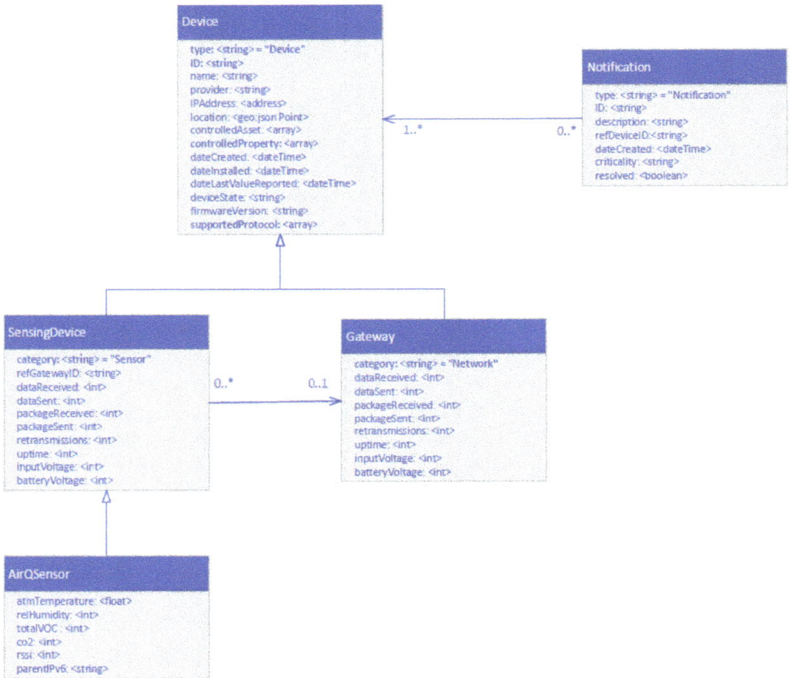


**Figure 2.** Contextual entities and their associations.

### *4.2. Functional Architecture*

In Section 3, we described the logical architecture of the proposed platform (Figure 1), while, in this section, we present its functional architecture, which comprises three layers, as depicted in Figure 3. The *Application Layer* encompasses user interfaces and dashboards used for visualizing all information collected by the MaaS platform. This layer can also feed third-party visualization platforms such as Kibana (https://www.elastic.co/kibana (accessed on 25 March 2024)), Grafana (https://grafana.com (accessed on 25 March 2024)), ThingSpeak (https://thingspeak.com (accessed on 25 March 2024)), etc. Currently, EPOPTIS feeds data with the visualization platform presented in [37].

The second layer, the *Service Layer*, involves various processes that provide critical functionalities, including identity and authorization management, contextual information management, historical data management, data integrity verification, and alert-based notification mechanisms. In this regard, FIWARE offers a set of specifications accessible through well-defined interfaces and supports a flexible architecture that facilitates the interconnection of devices with IoT applications. FIWARE's adoption fits our platform's needs as it simplifies development by providing a collection of extensible, scalable, and configurable components that can foster application development.

The third layer, the *Infrastructure Layer*, offers the necessary hardware and virtualized resources required to deploy the MaaS platform effectively. By adhering to the requirements as detailed in Section 3.1, the proposed architecture allows cloud services to interact seamlessly with IoT Devices, ensuring a robust and efficient MaaS platform.

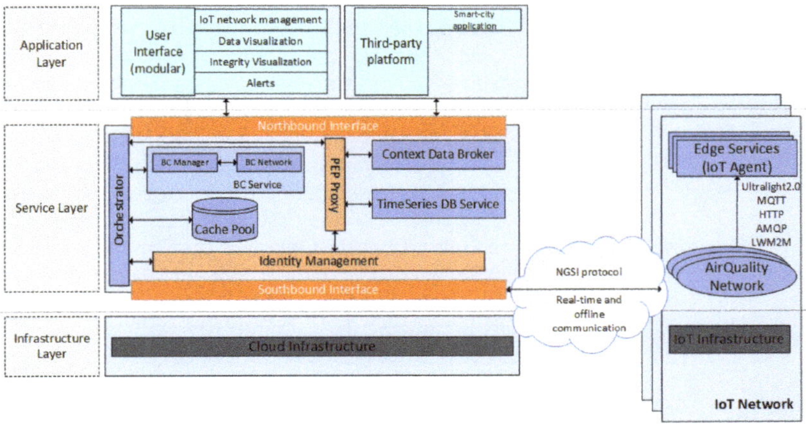


**Figure 3.** Functional architecture of the EPOPTIS platform.

4.2.1. Orchestrator

In order to construct a reliable, robust, and secure cloud-based platform, the monitoring, management, and orchestration of a variety of underlying heterogeneous technologies is required. In the core of the proposed architecture resides the *Orchestrator*, which serves as a gateway between our platform and third-party applications or IoT networks that wish to utilize the underlying provided services. The *Orchestrator* receives and verifies submitted requests to the platform, performs the corresponding actions, and finally generates the appropriate responses. It offers a RESTful API for communication and data exchange, consisting of two logically separated interfaces: (a) the *Northbound Interface*, which handles requests from user interfaces and third-party platforms, enabling the data retrieval of monitoring data and other provided services, and (b) the *Southbound Interface*, which handles requests from the IoT networks. The *Orchestrator* implements a number of software modules, named *Resource Modules*, which are responsible for managing the various heterogeneous system resources. Upon receiving a request, it communicates with the internal services, as dictated by the processing workflow, and finally returns the appropriate response. Below, we provide a description of the *Resource Modules* :

- ***Identity Module:*** This module is responsible for managing information about logical hierarchical entities on our platform, including virtual representation of human and non-human users (i.e., IoT users), ecosystems, roles, permissions, and their relevant connection graphs. The *Identity Module* utilizes the *Identity Management* component to store and associate this type of information in its internal database.
- ***Device Module:*** This module is responsible for retrieving information regarding physical devices and their association with IoT networks, supporting the retrieval of essential information about both types of physical devices, gateways, and sensing devices, including their attributes, as described in the contextual model. The *Device Module* utilizes the *Context Data Broker* component to retrieve this kind of information.
- ***Statistics Module:*** This module is responsible for retrieving and calculating statistics about logical entities and their associations. It provides detailed information about users, devices, and ecosystems, including statistics such as the number of users within an IoT ecosystem and the assigned role for each user within an IoT network.

The *Statistics Module* utilizes the *Context Data Broker* and the *Identity Management* components to provide this type of information.

- ***Notification Module:*** This module is responsible for the retrieval of information regarding generating notifications and organizing them in two categories: the *IoT Ecosystem* or the *Device*. The retrieved information includes criticality level, descriptions, and the attribute on which the notification was triggered. Additionally, this module allows authorized users to manage generated notifications (e.g., appropriately labeling them once they have been resolved). The *Notification Module* utilizes the *Context Data Broker* to support these operations.
- ***IoT Module:*** This module is responsible for registering a new device and updating its attributes. It receives NGSI payloads from IoT networks and implements mechanisms for: (i) triggering alert-based notifications according to the SC rules, (ii) caching the incoming NGSI payloads using a *Cache Pool*, calculating and storing the appropriate hash value required for the data integrity verification, and (iii) updating contextual information in order to support a seamless interaction with the IoT networks. This module utilizes the *Cache Pool*, the *Context Data Broker*, and the *BC Service* (described in Section 4.2.6) in order to support all the above operations.
- ***Query Module:*** This module is responsible for retrieving historical data and labeling them utilizing the data integrity verification mechanism. A detailed description of how this mechanism works is provided in Section 4.3. This module utilizes the *TimeSeries DB* service for retrieving historical data and the *BC Service* to obtain the appropriate hash value used to label them according to the outcome of the integrity protection mechanism (i.e., corrupted/not corrupted).

#### 4.2.2. Identity Management and PEP Proxy

These two components are responsible for the authentication and authorization operations of the platform, which encompass: (i) identity and user management and (ii) user authorization and access control. The *Identity Management* component provides the ability to create, update, and manage user accounts, storing user profile information, authentication credentials, roles, and permissions. When a request is made to access a protected resource, the *PEP Proxy* intercepts it and enforces the access control policies defined in the *Identity Management* component. Here, we utilize the FIWARE Keyrock (https://fiware-idm.readthedocs.io/en/latest (accessed on 25 March 2024)) for identity and user management and the Wilma PEP Proxy (https://fiware-pep-proxy.readthedocs.io/en/latest (accessed on 25 March 2024)) for authorization and access control.

#### 4.2.3. Cache Pool

This component acts as in-memory data storage for the NGSI payloads, which are grouped within specific time intervals before being fed into the caching mechanism. A detailed description of how this mechanism works is provided in Section 4.3. Here, we utilize Redis (https://github.com/redis/redis (accessed on 25 March 2024)) for the in-memory data storage.

#### 4.2.4. Context Data Broker

When connecting IoT devices to an IoT platform, the publish/subscribe pattern has proven to be a suitable method for messaging [38]. To this direction, the *Context Data Broker* component achieves the decoupling of producers and consumers of context information, implementing the Publish/Subscribe design pattern based on the FIWARE Orion Context Broker (https://fiware-orion.readthedocs.io/en/master (accessed on 25 March 2024)). Context producers (e.g., IoT ecosystems) publish their data to this component through the FIWARE NGSI API, without the requirement to know who the consumers of such data are. Context consumers (e.g., third-party applications, historical databases) do not need to know the origin of the data but are solely interested in the event itself which is consuming them. This decoupling mechanism allows context-aware IoT ecosystems to interact with the context information in a flexible and scalable manner, enabling seamless integration and fostering the development of a context-driven approach. The *Context Data Broker* supports the querying of the last value referring to the attributes of each context entity and also provides a subscription mechanism, enabling the persistent storage of the historical attributes.

4.2.5. TimeSeries DB Service

This is a component based on FIWARE QuantumLeap (https://quantumleap.readthedocs.io/en/latest (accessed on 25 March 2024)), designed to provide persistent storage of context information from the *Context Data Broker* in an external repository by converting the NGSI structured data into a tabular format and persistently storing context changes in a high-performance CrateDB database (https://crate.io (accessed on 25 March 2024)). It also provides an interface for performing complex queries on historical data (e.g., the latest *N* samples collected from a specific IoT device, the minimum value over a time period per hour or month, etc.).

4.2.6. Blockchain Service

Data integrity verification is one of the core functionalities of the proposed platform and this is utilized through the *BC Service*, which consists of two discrete components, namely (i) the *BC Network* and (ii) the *BC Manager*. The *BC Network* maintains a permissioned decentralized ledger based on HLF that facilitates the process of transaction recording and asset tracking in an immutable manner. The network structure and data flow are designed and implemented in order to achieve high service availability and to minimize the risk of being compromised. The utilization of a private HLF network offers several key advantages, particularly in terms of high availability and reliability. Using HLF, we have utilized a network of four organizations (https://hyperledger-fabric.readthedocs.io/en/latest/network/network.html (accessed on 25 March 2024)), three acting as *Peer* organizations and one as *Orderer*. Each of the *Peer* organizations operates with two *Peer* nodes, while the *Orderer* organization uses two *Orderer* nodes. The endorsement policy used assures a high availability, also requiring the majority of organizations to endorse transactions using at least one of their *Peer* nodes. This redundancy ensures that if a *Peer* node experiences downtime for any reason, the other *Peer* nodes within the same organization can seamlessly take over during transaction validation. Similarly, the presence of two *Orderer* nodes within the *Orderer* organization, guarantees fault tolerance in case a node becomes unavailable.

The *BC Manager* functions as a client for one of the three peer organizations. A client in the context of HLF refers to an authorized application that can interact with the *BC Network*, essentially acting as an interface (implemented as a REST API with Golang) that links the HLF network with the *Orchestrator*.

The *BC Service* provides two core functionalities:

- ***Data labeling***. SCs are crucial for the automated label compliance assignment of the collected IoT data. These enforce configurable rules that encompass various criteria, such as acceptable sensor data ranges, input voltage limits, etc. When data are submitted to the platform, SCs automatically assign a label based on these predefined criteria. Therefore, during the data retrieval processes, the historical data are labeled as compliant or non-compliant (based on the previously reported rules), also feeding a suitable API for visualization purposes.
- ***Hash computation and storage***: The *BC Service* is responsible for storing and retrieving hash values, which are computed by the *Orchestrator* using the SHA-3 (https://csrc.nist.gov/pubs/fips/202/final (accessed on 25 March 2024)) hash algorithm. The corresponding hashed values are securely stored in the immutable BC ledger in a key-value format, significantly reducing retrieval times and conserving storage space within the ledger, while the actual data records are stored in the *TimeSeries DB*. The hashed values can be retrieved by the *Orchestrator* in order to support the data integrity verification mechanism during historical data retrieval.

Overall, the *BC Service*, with its decentralized nodes, automated recovery mechanisms, and data validation and storage processes, offers a resilient and reliable foundation for managing and securing IoT data.

### *4.3. Interactions of the Functional Components*

#### 4.3.1. NGSI Data Persistent Storage

Prior to sending an NGSI payload to the *Orchestrator*, the *IoT Agent* (Figure 1) authenticates and obtains an access token with permissions that allow access to the platform's internal services. In general, an NGSI request can be of two types: (i) a registration request, indicating that the NGSI payload is associated with a device that has not been already defined in the contextual entities (Figure 2), or (ii) an update request, signifying that the NGSI payload should update an existing contextual entity. After receiving an NGSI request from an *IoT Agent*, the *Orchestrator* manages and orchestrates all internal services for the (i) creation of the appropriate label for the payload; (ii) update of the relevant *Notification* entity to the *Context Data Broker*, if required; (iii) update of the relevant *Device* entity to the *Context Data Broker*; (iv) trigger of the appropriate broker notification to store the related attributes persistently in the *TimeSeries DB*; and (v) support of the mechanism that caches the incoming payloads and computes the corresponding hash values that are finally stored in the BC ledger.

In more detail and as shown in Figure 4, the *Orchestrator* sends a request for the validation of an NGSI payload (1). The *PEP Proxy* monitors the request flow and checks whether or not the request has the permission to access the *BC Service*'s resources for validating the payload's content using the SC-based rules. The *Identity Management* service acts as the policy decision point (PDP) and infers if the request should access the specific resource or not (2, 3). The *PEP Proxy* enforces PDP's decision to the *BC Service*, enabling or not the validation of the NGSI payload (4). The *BC Service* responds with an appropriate label that characterizes the payload's validity (5).

Subsequently, if the payload does not validate the SC-based rules, the *Orchestrator* creates (or updates) the appropriate *Notification* entity in the *Context Data Broker*. The *PEP Proxy* monitors the request flow and checks whether or not the request has the access permission to create or update a *Notification* entity (6) and the PDP infers if the request has the permission to access the specific resource or not (7, 8). The *PEP Proxy* enforces PDP's decision to *Context Data Broker*, enabling or not the creation or update of the *Notification* entity (9). Once the *Notification* entity is created or updated, the *Context Data Broker* stores the latest value of a notification and then sends the *Orchestrator* a suitable response message (10).

Additionally, the *Orchestrator* creates (or updates) the *Device* entity (11), ensuring that the contextual entities are kept updated to the latest state. The *PEP Proxy* enforces access control to the *Context Data Broker* (12, 13), triggering the appropriate contextual notifications (14), persistently storing the historical attributes of the *Device* entity (15). Once the payload's attributes are successfully stored in the *TimeSeries DB*, the appropriate responses are generated (16, 17). Moreover, the *Orchestrator* implements a caching mechanism that facilitates the calculation of the hash value that for a group of payloads. Upon receiving NGSI payloads, the *Orchestrator* caches them in the *Cache Pool* (18) and awaits the appropriate response (19). Once a predefined volume of payloads has been cached, the *Orchestrator* calculates the hash value of this payload group (20) and stores it in the BC ledger (24). The access to the *BC Service* requires the access control process enforced by the *PEP Proxy* on the *BC Service*'s resources (21, 22, 23). The hash value is generated using the SHA-3 secure hash algorithm, enabling the immutable storage of an imprint of the data instead of storing the actual data themselves. This approach provides a more efficient and secure way to represent the data in the *BC Service*, reducing storage requirements and ensuring data integrity.

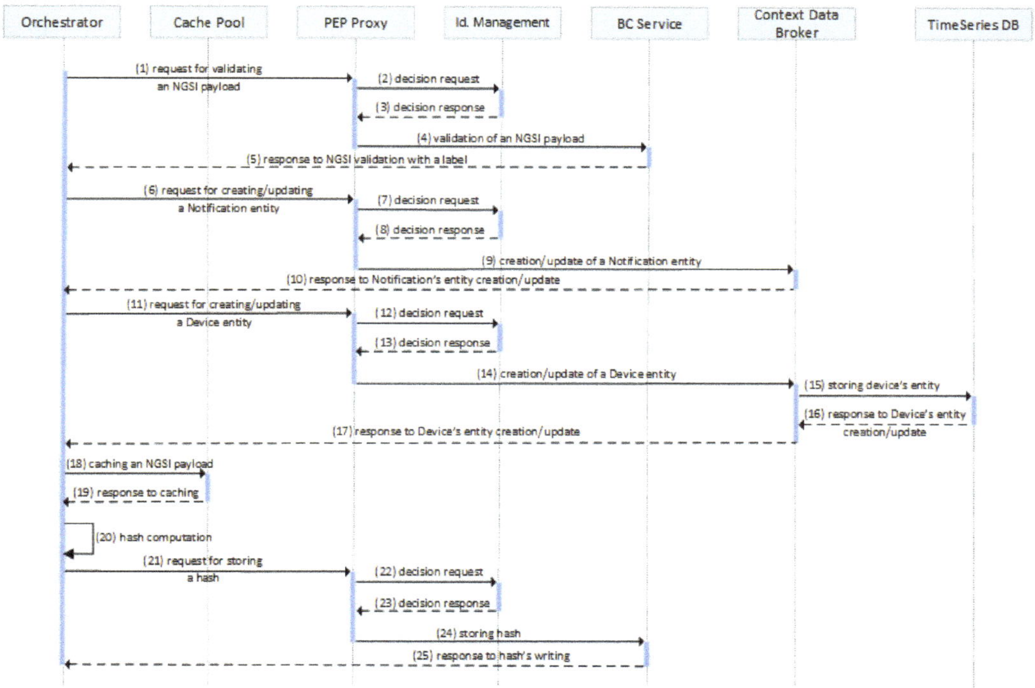


**Figure 4.** Functional component interactions for storing NGSI data.

#### 4.3.2. Retrieval of Historical Data

After receiving a retrieval request, the *Orchestrator* manages and orchestrates all internal services for (i) retrieving historical data from the *TimeSeries DB*, (ii) calculating the hash value of the retrieved data, (iii) fetching the hash value through the *BC Service* (which was previously computed during storage), and (iv) comparing the hashes and labeling the retrieved data, inferring its integrity. All these interactions are depicted in Figure 5. Initially, the *Orchestrator* sends a request in order to retrieve historical data (1). The *PEP Proxy* checks if access to the resources of the *TimeSeries DB Service* for retrieving the data is permitted. The *Identity Management* service acts as the PDP and infers if the request should access the specific resource or not (2,3). The *PEP Proxy* enforces the PDP's decision to the *TimeSeries DB Service*, enabling or not the retrieval of the historical data (4). The *TimeSeries DB Service* provides the requested data (5), and then, the *Orchestrator* calculates the corresponding hash value (6). Additionally, the *Orchestrator* retrieves the hash value that was previously stored by sending a request to the *BC Service* (7). The access to this service requires the access control process enforced by the *PEP Proxy* on the *BC Service*'s resources (8, 9, 10). The *BC Service* responds with the hash value that was previously stored (11). By comparing this with the hash value calculated for the recently retrieved data (12), any (unauthorized) modifications of the initial data received from *Orchestrator* during the storing process can be detected.

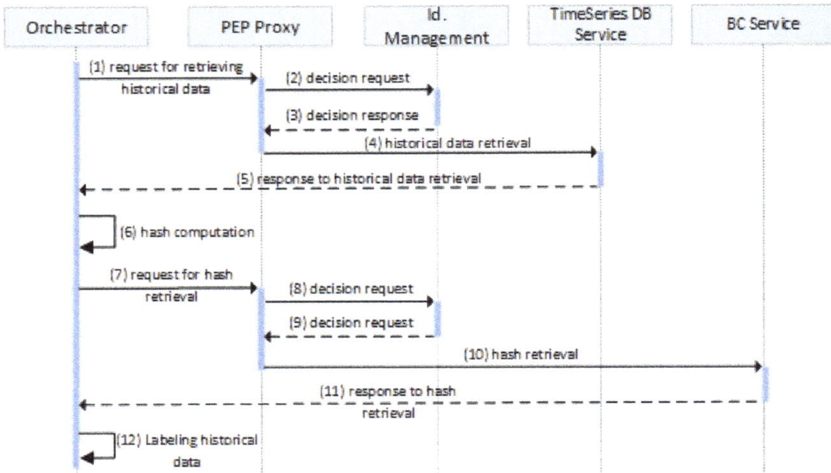


**Figure 5.** Functional component interactions for retrieving historical data.

## 5. Performance Evaluation

With the intended goal to act as a multi-tenant solution offering unified monitoring and management services for heterogeneous IoT networks, the cloud platform presented in this work needs to efficiently handle a substantially large number of IoT devices pushing data updates, either directly or through appropriate gateways. Therefore, in this section, we provide an extensive performance evaluation of a Proof-of-Concept (PoC) deployment of the proposed platform, under variable workload, and further present and discuss the results of the evaluation.

### *5.1. Testbed and Deployment*

Our testbed essentially comprises two distinct entities, namely (i) the *Load Generator infrastructure* and (ii) the *Cloud Infrastructure*. The *Load Generator* (LG) is responsible for generating configurable workloads (IoT device data updates) for the cloud backend and collecting appropriate performance metrics. We used *Apache JMeter* (https://jmeter.apache.org (accessed on 25 March 2024)) as the specialized software for workload generation and performance metric collection, which is a mature and industry-grade modular open-source tool used for the automated performance evaluation of web-based systems through load and regression tests. Developed in Java, it integrates support for various protocols and applications (e.g., HTTP(s), REST Webservices, FTP, LDAP, etc.) and supports the creation and execution of configurable and complex workloads on a web server, collecting and storing replies, as well as analyzing and presenting aggregate test statistics (e.g., latency, throughput, response time, etc.).

On the other hand, the *Cloud Infrastructure* (CI) hosts a complete deployment of our platform, as depicted in Figure 3. The cloud backend leverages several Generic Enablers of the FIWARE ecosystem, as described in Section 4.2. Specifically, the *Orchestrator* is a web application developed in Python, which is responsible for orchestrating the rest of the backend services and exposing a unified and comprehensive RESTful API for third-party applications to consume. Additionally, we developed the *BC Manager*, named here as *Pearl*, which is responsible for the deployment and maintenance of the *BC Service*, necessary for the IoT data validation and integrity verification functionalities offered by the platform. Essentially, it is an application written in the Go programming language that offers a rich RESTful API for the interaction between the *Orchestrator* and the *BC Service*.

Moreover, we used a microservices architecture [39] for the cloud-based deployment of the platform. In particular, we employed Docker containers operated by the Container Orchestration Engine, Kubernetes (https://kubernetes.io (accessed on 25 March 2024)), which conveniently enables the management of distributed and horizontally scalable cloud-native applications, with an emphasis on high availability and fault tolerance. We deployed Kubernetes in a self-managed private cluster that consists of six virtual machines (VMs): one master node and five worker nodes, one of which exclusively hosts the *BC Service*, i.e., both the *BC Manager* and the *BC Network*. Table 3 summarizes the VM specifications used as LG infrastructure and CI infrastructure for the platform deployment.

**Table 3.** Specifications of VMs deployed for performance evaluation.

| | vCPUs | RAM | Storage | Operating System |
|---|---|---|---|---|
| **LG infrastructure** | 4 | 8 GB | 20 GB HDD | Ubuntu 20.04LTS |
| **CI infrastructure—master node** | 4 | 10 GB | 30 GB HDD | Ubuntu 20.04LTS |
| **CI infrastructure—worker nodes (4×)** | 4 | 10 GB | 30 GB HDD/SSD | Ubuntu 20.04LTS |
| **CI infrastructure—worker nodes (BC)** | 4 | 16 GB | 50 GB SSD | Ubuntu 20.04LTS |

From a performance point of view, we enabled CPU pinning (tying virtual CPUs to real physical CPUs of the host) for the VM hosting the CI, with KVM hypervisor set in "host-passthrough" CPU mode. Preliminary experimentation showed that this configuration offers substantial performance gains compared to different configuration choices. In addition, we tuned the operating system limits so that our measurements reflect the capabilities of the applications, instead of the limits enforced by the operating system. Thus, the user limits for file size, max memory, and CPU time were set to unlimited, and the open file limit was increased to 64,000.

We used the *MicroK8s* (https://microk8s.io (accessed on 25 March 2024)) distribution for installing the Kubernetes cluster, which is maintained by Canonical (https://canonical.com (accessed on 25 March 2024)) and is provided through the *snap* package manager. It offers a lightweight *Certified Kubernetes Software Conformance* (https://www.cncf.io/training/certification/software-conformance (accessed on 25 March 2024)) distribution, which simplifies the selection of several Kubernetes functionalities through easy (de-)activation of add-ons (e.g., DNS, ingress, metrics-server, etc.). In addition, we utilized the *Helm* (https://helm.sh (accessed on 25 March 2024)) package manager for installing and maintaining the platform services after developing the necessary Helm charts. The interactions between the deployed components are depicted in Figure 6. It is noted that we used the Nginx Kubernetes Ingress Controller as a reverse proxy to route incoming traffic towards the appropriate backend service. Table 4 summarizes the software version as well as the best-performing configuration for each software component according to our extensive preliminary experimentation.

**Table 4.** Software versions and best-performing configuration for each software component.

| Software Component | Software Version | Best Performing Configuration |
|---|---|---|
| MicroK8s | 1.26.1 | - |
| Orchestrator | 0.1.0 | • Gunicorn worker type: sync<br>• Gunicorn workers: 9 |
| Redis | 7.0.7 | - |
| Keyrock | 8.0.0 | - |
| MySQL | 8.0.32 | - |
| Orion | 3.6.0 | • reqMutexPolicy: none<br>• reqPool: 4 |

**Table 4.** *Cont.*

| Software Component | Software Version | Best Performing Configuration |
|---|---|---|
| MongoDB | 4.4.11 | - |
| QuantumLeap | 0.8.0 | • Gunicorn worker type: gthread<br>• Gunicorn workers: 9 |
| CrateDB | 4.6.7 | • Heap size: 2 GB |
| Wilma | 8.0.0 | - |
| Pearl | 0.1.0 | - |
| Hyperledger Fabric | 2.4 | - |

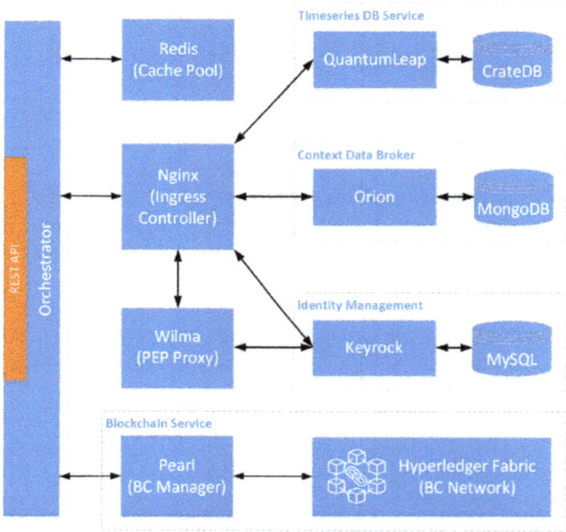


**Figure 6.** Interactions between the deployed platform components.

## 5.2. Test Plan

In order to quantify the performance of our platform, we measured (i) the latency of the device data update requests (HTTP PATCH requests—they carry NGSI-based payloads) towards the REST API of the *Orchestrator* and (ii) the resource utilization (CPU and RAM utilization) for all platform services under a variable workload. For orchestrating the experiments, we created 10 different ecosystems and registered 10 IoT devices per ecosystem. Two different cases were examined, namely (i) ecosystems without IoT data integrity verification (no use of the *BC Service*) and (ii) ecosystems with IoT data integrity verification (use of the *BC Service*). Each data update request has a payload of 190 bytes. Furthermore, in order to evaluate the horizontal scalability of the platform, we leveraged the native Kubernetes horizontal scaling mechanism and created load-balanced replicas for the most resource-intensive services (*Orchestrator*, QuantumLeap, and Orion Context Data Broker), as observed during preliminary experimentation.

### 5.2.1. Ecosystems without Data Integrity Verification

We initially increased the number of concurrent users (i.e., active JMeter threads) in a step-wise fashion in order to empirically quantify the maximum rate of requests/s the corresponding endpoint can successfully serve. This process was repeated independently for each number of replicas. Subsequently, we configured JMeter so as to generate asynchronous update requests and increase load in a step-wise fashion (step = 10 requests/s, up to the empirically measured maximum rate. Figure 7 depicts the corresponding test plan for a single replica per service. We summarize the parameters used in our experiments in Table 5.

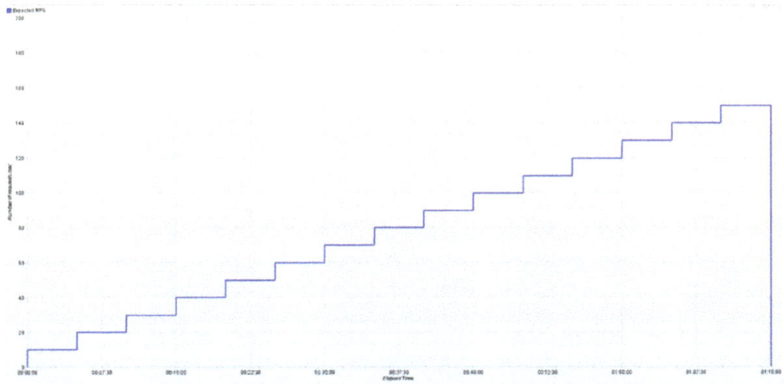

**Figure 7.** Step-wise increase in IoT data update load (no data integrity verification).

**Table 5.** Experimental parameters (no data integrity verification).

| Parameter | Values |
|---|---|
| Average data update load (Basic scenario—1 replica) | 10–150 requests/s (step = 10) |
| Average data update load (2 replicas) | 10–190 requests/s (step = 10) |
| Average data update load (3 replicas) | 10–240 requests/s (step = 10) |
| Average data update load (4 replicas) | 10–260 requests/s (step = 10) |
| Payload size | 190 bytes |

5.2.2. Ecosystems with Data Integrity Verification

In this scenario, we utilized the *BC Service* through the Pearl BC Manager for implementing the IoT data integrity verification mechanism. Following the same strategy as described above, we initially increased the number of active JMeter threads in a step-wise fashion in order to empirically quantify the maximum rate of requests/s the corresponding endpoint can successfully serve. In this case, data update requests are served at a lower rate, mainly due to the consensus mechanism and the transaction processing overhead the *BC Service* introduces. As a result, we configured JMeter to generate asynchronous update requests by applying increasing load at levels (5–30 requests/s, step = 5), as shown in Figure 8. Having identified the communication with the *BC Service* as the bottleneck of this scenario, we solely use a single replica of any backend service.

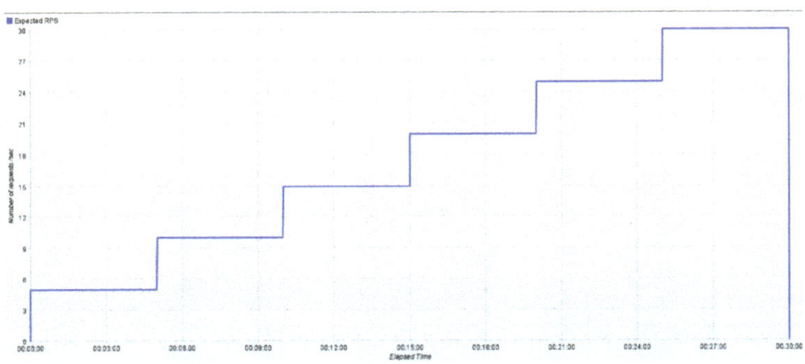


**Figure 8.** Step-wise increase in IoT data update load (with data integrity verification).

### 5.3. Evaluation Results

In this section, we present the evaluation results for different IoT data update load levels. We provide the average and standard error (in the form of error bars) for CPU and RAM utilization of all backend platform services, each one running in separate (possibly replicated) Kubernetes pods, as well as the latency of IoT data update requests.

#### 5.3.1. Ecosystems without Data Integrity Verification

The average CPU and RAM utilizations of the backend services (each one corresponding to a unique Kubernetes pod replica) when no data integrity verification is applied are depicted in Figure 9 and Figure 10, respectively. As expected, average CPU utilization increases as data update rate increases for all backend services. Keyrock and MySQL services have the lowest CPU utilization, which is almost constant irrespective of the number of requests/s. This happens mainly due to the Wilma PEP Proxy caching of access control decisions that significantly reduces the number of requests towards the PDP resting in Keyrock. The two backend services with the highest average CPU utilization are the *Orchestrator* (3000 millicores for 150 requests/s) and QuantumLeap (1700 millicores fro 150 requests/s), both of them developed as web services using the Python Flask framework and Gunicorn WSGI HTTP server. For any service apart from the *Orchestrator*, CPU utilization increases less than linearly to the load (in requests/s), illustrating the scalability of the platform.

By carefully inspecting Figure 10, we conclude that most of the backend services have relatively low RAM requirements. CrateDB is the only service that requires almost 1.8 GB RAM, allocated as Java heap memory. In addition, there is a negligible increase in RAM utilization, as load increases. Figure 11 depicts the latency for IoT data update requests under varying load. We observe that the latency median remains almost constant and lower than 40ms in any case, except for 150 requests/s, where it is almost 55 ms.

Figure 12 illustrates the average CPU utilization per node of the Kubernetes cluster the backend services are deployed on for the basic scenario (one replica per service). Observe that node "worker-ssd-1" utilizes almost 90% of the available CPU, while other worker nodes are underutilized. Based on this observation, as mentioned before, we leverage the Kubernetes horizontal pod scaling functionality for horizontally scaling the three most intensive backend services (*Orchestrator*, QuantumLeap, and Orion Context Data Broker) in order to improve utilization of available cluster resources. As shown in Table 5, two replicas may serve at most 190 requests/s (+27% compared to basic scenario), three replicas at most 240 requests/s (+60% compared to basic scenario), and four replicas at most 260 requests/s (+73% compared to basic scenario).

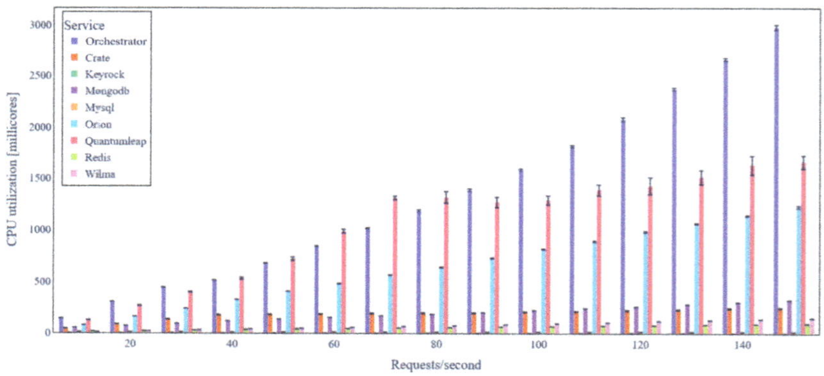


**Figure 9.** Backend service average CPU utilization under variable IoT data update request rate (basic scenario).

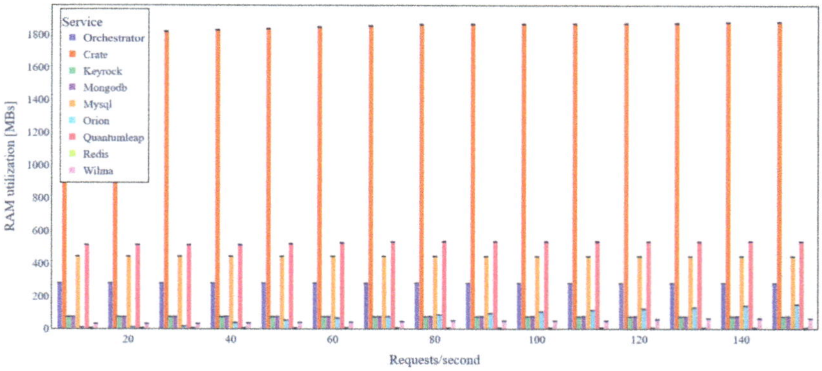


**Figure 10.** Backend service average RAM utilization under variable IoT data update request rate (basic scenario).

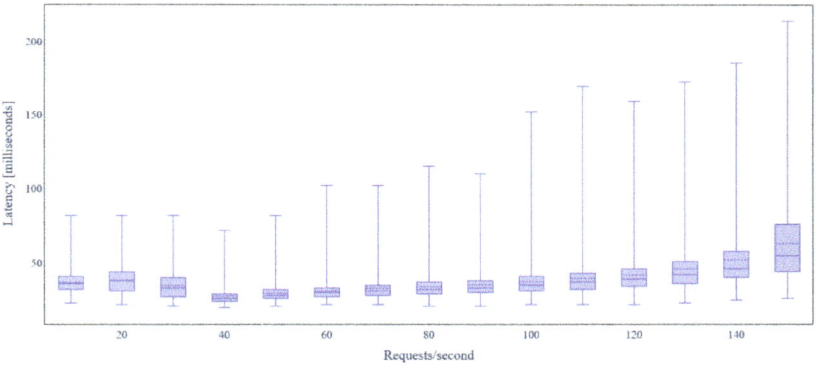


**Figure 11.** Latency under variable IoT data update request rate (basic scenario).

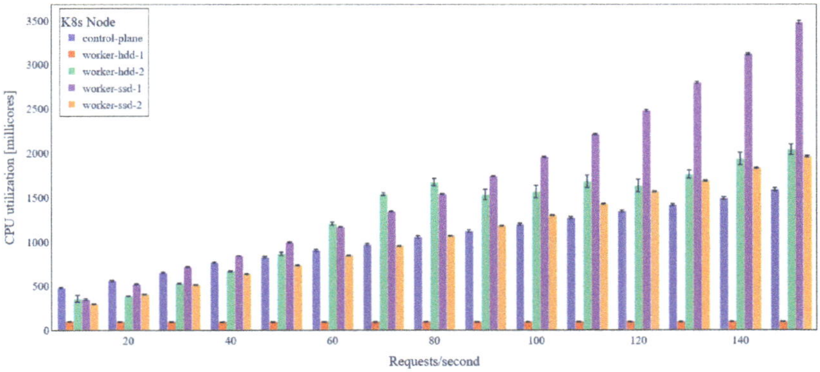


**Figure 12.** Average node CPU utilization under variable IoT data update request rate (basic scenario).

Figure 13 illustrates the average CPU utilization of all backend services when the three most intensive services, namely *Orchestrator*, QuantumLeap, and Orion Context Data Broker, are horizontally scaled (two, three, and four pod replicas per service). For the sake of clarity, we omit rates lower than 100 requests/s, since average CPU utilization exhibits no significant differences compared to the basic scenario for these rates. We note that we report aggregate average CPU utilization per service by calculating the sum of the CPU utilization of all replicas. As in the basic scenario, average CPU utilization increases with the increase in the IoT data update rate. This happens in a sub-proportional manner for all services apart from the *Orchestrator*, possibly due to the worker model employed by the *Gunicorn* (https://gunicorn.org (accessed on 25 March 2024)) WSGI HTTP Server. In any case, the three backend services that are horizontally scaled remain the most CPU-intensive ones.

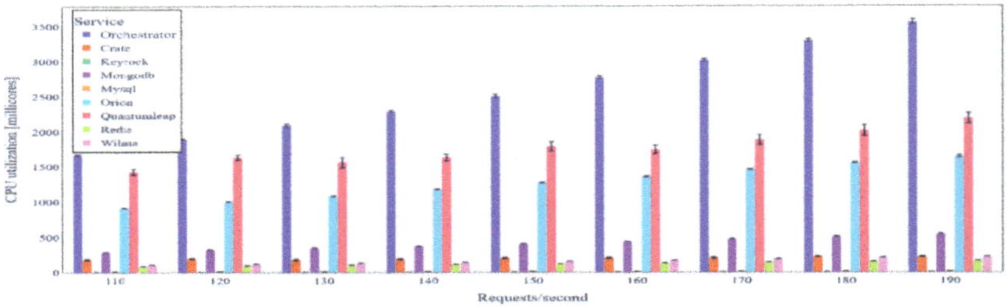


(**a**) Two-pod replicas.

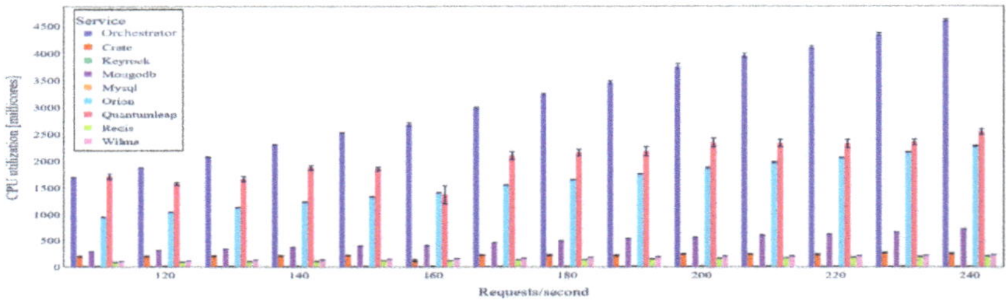


(**b**) Three-pod replicas.

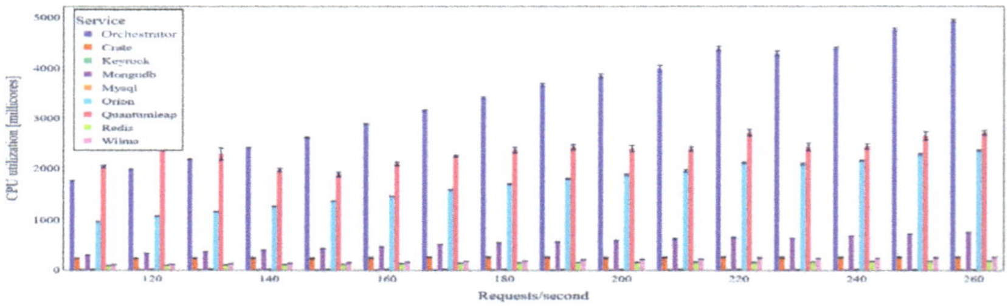


(**c**) Four-pod replicas.

**Figure 13.** Backend services average CPU utilization under variable IoT data update request rate and horizontal scaling of three most intensive services.

The average RAM utilization for two, three, and four pod replicas of the three most intensive services is depicted in Figure 14. Similar to the basic scenario, CrateDB exhibits the highest RAM utilization (around 2.2 GB). RAM utilization for the *Orchestrator* and QuantumLeap services increases almost proportionally to the number of replicas. On the contrary, the Orion Context Data Broker's RAM utilization is sub-proportional to the load, possibly due to more efficient memory management (application written in C++).

Table 6 summarizes the results in terms of resource utilization for a data update request rate of 150 requests/s, which is the maximum rate achieved for the baseline scenario. We observe small variations in terms of the aggregate CPU utilization for the three most CPU-intensive backend services. We still observe that the RAM utilization of the Orchestrator and QuantumLeap services rises nearly proportionally with the number of replicas, while that of the Orion Context Data Broker does not increase as much with the load, indicating a more efficient implementation of the latest. As expected, horizontal scaling of the three aforementioned backend services does not affect the utilization of any other backend service. Once more, the results indicate the fact that our system can scale efficiently with regard to the number of backend service instances.

**Table 6.** Average CPU (milliCores) and RAM (MBs) utilization for data update request rate of 150 requests/s.

| Service | Baseline | | Two Pods | | Three Pods | | Four Pods | |
|---|---|---|---|---|---|---|---|---|
| | CPU | RAM | CPU | RAM | CPU | RAM | CPU | RAM |
| Orchestrator | 2999 | 281 | 2497 | 572 | 2510 | 850 | 2593 | 1144 |
| QuantumLeap | 1660 | 537 | 1650 | 1058 | 1720 | 1583 | 1750 | 2102 |
| Orion | 1235 | 154 | 1230 | 167 | 1320 | 190 | 1361 | 212 |
| MongoDB | 352 | 75 | 346 | 80 | 353 | 85 | 367 | 89 |
| Crate | 266 | 1808 | 201 | 2130 | 215 | 2201 | 235 | 2230 |
| Wilma | 147 | 62 | 150 | 62 | 142 | 62 | 146 | 58 |
| Redis | 114 | 6 | 119 | 6 | 115 | 7 | 116 | 6 |
| MySQL | 15 | 446 | 14 | 445 | 15 | 446 | 15 | 446 |
| Keyrock | 2 | 74 | 2 | 75 | 3 | 75 | 3 | 74 |

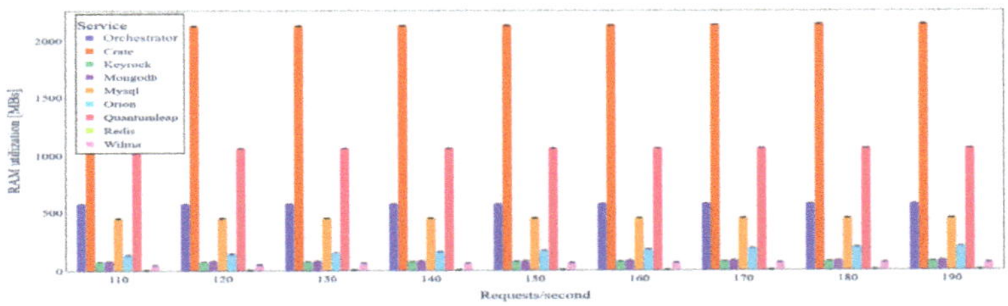


(**a**) Two-pod replicas.

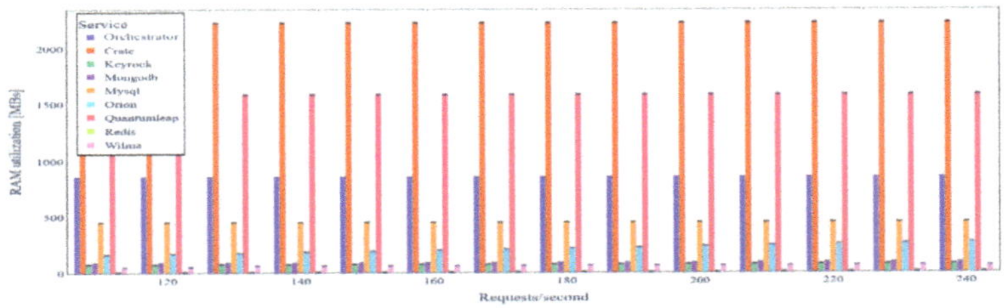


(**b**) Three-pod replicas.

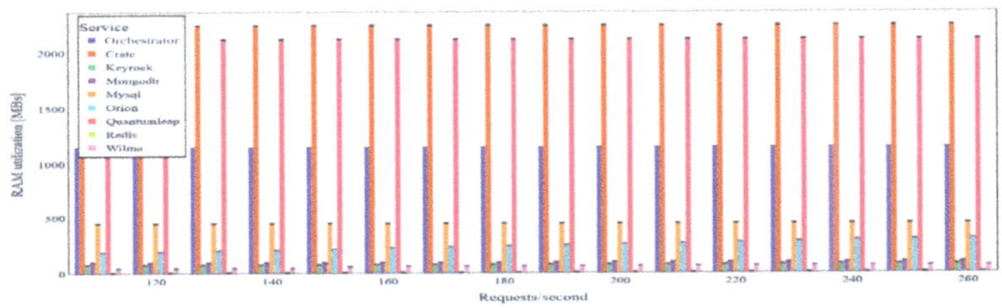


(**c**) Four-pod replicas.

**Figure 14.** Backend service average RAM utilization under variable IoT data update request rate and horizontal scaling of three most intensive services.

Finally, Figure 15 illustrates the latency of the IoT data update requests for horizontal scaling of the three most intensive services. The median latency is lower than 45 ms up to 150 requests/s but then increases more profoundly with the load increase, indicating increased pressure on the platform's endpoint. In any case, the median latency is lower than 100 ms.

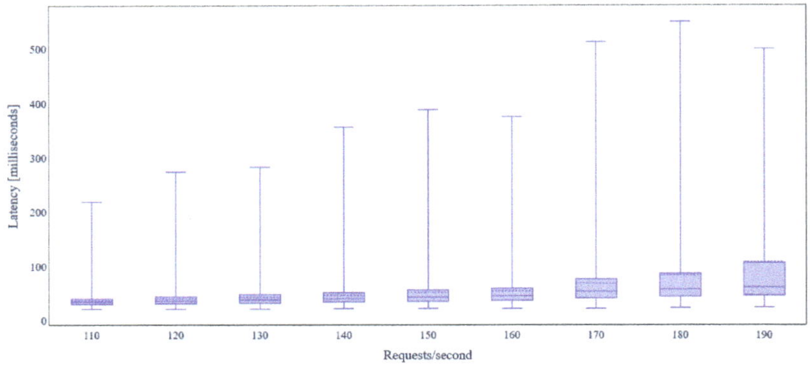


**(a)** Two-pod replicas.

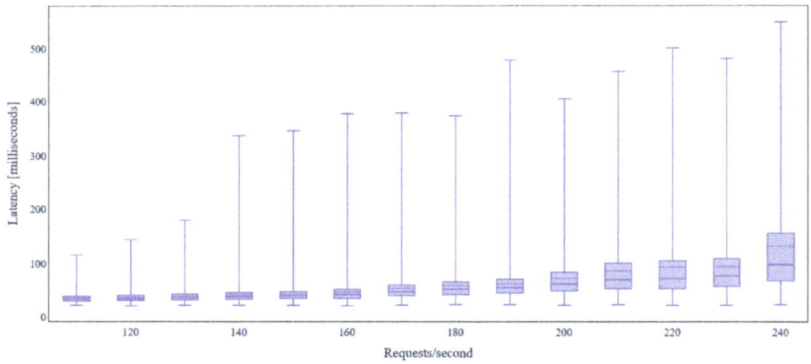


**(b)** Three-pod replicas.

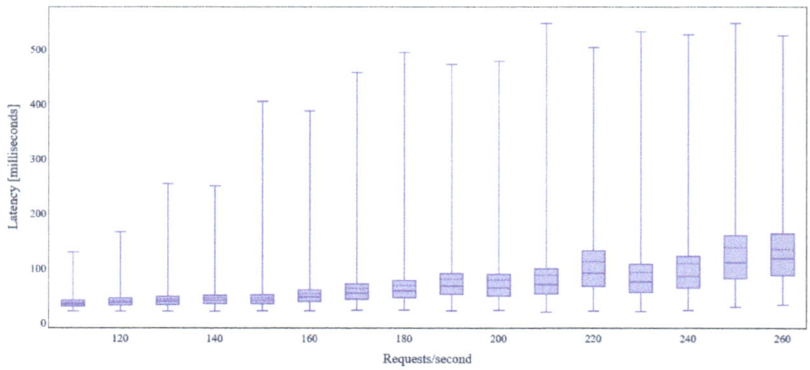


**(c)** Four-pod replicas.

**Figure 15.** Latency under variable IoT data update request rate and horizontal scaling of the three most intensive services.

5.3.2. Ecosystems with Data Integrity Verification

Here, we utilize the *BC Service* for the data integrity verification mechanism. Figure 16 illustrates the average CPU utilization of all backend services under variable IoT data update load in the case of data integrity verification. CPU utilization increases as the data update rate increases, up to the value of 20 requests/s. Then, CPU utilization remains almost constant due to the bottleneck introduced by the consensus mechanism of the *BC Service*. As previously, the services with the highest CPU utilization are the *Orchestrator* (600 millicores for 30 requests/s), QuantumLeap (170 millicores for 30 requests/s), and Orion Context Data Broker (105 millicores for 30 requests/s), but CPU utilization is in general considerably lower when compared to the one without data integrity verification.

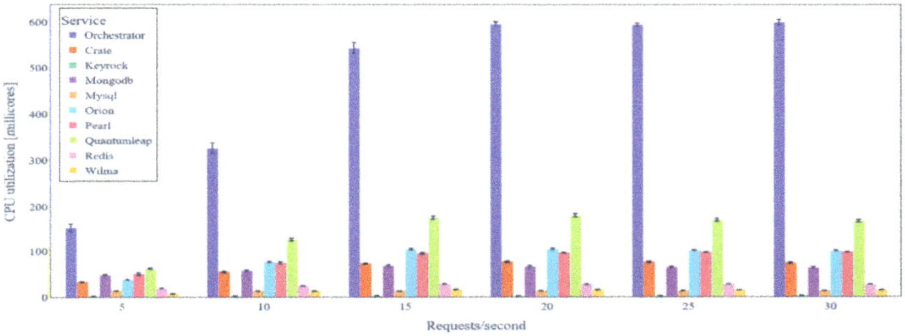


**Figure 16.** Backend service average CPU utilization under variable IoT data update request rate (with data integrity verification).

The average RAM utilization of the backend services when data integrity verification takes place is shown in Figure 17. No significant change in RAM utilization is observed as load varies. CrateDB still has the highest RAM demand (around 1.6 GB). Finally, Figure 18 depicts the latency for IoT data update requests under varying load when data integrity verification is performed. It is obvious that the median of the latency is significantly higher compared to the case where no data integrity verification takes place due to the delay of the consensus and transaction commit mechanisms introduced by the *BC Service*.

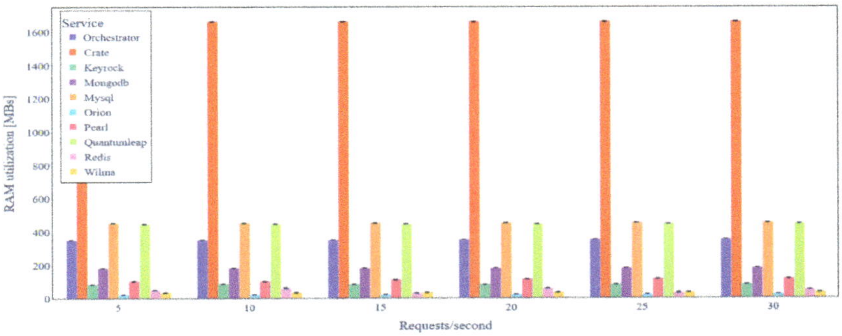


**Figure 17.** Backend service average RAM utilization under variable IoT data update request rate (with data integrity verification).

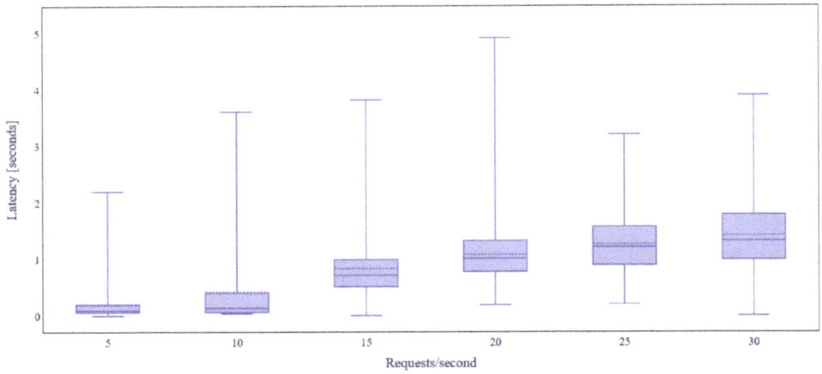


**Figure 18.** Latency under variable IoT data update request rate (with data integrity verification).

## 6. Conclusions and Further Work

In this paper, we proposed a Monitoring-as-a-Service platform for IoT applications based on FIWARE, which utilizes BC and SC technologies for data integrity verification. We presented the system architecture and thoroughly described the implementation details of our platform as well as the interactions between the system components. Additionally, we extensively evaluated a Proof-of-Concept Kubernetes-based deployment of the platform in terms of resource utilization (CPU, RAM) and latency under a variable rate of incoming IoT data. Most backend services, deployed as separate pods, have low computational requirements apart from three services that are more CPU intensive. By leveraging the native Kubernetes horizontal scaling functionality for the most intensive services, we achieve higher system throughput with a low expense in terms of RAM utilization. The evaluation shows that our platform enjoys scalability, provided that sufficient computational and memory resources are available. The incorporation of a BC-based data verification mechanism upheld the integrity of stored IoT data with no significant penalty on CPU and RAM utilization but at a discernible expense to the overall system throughput. Further work includes the investigation of Directly Acyclic Graph (DAG) ledgers such as IOTA (https://www.iota.org (accessed on 25 March 2024)) and NANO (https://nano.org (accessed on 25 March 2024)), which could support a much higher throughput in decentralized systems because the transactions can be sent and confirmed in parallel without the necessity to be grouped into sequential blocks as in traditional BC systems. Moreover, we aim to decentralize the identity management, authentication, and authorization mechanisms by substituting FIWARE Keyrock functionalities for corresponding processes that execute within SCs and use decentralized identifiers (https://www.w3.org/TR/did-core (accessed on 25 March 2024)) for enhanced privacy.

**Author Contributions:** Conceptualization, A.F. and P.C.; methodology, P.C., P.Z., E.P. and N.K.; software, P.Z., E.P., N.K. and P.C.; validation, A.F., P.C., P.Z. and N.K.; formal analysis, P.C., P.Z., E.P. and N.K.; resources, A.F., P.C. P.Z., E.P. and N.K.; writing—original draft preparation, P.C., P.Z., N.K. and E.P.; writing—review and editing, A.F. and P.C.; supervision, A.F.; project administration, A.F. and P.C.; funding acquisition, A.F. All authors have read and agreed to the published version of the manuscript.

**Funding:** This research has been financed by the European Union and Greek national funds through the Operational Program Competitiveness, Entrepreneurship and Innovation, under the call RESEARCH – CREATE – INNOVATE (project code: T1EDK-00070).

**Institutional Review Board Statement:** Not applicable.

**Data Availability Statement:** The data presented in this study are available on request from the corresponding author.

**Conflicts of Interest:** The authors declare no conflicts of interest. The funders had no role in the design of the study; in the collection, analyses, or interpretation of data; in the writing of the manuscript; or in the decision to publish the results

## Abbreviations

The following abbreviations are used in this manuscript:

| | |
|---|---|
| API | Application Programming Interface |
| BC | Blockchain |
| CI | Cloud Infrastructure |
| CIM | Context Information Model |
| CPU | Central Processing Unit |
| CSS | Cloud Storage Service |
| DNS | Domain Name System |
| FTP | File Transfer Protocol |
| HLF | HyperLedger Fabric |
| HTTP | Hypertext Transfer Protocol |
| IoT | Internet of Things |
| LG | Load Generator |
| MaaS | Monitoring as a Service |
| MQTT | Message Queuing Telemetry Transport |
| NGSI | Next-Generation Service Interface |
| PDP | Policy Decision Point |
| PEP | Policy Enforcement Point |
| RAM | Random Access Memory |
| REST | Representational State Transfer |
| SC | Smart Contract |
| TPA | Third-Party Auditor |
| VM | Virtual Machine |

## References

1. Kalaitzakis, M.; Bouloukakis, M.; Charalampidis, P.; Dimitrakis, M.; Drossis, G.; Fragkiadakis, A.; Fundulaki, I.; Karagiannaki, K.; Makrogiannakis, A.; Margetis, G.; et al. Building a Smart City Ecosystem for Third Party Innovation in the City of Heraklion. In *Mediterranean Cities and Island Communities: Smart, Sustainable, Inclusive and Resilient*; Springer International Publishing: Berlin/Heidelberg, Germany, 2019; pp. 19–56. [CrossRef]
2. Afonso, J.A.; Monteiro, V.; Afonso, J.L. Internet of Things Systems and Applications for Smart Buildings. *Energies* **2023**, *16*, 2757. [CrossRef]
3. Atalla, S.; Tarapiah, S.; Gawanmeh, A.; Daradkeh, M.; Mukhtar, H.; Himeur, Y.; Mansoor, W.; Hashim, K.F.B.; Daadoo, M. IoT-Enabled Precision Agriculture: Developing an Ecosystem for Optimized Crop Management. *Information* **2023**, *14*, 205. [CrossRef]
4. Ejaz, W.; Naeem, M.; Shahid, A.; Anpalagan, A.; Jo, M. Efficient Energy Management for the Internet of Things in Smart Cities. *IEEE Commun. Mag.* **2017**, *55*, 84–91. [CrossRef]
5. Folgado, F.J.; González, I.; Calderón, A.J. Data acquisition and monitoring system framed in Industrial Internet of Things for PEM hydrogen generators. *Internet Things* **2023**, *22*, 100795. [CrossRef]
6. Tarrés-Puertas, M.I.; Brosa, L.; Comerma, A.; Rossell, J.M.; Dorado, A.D. Architecting an Open-Source IIoT Framework for Real-Time Control and Monitoring in the Bioleaching Industry. *Appl. Sci.* **2024**, *14*, 350. [CrossRef]
7. Lampropoulos, G.; Siakas, K.; Anastasiadis, T. Internet of Things in the Context of Industry 4.0: An Overview. *Int. J. Entrep. Knowl.* **2019**, *7*, 4–19 . [CrossRef]
8. Chatterjee, A.; Ahmed, B.S. IoT anomaly detection methods and applications: A survey. *Internet Things* **2022**, *19*, 100568. [CrossRef]
9. Xenakis, A.; Karageorgos, A.; Lallas, E.; Chis, A.E.; Gonzalez-Velez, H. Towards Distributed IoT/Cloud based Fault Detection and Maintenance in Industrial Automation. *Procedia Comput. Sci.* **2019**, *151*, 683–690. [CrossRef]
10. Passlick, J.; Dreyer, S.; Olivotti, D.; Grutzner, L.; Eilers, D.; Breitner, M. Predictive maintenance as an internet of things enabled business model: A taxonomy. *Electron. Mark.* **2021**, *31*, 67–87. [CrossRef]

11. Psychoula, I.; Singh, D.; Chen, L.; Chen, F.; Holzinger, A.; Ning, H. Users' Privacy Concerns in IoT Based Applications. In Proceedings of the 2018 IEEE SmartWorld, Ubiquitous Intelligence & Computing, Advanced & Trusted Computing, Scalable Computing & Communications, Cloud & Big Data Computing, Internet of People and Smart City Innovation (SmartWorld/SCALCOM/UIC/ATC/CBDCom/IOP/SCI), Guangzhou, China, 8–12 October 2018; pp. 1887–1894. [CrossRef]
12. Jeon, H.; Lee, C. Internet of Things Technology: Balancing privacy concerns with convenience. *Telemat. Inform.* **2022**, *70*, 101816. [CrossRef]
13. Babun, L.; Denney, K.; Celik, Z.B.; McDaniel, P.; Uluagac, A.S. A survey on IoT platforms: Communication, security, and privacy perspectives. *Comput. Netw.* **2021**, *192*, 108040. [CrossRef]
14. Shelby, Z.; Hartke, K.; Bormann, C. RFC 7252—The Constrained Application Protocol (CoAP). Available online: https://datatracker.ietf.org/doc/html/rfc7252 (accessed on 25 March 2024).
15. Fedor, M.; Schoffstall, M.; Davin, J.; Case, J. RFC 1157—Simple Network Management Protocol (SNMP). Available online: https://datatracker.ietf.org/doc/html/rfc1157 (accessed on 25 March 2024).
16. Enns, R.; Bjorklund, M.; Schoenwaelder, J.; Bierman, A. RFC 6241—Network Configuration Protocol (NETCONF). Available online: https://datatracker.ietf.org/doc/html/rfc6241 (accessed on 25 March 2024).
17. Bierman, A.; Bjorklund, M.; Watsen, K. RFC 8040—RESTCONF Protocol. Available online: https://datatracker.ietf.org/doc/html/rfc8040 (accessed on 25 March 2024).
18. Veillette, M.; Stok, P.; Pelov, A.; Bierman, A.; Bormann, C. CoAP Management Interface (CORECONF). Available online: https://datatracker.ietf.org/doc/draft-ietf-core-comi/ (accessed on 25 March 2024).
19. Li, Z.; Xie, Z.; Liu, L.; Wu, Y. Design and Implementation of an Integrated City-Level IoT Platform Based on Edge Computing and Cloud Native. In Proceedings of the 2022 IEEE 6th Advanced Information Technology, Electronic and Automation Control Conference (IAEAC ), Beijing, China, 3–5 October 2022; pp. 463–467. [CrossRef]
20. Lopez-Riquelme, J.; Pavon-Pulido, N.; Navarro-Hellin, H.; Soto-Valles, F.; Torres-Sanchez, R. A software architecture based on FIWARE cloud for Precision Agriculture. *Agric. Water Manag.* **2017**, *183*, 123–135. [CrossRef]
21. Neagu, G.; Preda, S.; Stanciu, A.; Florian, V. A Cloud-IoT based sensing service for health monitoring. In Proceedings of the 2017 E-Health and Bioengineering Conference (EHB), Sinaia, Romania, 22–24 June 2017, pp. 53–56. [CrossRef]
22. Galán, F.; Fazio, M.; Celesti, A.; Glikson, A.; Villari, M. Exploiting the FIWARE Cloud Platform to Develop a Remote Patient Monitoring System. In Proceedings of the 2015 IEEE Symposium on Computers and Communication (ISCC), Larnaca, Cyprus, 6–9 July 2015. [CrossRef]
23. Fernández, P.; Santana, J.M.; Ortega, S.; Trujillo, A.; Suárez, J.P.; Domínguez, C.; Santana, J.; Sánchez, A. SmartPort: A Platform for Sensor Data Monitoring in a Seaport Based on FIWARE. *Sensors* **2016**, *16*, 417. [CrossRef] [PubMed]
24. Hui, L.; Gui-rong, W.; Jian-ping, W.; Peiyong, D. Monitoring platform of energy management system for smart community. In Proceedings of the 2017 29th Chinese Control And Decision Conference (CCDC), Chongqing, China, 28–30 May 2017; pp. 1832–1836. [CrossRef]
25. Openhab Smart Home Platform. Available online: https://www.openhab.org (accessed on 22 March 2024).
26. Samsung SmartThings Platform. Available online: https://www.samsung.com/us/smartthings (accessed on 22 March 2024).
27. Apple HomeKit. Available online: https://www.apple.com/shop/accessories/all/homekit (accessed on 22 March 2024).
28. Amazon Web Services IoT. Available online: https://aws.amazon.com/iot (accessed on 22 March 2024).
29. IBM Watson. Available online: https://www.ibm.com/watson (accessed on 22 March 2024).
30. Liu, B.; Yu, X.L.; Chen, S.; Xu, X.; Zhu, L. Blockchain Based Data Integrity Service Framework for IoT Data. In Proceedings of the 2017 IEEE International Conference on Web Services (ICWS), Honolulu, HI, USA, 25–30 June 2017; pp. 468–475. [CrossRef]
31. Wu, X.; Kong, F.; Shi, J.; Bao, L.; Gao, F.; Li, J. A blockchain internet of things data integrity detection model. In Proceedings of the 1st International Conference on Advanced Information Science and System; Association for Computing Machinery, New York, NY, USA, 15–17 November 2019; AISS '19. [CrossRef]
32. Eghmazi, A.; Ataei, M.; Landry, R.J.; Chevrette, G. Enhancing IoT Data Security: Using the Blockchain to Boost Data Integrity and Privacy. *IoT* **2024**, *5*, 20–34. [CrossRef]
33. Chanai, P.; Kakkasageri, M. Blockchain-based data integrity framework for Internet of Things. *Int. J. Inf. Secur.* **2024**, *23*, 519–532. [CrossRef]
34. Zhang, K.; Xiao, H.; Liu, Q. Data Integrity Verification Scheme Based on Blockchain Smart Contract. In Proceedings of the 2022 IEEE International Conference on Trust, Security and Privacy in Computing and Communications (TrustCom), Wuhan, China, 9–11 December 2022; pp. 857–863. [CrossRef]
35. Chen, C.; Wang, L.; Long, Y.; Luo, Y.; Chen, K. A blockchain-based dynamic and traceable data integrity verification scheme for smart homes. *J. Syst. Archit.* **2022**, *130*, 102677. [CrossRef]
36. Rahman, M.S.; Chamikara, M.; Khalil, I.; Bouras, A. Blockchain-of-blockchains: An interoperable blockchain platform for ensuring IoT data integrity in smart city. *J. Ind. Inf. Integr.* **2022**, *30*, 100408. [CrossRef]
37. Doulgeraki, P.; Karuzaki, E.; Sykianaki, E.; Partarakis, N.; Bouhli, M.; Ntoa, S.; Stephanidis, C. Web-Based Management for Internet of Things Ecosystems. In Proceedings of the HCI International 2023 Posters, Copenhagen, Denmark, 23–28 July 2023; Stephanidis, C., Antona, M., Ntoa, S., Salvendy, G., Eds.; Springer: Cham, Switzerland, 2023; pp. 475–482.

38. Domínguez-Bolaño, T.; Campos, O.; Barral, V.; Escudero, C.J.; García-Naya, J.A. An overview of IoT architectures, technologies, and existing open-source projects. *Internet Things* **2022**, *20*, 100626. [CrossRef]
39. Dragoni, N.; Giallorenzo, S.; Lafuente, A.L.; Mazzara, M.; Montesi, F.; Mustafin, R.; Safina, L., Microservices: Yesterday, Today, and Tomorrow. In *Present and Ulterior Software Engineering*; Springer International Publishing: Berlin/Heidelberg, Germany, 2017; pp. 195–216. [CrossRef]

*sensors*

MDPI

*Article*

# FLEXTORY: Flexible Software Factory of IoT Data Consumers

**Rafael López-Gómez *, Laura Panizo and María-del-Mar Gallardo**

ITIS Software, Andalucía Tech, Universidad de Málaga, 29071 Malaga, Spain; laurapanizo@uma.es (L.P.); mdgallardo@uma.es (M.-d.-M.G.)

* Correspondence: rafaellopez@uma.es

**Abstract:** The success of the Internet of Things (IoT) has driven the development, among others, of many different software architectures for producing, processing, and analyzing heterogeneous data. In many cases, IoT applications share common features, such as the use of a platform or middleware, also known as *message broker*, that collects and manages data traffic between endpoints. However, in general, data processing is very dependent on the case study (sensors that send temperature data, drones that send images, etc.). Thus, the applications responsible for receiving and processing data, which we call *consumers*, have to be built ad hoc, since some of their elements have to be specially configured to solve specific needs of the case study. This paper presents FLEXTORY, a *software factory tool* to make it easier for IoT developers to *automatically* construct configurable *consumer* applications, which we call *FLEX-consumers*. FLEXTORY guides developers through the process of generating Java *consumers* by selecting some desired features such as, for instance, the particular communication protocol to be used. This way, the developer only has to concentrate on designing the algorithm to process the data. In short, the use of FLEXTORY will result in *consumer* applications with configurable behavior, namely *FLEX-consumers*, that can connect to a messaging server (for example RabbitMQ) and process the received messages.

**Keywords:** Internet of Things; software factory; message broker

**Citation:** López-Gómez, R.; Panizo, L.; Gallardo, M.-d.-M. FLEXTORY: Flexible Software Factory of IoT Data Consumers. *Sensors* **2024**, *24*, 2550. https://doi.org/10.3390/s24082550

Academic Editors: Allel Hadjali, Behnam Mobaraki and Jose Turmo

Received: 5 March 2024
Revised: 30 March 2024
Accepted: 13 April 2024
Published: 16 April 2024



## 1. Introduction

The Internet of Things (IoT) can be defined as the interconnection of heterogeneous devices through the Internet. With the evolution of wireless networks, in particular the fifth generation of mobile networks (5G), IoT has become an enabling technology for a large set of applications from many different domains, such as smart cities, smart farming, Industry 4.0, and e-Health [1]. In IoT applications, at least three main actors can be identified: the source of data (the *producers/publishers*), such as sensors that produce information, the processors (the *consumers*) that analyze and transform data following some criteria, and the *actuators* that respond properly according to the information registered.

Despite the wide variety of IoT architectures, most of them rely on an intermediate platform (middleware), called *message broker*, that abstracts the data transmission between IoT peer devices. Brokers support reliable communication and, additionally, can hide the existence of devices connected at any given time. Thus, the communication endpoints (the *producer* and *consumer* nodes) are only self-aware. Thanks to this feature, it is possible to develop loosely coupled and scalable IoT applications. Traditionally, when a developer faces the task of constructing an IoT application for a given domain following this architecture, he/she has to select the intermediate platform that will connect the source and target entities, implement software modules that carry out this communication, and design the software component that processes the data produced by the data source.

Currently, there is a wide variety of general purpose brokers, such as RabbitMQ, Apache Kafka, and Mosquito. One of the main differences between them is the communication protocols used to connect the broker with the *producer* and *consumer* applications. In the IoT domain, brokers usually make use of the two well-known communication protocols

AMQP [2] and MQTT [3] that take into account the limited resources of many IoT devices. It is worth mentioning that some broker systems support both communication protocols.

In this work, we have followed the ideas of *Software Product Line Engineering* (SPL) [4], a research area whose aim is to model families of software products that can be used to generate software adapted to the final user's needs. The SPL community has addressed multiple problems of the IoT domain, such as the design of App Stores that support the deployment of customized applications [5] and the self-adaptation of applications running in heterogeneous IoT devices [6].

The term *software factory* is not new [7] and refers to the set of techniques and tools that automate and simplify software creation [8]. They have been traditionally used to develop industrial applications. Nowadays, they are part of the techniques of SPL Engineering and are broadly used to customize software products of different domains [9]. In particular, we have constructed the tool FLEXTORY that allows the development of IoT *consumer* applications with several facilities. The tool is able to *automatically* construct the software components to read the data from the producer. In addition, FLEXTORY allows the easy integration of any data processing algorithm in the IoT application. This characteristic is very interesting, since it means that the developer only has to focus on the design of the algorithm suitable for the application being implemented. The tasks related to the integration of the algorithm with the rest of the application's software modules are already provided by the tool. To the best of our knowledge, in the literature, there are no tools similar to FLEXTORY that generate *consumer* applications with flexible behavior in the IoT domain.

The objective of this paper is to describe FLEXTORY in detail. FLEXTORY generates *consumer* applications able to read and process data received from a message broker system. To avoid confusion due to names, we use the term *FLEX-consumer* for the *consumer* applications built with FLEXTORY. As commented above, *FLEX-consumers* have a flexible behavior with many configurable options to ease their deployment. All *FLEX-consumers* have been developed in Java, which is well known for its portability. In addition, Java is also supported by some devices with low computing power [10].

Figure 1 depicts the interaction between the tool FLEXTORY and *FLEX-consumers* and their users. As shown, two different users appear in the diagram. On the one hand, the FLEXTORY user (called developer) makes use of FLEXTORY to construct *FLEX-consumers*. The developer introduces some parameters needed to construct the desired *FLEX-consumer*, such as, for instance, the communication protocol to be used and the algorithm that will process data. On the other hand, the *FLEX-consumer* user configures the application in order to adapt its execution to the particular expected behavior. Usually, the *FLEX-consumer* configuration is related to how and when the data processing algorithm must be executed.

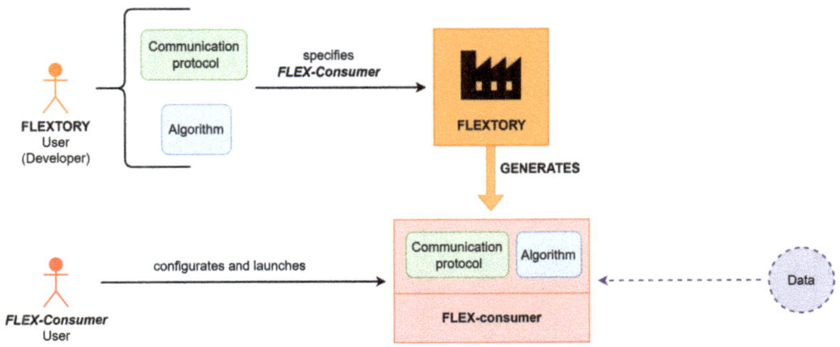


**Figure 1.** Overview of FLEXTORY and *FLEX-consumer* proposal.

To demonstrate the utility of FLEXTORY and *FLEX-consumers* in the development and deployment of IoT *consumer* applications, we present a non-trivial case study related to the learning technique of black-box systems [11]. FLEXTORY is used to produce a *FLEX-*

*consumer* that reads a sequence of observations and runs a learning algorithm that is able to produce models of the system. In particular, in the case study, the *FLEX-consumer* is used to learn models of the DASH protocols [12]. This example shows how FLEXTORY facilitates the construction of complex IoT applications. As commented above, the developer only has to focus on the construction of Java classes that carry out the learning process. Finally, we have evaluated the tool with a user study in which a group of post-graduate students have performed a task with FLEXTORY and have filled out a questionnaire.

The tool and the documentation is publicly available at https://gitlab.com/morse-uma/formal-methods/flextory/, accessed on 28 March 2024.

The rest of the paper is organized as follows. First, Section 2 introduces some of the most common IoT communication protocols and message brokers. Then, Section 3 summarizes related work. In Section 4, we describe in detail the design and implementation of FLEXTORY and the generated *FLEX-consumers*. Sections 5 and 6 present two case studies and the user study results, respectively. Finally, in Sections 7 and 8, we discuss the strengths and weaknesses of our proposal and summarize conclusions and future work. Additionally, we provide supplementary material in Appendix A, expanding on Sections 4 and 5 to provide more information on the development and application of FLEXTORY.

## 2. Background

In this section, we first present and compare some of the most widely used IoT communication protocols. Then, we introduce the IoT architecture based on message brokers, which is the basis of the design of the tool FLEXTORY. We recommend that the reader consult a state-of-the-art survey on IoT, such as [1], to gain a broader view of this technology.

### 2.1. IoT Communication Protocols

In general, communication protocols establish a set of rules so that different devices can communicate, as well as the format of the messages exchanged. In the case of the IoT domain, communication protocols have to deal with some specific features. Firstly, they must take into account the large number of devices that can be connected, the different tasks that they may perform, and the disparity in devices' computation capability. In addition, the IoT architecture should be scalable and flexible; that is, adding or removing devices should not produce noticeable changes in the IoT solution. This requirement entails adopting low coupling between devices. Finally, the security of communications must also be ensured.

IoT protocols may be based on different communication patterns, Publish/Subscribe being one of the most widely used. In this communication pattern, there are two entities that interchange data via a middleware broker. On the one hand, the *subscriber* tells a broker the topic of the messages that it wants to receive. On the other hand, the *publisher* sends data about a certain topic to the broker. As shown in Figure 2, the message broker is in charge of distributing the messages to the *subscribers* subscribed to each topic.

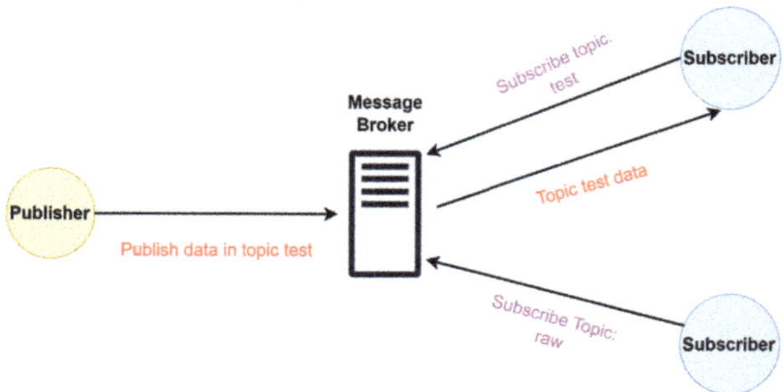


**Figure 2.** Example of Publish/Subscribe communication pattern.

We will now describe some IoT protocols based on the Publish/Subscribe pattern that work over the application layer of the OSI network model. Since AMQP and MQTT are, by far, the most widely used protocols, we have decided that developers can configure FLEXTORY to allow the resulting *FLEX-consumer* to make use of one of these two protocols.

- MQ Telemetry Transport [13,14] (MQTT) is an OASIS open standard that defines a Machine to Machine (M2M) communication protocol for IoT environments. MQTT typically works over TCP and transports binary data. MQTT is designed to be lightweight so that it can be used by devices with low computing capacity. In addition, message headers are small to accommodate low bandwidth networks. Finally, regarding security, different security mechanisms are available, such as encrypting connections using SSL/TLS and authentication.
- Advanced Message Queuing Protocol [15] (AMQP) is also an OASIS open standard. AMQP is designed to support a wide variety of communication patterns. Messages are distributed between the endpoint devices by means of a flexible and complex mechanism based on *exchanges*, which are abstract entities declared by users to which messages are sent. *Exchanges* take a message and route it into queues. Users have to create queues if they do not exist and bind them to a specific *exchange*. The routing algorithm that distributes messages from *exchanges* to queues depends on the *exchange* type and the binding rules. There are many *exchange* types, but in this project, we have used the default *exchange* that is pre-declared by the broker. This *exchange* offers a Publish/Subscribe pattern with one special property that makes it very useful for applications: every queue that is created is automatically bound to the *exchange* with a "topic name" (routing key in AMQP nomenclature) that is the same as the queue name. For example, if a queue with the name "test" is declared, the message broker will bind it to the default exchange using "test" as the topic. AMQP provides authentication and encryption based on SASL and TLS.
- Constrained Application Protocol [16] (CoAP) is a protocol for devices with limited resources that provides a REST model between application endpoints with messages in binary format. Targeting efficiency, CoAP runs over the UDP transport layer protocol to distribute messages in an asynchronous manner. Regarding security, it supports DTLS.
- Extensible Messaging and Presence Protocol [17] (XMPP) is an IETF open standard that uses the Extensible Markup Language (XML) as the data format. XMPP was initially designed for instant messaging services but was later extended to cover different communication patterns, such as Request/Response, Asynchronous Messaging, Publish/Subscribe, event subscription (Observe) and delayed delivery. In terms of security, XMPP supports SASL and TLS. However, the support of end-to-end encryption is a work in progress.

### 2.2. Message Brokers

A message broker, or a messaging server, is a middleware between applications or devices, both senders and receivers, that exchange messages. In the IoT context, a message broker consists of a centralized notification service that is a main server with a fixed IP address known by all devices. As shown in Figure 3, the server is responsible for receiving messages from all sending devices (*publishers*) and distributing them to receivers (*consumers*). The connected devices are only aware of themselves. *Consumers* do not know the true origin of the data, and publishers do not know by who or how the data will be processed. This isolation is achieved thanks to the message broker management and provides scalability and low coupling.

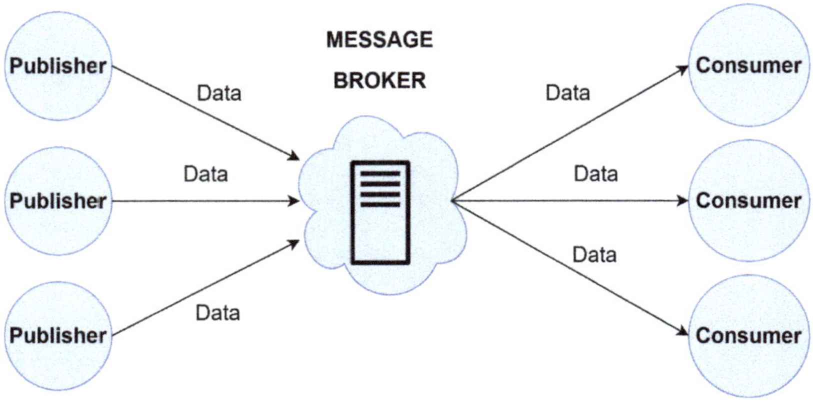


**Figure 3.** Message broker architecture.

Nowadays, a wide variety of message brokers are available, both self-hosted and cloud-hosted. In particular, we have tested the *FLEX-consumers* using RabbitMQ, since it is one of the message brokers that supports MQTT and AMQP. Table 1 shows a comparison between some of the most popular message brokers (despite the fact that all solutions can be deployed in a cloud service provider, we have only considered native cloud hosting). In addition to the type of business model followed, either proprietary or open source, some other differences can be highlighted. For example, the RabbitMQ model has commercial features to gain access to a virtualized version of the message broker and even cloud hosting and technical assistance. Regarding the protocols supported, the brokers accept AMQP and MQTT natively or via plugin, with the exception of Apache Kafka that uses a custom protocol. Finally, Microsoft's and Amazon's options also cover the integration and communication with other services provided by these companies.

**Table 1.** Features of different Message brokers.

| Name | Business Model | Hosting | Native Protocols | Additional Information |
|---|---|---|---|---|
| RabbitMQ [18] | Open Source | Self-hosting | AMQP | Plugin support including MQTT extension |
| Azure IoT Hub [19] | Proprietary | Cloud hosting | HTTP, AMQP, and MQTT | Integration with Azure services |
| Apache Kafka [20] | Open Source | Self-hosting | TCP-based protocol | Supports AMQP and MQTT via plugin |
| Eclipse Mosquitto [21] | Open Source | Self-hosting | MQTT | - |
| AWS IoT Core [22] | Proprietary | Cloud hosting | MQTT and HTTPS | Integration with AWS and other Amazon services |

## 3. Related Work

In the IoT domain, it is common to use the three-layered architecture that distinguishes between the roles of *producers*, message brokers, and *consumers*. One of the main challenges to be addressed when using this architecture is how to adapt it to changes in the data format and the processing algorithm. There exist many proposals that integrate all the components of this three-layered architecture into a custom solution. For example, JCL [23] is a middleware whose purpose is to integrate IoT with High Performance Computing (HPC). It also has an API in which different categories of devices can be programmed. Since JCL requires that its own components be installed in all the systems belonging to the IoT ecosystem, its use is limited to devices that are JCL compatible. Our proposal is focused on end devices that process data (*consumers*) and relies on the portability paradigm of the Java language. In contrast to JCL, *FLEX-consumers* are automatically generated.

D-LITe [24] is another all-in-one solution that uses a choreography approach. Programming is based on cooperation between nodes, each one performing a small part of the total application. To program a node, D-LITe uses finite state machines with an output alphabet, called Finite State Transducers (FSTs), to describe the logic of the application. When a user wants to describe the application, he/she does it using a specific format called SALT. Subsequently, the rules are transformed into a set of FSTs (one per node) that are sent through the network. Finally, each node's rule analyzer is responsible for executing its FST. The use of D-LITe is limited both by the compatible devices and by programming options offered by the SALT format. As mentioned before, our proposal offers *consumers* with flexible behavior that can be changed in every execution. In addition, there are no programming restrictions because they are built using Java.

IoTSuite [25] also offers a tool suite that covers all the layers that constitute an IoT infrastructure. These tools automatize tasks in different phases of developing an IoT application. For example, programmers can write high-level textual specifications that can be analyzed and transformed into code by the compiler tool. There is also an execution system that incorporates a middleware to coordinate nodes. IoTSuite requires that it be compiled and installed on all devices that will be used, which limits its application.

The platform SYNAISTHISI [26] is another approach to support multiple communication protocols, such as MQTT and AMQP, by integrating different open-source frameworks including, among others, RabbitMQ. This work does not focus on how to support the development of IoT applications. In contrast, its objective is to support the interoperability of these different frameworks and provide a unified user-access control on IoT data and services. The platform is available as a set of dockerized containers; thus, it is easily deployable. Both tools, the SYNAISTHISI framework and FLEXTORY, aim to support the fragmented IoT ecosystem from different approaches. Clearly, the SYNAISTHISI platform can help us to test *FLEX-consumers* generated by FLEXTORY in different scenarios (protocols and brokers) and, alternatively, FLEXTORY can easily produce *FLEX-consumers* with different communication protocols to test the interoperability of SYNAISTHISI.

There are also proposals that only focus on endpoint devices, either *publishers* or *consumers*. For instance, FRASAD [27] is a framework that facilitates the development of programs for sensor nodes (devices physically connected to sensors). FRASAD has been built following a software architecture centered on nodes and a programming model based on rules that allows applications to be described using a language specific to the sensor domain. The application code is generated from models built with the language through an automatic transformation process. FRASAD focuses on the *publisher* part, while our proposal centers on the *consumer* applications.

On the *consumers* side, Midas [28] is a framework to help researchers create and manage IoT applications with heterogeneous data sources. Midas has a module to process the data features of interest by means of the so-called analysis functions that make use of machine learning techniques. The main characteristic of Midas is its modularity, making it easy to incorporate new components in order to add new data streams or analysis functions. In addition, it is implemented as a distributed architecture to assure scalability. Compared with Midas, FLEXTORY's goal is different, since it is conceived as a meta-tool to create new configurable tools with respect to the structure of input data, the type of message brokers to be used, and the algorithms to process data, among other features.

## 4. Software Description

Figure 4 presents a general overview of FLEXTORY and *FLEX-consumers* developed in this work. The top part of the figure shows FLEXTORY's input and output. Thus, the user of FLEXTORY (a developer of IoT *consumer* applications) introduces parameters that FLEXTORY needs to build a *FLEX-consumer*. The mandatory inputs are the type and structure of the data (in JSON format) that the *FLEX-consumer* will receive from the message broker, the algorithm for data processing, and the communication protocol (AMQP or MQTT) to be used to connect with the message broker. With this information, FLEXTORY automatically generates a *FLEX-consumer*.

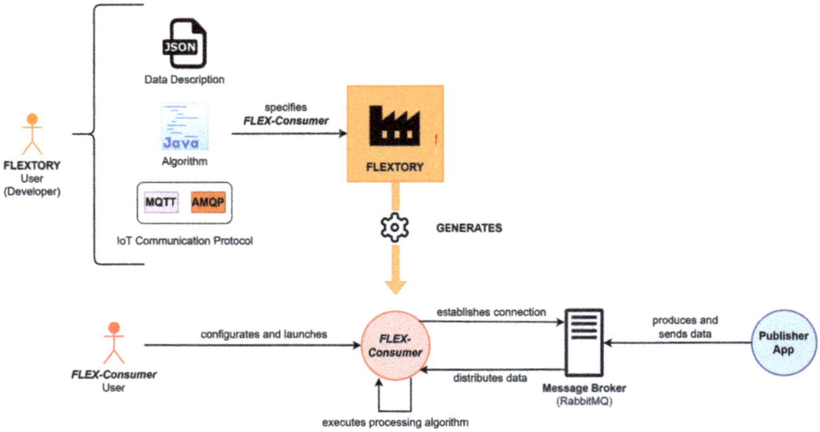


**Figure 4.** General overview of the inputs and outputs of FLEXTORY and *FLEX-consumers*.

The lower part of Figure 4 shows how the resulting *FLEX-consumer* can be configured and executed. Among other functionalities, the *FLEX-consumer* can be configured by the user to connect to a specific message broker and to subscribe to different types of data topics (as described in Section 2.1). Furthermore, there are some other customizable parameters used to establish some execution conditions of the processing algorithm and to decide when the *FLEX-consumer* should stop its execution. We have tested the *FLEX-consumers* generated with FLEXTORY using RabbitMQ as a message broker, but *FLEX-consumers* can connect to any other messaging server supporting AMQP or MQTT protocols.

The rest of the section describes the design of FLEXTORY and the *FLEX-consumers* shown in Figure 4. We first present the design of FLEXTORY, including its main functional and non-functional requirements and some implementation details. Then, we introduce the *FLEX-consumer* requirements related to their configuration.

### 4.1. *Design and Implementation of* FLEXTORY

As commented before, FLEXTORY automatically creates *consumer* applications with configurable behavior (the *FLEX-consumers*). On the one hand, FLEXTORY must be able to generate *FLEX-consumers* adapted to different IoT domains. This implies that FLEXTORY must allow both different formats for the data received by *FLEX-consumers* and also different processing algorithms to be applied to data. On the other hand, FLEXTORY must construct *FLEX-consumers* able to change the message broker as well as the conditions to trigger the processing algorithm in each execution.

Table 2 contains the list of functional and non-functional requirements that have guided the construction of FLEXTORY. These requirements have been selected to offer developers maximum flexibility when choosing how to build a *FLEX-consumer*. The most relevant requirements are FR-1 to FR-5. FR-1 establishes the need to provide the format of the data received by the *FLEX-consumer*. FR-2 to FR-4 describe the requirements related to the processing algorithm used by the *consumer* applications produced by FLEXTORY. FR-5 is related to the communication protocols to be included in the *FLEX-consumers*.

**Table 2.** Functional and non-functional requirements of FLEXTORY.

| Id | Description |
|---|---|
| FR-1 | Developers have to specify the format of the data received by the *FLEX-consumer*. |
| FR-2 | FLEXTORY will have a way to enter the algorithm that the *FLEX-consumer* will execute to process the received data. |
| FR-3 | FLEXTORY should offer a template of the processing algorithm that users have to complete. |
| FR-4 | FLEXTORY will allow the user to add external dependencies of the *FLEX-consumer*'s processing algorithm in JAR format. |
| FR-5 | Developers have to select either MQTT or AMQP as the *consumer*'s protocol, but not both. |
| FR-6 | The name of the resulting *FLEX-consumer* has to be configured by the user. |
| FR-7 | FLEXTORY has to give warning messages to users when an error occurs. |
| NFR-1 | FLEXTORY must have an intuitive and easy-to-use GUI. |
| NFR-2 | FLEXTORY should be offered as a stand-alone application in JAR format. |
| NFR-3 | FLEXTORY should be compatible with several operating systems. |
| NFR-4 | FLEXTORY must be executable on systems that have the Java Development Kit (Java JDK). |

Figure 5 shows the use case diagram that describes FLEXTORY's main capabilities. The main actor is the FLEXTORY user interacting with FLEXTORY to generate a *FLEX-consumer*. To this end, the user uploads the format of the messages to be processed, currently using a JSON schema. In addition, the user defines the processing algorithm. This requires uploading, at a minimum, the implementation of the *Algorithm* class. To ease the process, the user can download a template to be completed. Optionally, if needed, the user can upload external dependencies packaged in JAR format. Moreover, the user must select the communication protocol between the two that are currently available (AMQP and MQTT). Finally, FLEXTORY uses the Java compiler to generate the *FLEX-consumer*, so it is essential that it be installed.

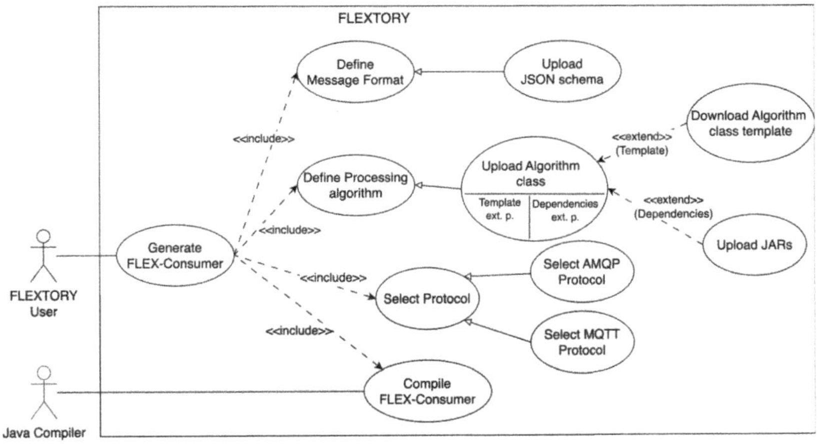


**Figure 5.** FLEXTORY use case diagram.

Figure 6 shows FLEXTORY's architecture. In Appendix A, we have included supplementary material, such as the class diagram. FLEXTORY follows the classical *Model/View/Controller* design pattern. The *View* module includes the visual components used by the Graphical User Interface (GUI) to guide the user through the configuration and generation of the *FLEX-consumer* application. There are two main visual components: the MainFrame, which provides the skeleton of the FLEXTORY GUI, along with different panels that remain visible during the creation of the *FLEX-consumer* in order to ease the interaction with the user. The *Model* module is in charge of generating the *FLEX-consumer*and is composed of two sub-components, the *Consumer Templates* and the *Compiler*. The former contains the code templates of different *consumer* components, such as the *User Interaction* module or different versions of the *Connection Management* module. The *Compiler* is in charge of integrating the templates with the data provided by the user in order to generate the *FLEX-consumer* executable. Finally, as usual, the *Controller* is the link between the *View* and the *Model*, reacting to user inputs and performing interactions on the *Model*. In addition, it can also react to *Compiler* events to properly update the *View*.

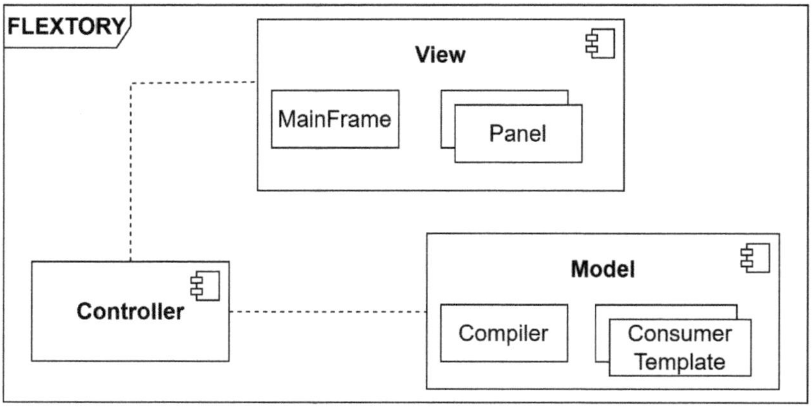


**Figure 6.** FLEXTORY components diagram implementing Model/View/Controller design pattern.

The current version of FLEXTORY is a Java application with a Java Swing GUI, packed in an executable JAR file. To generate *FLEX-consumers*, FLEXTORY guides users through a sequence of steps shown in the flow diagram in Figure 7. Although it is not explicit in the diagram, the user cannot advance to the next step if the selection made in the current step is wrong. We will use this diagram to present FLEXTORY's main implementation decisions.

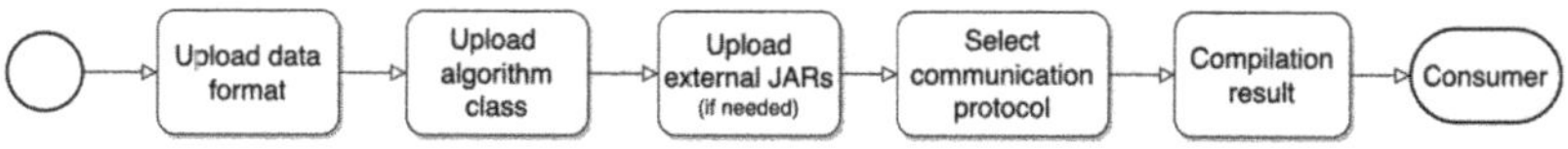


**Figure 7.** Simplified activity diagram of FLEXTORY.

In the first step, the user has to provide a JSON schema describing the format of the data distributed by the message broker. Then, the user uploads the processing algorithm to be used by the *FLEX-consumer*. The algorithm has to be coded in a special Java class called *Algorithm* that implements the Java *Runnable* interface, so that the user only has to implement the *run* method. To make this step easier, FLEXTORY provides a downloadable template of the *Algorithm* class. In addition, if the *Algorithm* class has some dependencies, they have to be provided as JAR files. In the fourth step, the user selects the communication protocol supported by the broker (AMQP or MQTT). With all the necessary files, the *Compiler* module transforms the data description (JSON schema) into a set of Plain Old Java Objects (POJOs) classes that will be part of the *FLEX-consumer*, and will support the deserialization of the data received from the message broker. Finally, the *FLEX-consumer* is built as a Java application that integrates the POJOs classes and the *Algorithm* class with its dependencies and the templates.

To illustrate the use of FLEXTORY, we show how to build a simple application that receives data from the well-known Iris dataset [29] and counts the number of flowers of each species. The dataset has samples of different iris species. Four traits (the length and width of the sepal and petal) are associated with each species. The FLEXTORY user (the developer) has to provide the file with this JSON schema in the first step (see the specific format in Appendix A). Next, the user has to complete the *Algorithm* class to count the number of samples of each species, as show in Listing 1. This algorithm has no external dependencies, so the user, after providing this class, can jump directly to the fourth step and select the communication protocol to generate the *FLEX-consumer* binaries (this example and the generated *FLEX-consumers* are available in gitlab).

**Listing 1.** Implementation of the Algorithm class of the Iris dataset example.

```
public class Algorithm implements Runnable {
  LinkedBlockingQueue<IrisSchema> data;

  public Algorithm(LinkedBlockingQueue<IrisSchema> data){
    this.data = data;
  }

  @Override
  public void run() {
    float vir= 0, ver= 0, set= 0, other= 0;
    for (IrisSchema flower : data) {
      switch (flower.getSpecies()) {
        case "setosa": set++;
          break;
        case "versicolor": ver++;
          break;
        case "virginica": vir++;
          break;
        default: other++;
      }
    }
    /* Log % of each specie*/
  }
}
```

### 4.2. *Design and Implementation of FLEX-Consumers*

The main purpose of a *consumer* application is to connect to a message broker and process the messages received by applying a processing algorithm. The design of the *FLEX-consumers* takes into account the following aspects. On the one hand, *FLEX-consumers* have to use a standard protocol to communicate with a message broker, such as the AMQP and MQTT protocols introduced in Section 2. To simplify the design, we assume that a *FLEX-consumer* uses only one of these protocols. On the other hand, *FLEX-consumers* can be configured in a *persistent* or *repetition* mode, since the internal behavior of the algorithm is unknown, i.e., it can be designed to process all incoming data in a persistent manner or, on the contrary, to process data batch. In particular, given a processing algorithm, we could define different repetition conditions that establish when the algorithm has to iterate again: each time a new message arrives, when a fixed number of messages are received, or after a specific time has elapsed. In consequence, the design of *FLEX-consumers* allows the configuration of different execution modes for the same algorithm. Moreover, *FLEX-consumers* can be configured with different *halting conditions* that define when the *FLEX-consumer* must close connections and stop the execution. For instance, data processing could stop when a *FLEX-consumer* reaches a maximum number of messages received or when a given time without receiving messages has elapsed.

Considering the foregoing, we have identified the main functional (FR) and non-functional (NFR) requirements of *FLEX-consumers*, which are listed in Table 3. We now describe the most relevant ones. FR-1 to FR-4 define the necessary parameters to establish a connection with a message broker, such as the communication protocol and the topic to subscribe. FR-5 to FR-7 focus on the behavior of the processing algorithm. For example, there will be repetition conditions due to elapsed time or number of messages. FR-8 mentions the need to include options to define conditions of when a *FLEX-consumer* should close connections and end.

**Table 3.** Functional and non-functional requirements of *FLEX-consumers*.

| Id | Description |
|---|---|
| FR-1 | A *FLEX-consumer* has to connect to a message broker using AMQP or MQTT protocols. |
| FR-2 | The IP address and port of the message broker has to be configurable. |
| FR-3 | *FLEX-consumers* should be able to subscribe to a specific queue or topic depending on the protocol. |
| FR-4 | The topic or the queue name to be subscribed has to be configurable. |
| FR-5 | *FLEX-consumers* must include configurable trigger conditions to control the processing algorithm execution. |
| FR-6 | A *FLEX-consumer* user can configure whether messages are discarded once processed or they continue to be processed in subsequent calls to the algorithm. |
| FR-7 | There should be an option to decide if the processing algorithm can be executed one last time before closing the connection with the message broker. |
| FR-8 | *FLEX-consumers* will offer options to configure when their execution stops. |
| FR-9 | *FLEX-consumers* must include an option to invoke the processing algorithm at the beginning of the lifecycle. |
| FR-10 | *FLEX-consumers* should include an error reporting system. |
| FR-11 | *FLEX-consumers* should provide help or usage information to users. |
| NFR-1 | *FLEX-consumers* must be executable on a system that has the Java Development Kit (Java JDK). |
| NFR-2 | The *FLEX-consumer*'s user interface should be user friendly. |

Figure 8 shows the use case diagram of a *FLEX-consumer*. The main actor is the *FLEX-consumer* user that launches the consumer in order to process data coming from the message broker. To this end, the user has to configure some mandatory parameters, such as the message broker IP address as well as the queue or topic depending on whether the *FLEX-consumer* uses AMQP or MQTT, and the trigger and halting conditions of the processing algorithm. Additionally, the user can configure connection credentials.

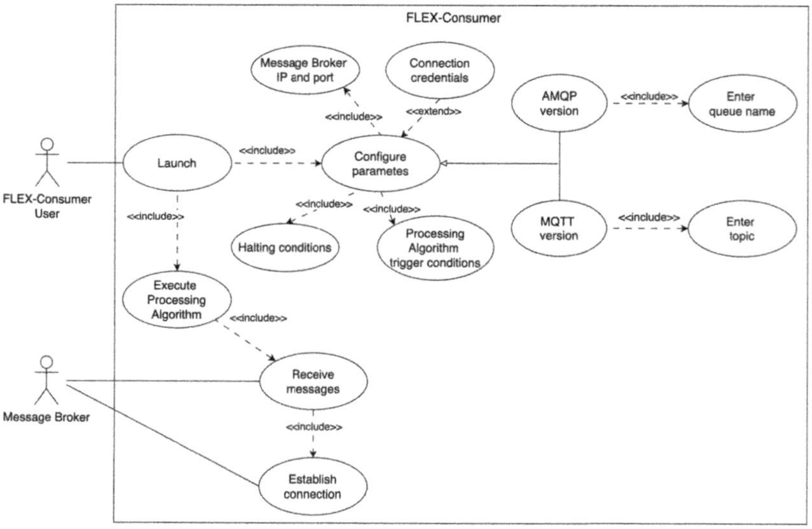


**Figure 8.** *FLEX-consumer* use case diagram.

These requirements lead us to the *FLEX-consumer* architecture shown in Figure 9. A *FLEX-consumer* comprises three main components. The *"User Interaction"* module is responsible for interacting with the user through the command line terminal, mainly to read the configuration parameters and display the results of the processing algorithm, including the errors, if they occur. The *"Connection Management"* module is in charge of establishing and managing the communication with the message broker. Finally, the *"Data Processing"* module deals with the execution of the processing algorithm following the entered configuration. Since this algorithm, which is specific to each *FLEX-consumer*, can have different internal sub-modules, this module can be conceived as a wrapper that controls the algorithm's execution and stop conditions.

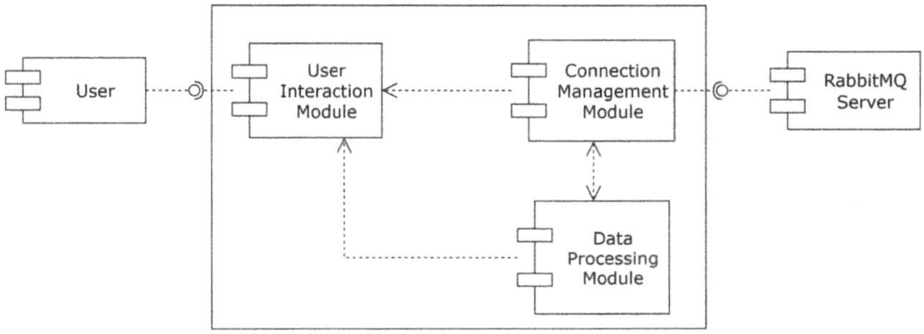


**Figure 9.** *FLEX-consumer* components diagram.

Regarding implementation, *FLEX-consumers* are Java applications in JAR format that are invoked by users using a command line. When *FLEX-consumers* are executed, they receive arguments that define how they must behave. For example, there exist parameters to state different connection options, such as the IP address of the message brokers, their listening port, and the topic to subscribe. For instance, for the Iris example introduced in Section 4.1, the *FLEX-consumer* produced by FLEXTORY can be run following different execution modes. All the invocations have the same structure: "java -jar <*FLEX-consumer* name> -ip <broker address> -t <topic name> <optional arguments>". The optional arguments offer very different customization options. For example, with "-mr 300 -d", the *FLEX-consumer* of Iris will count the number of flowers of each species every 300 received messages, deleting the current messages after they have been processed, i.e., it will only count the new data that have arrived in the last 300 messages. Another possibility is "-tr 4 -w 8", which indicates "count the number of each Iris species every 4 min and stop the *FLEX-consumer* execution if there are no new messages after 8 min since the last one received". Note that, in this case, the messages will not be deleted after being processed, meaning that all the received messages will be processed each time.

Finally, Figure 10 describes the lifecycle of a *FLEX-consumer*. In order to connect to the message broker (in the example RabbitMQ), the user provides the networking configuration (e.g., IP address and port of message broker, the topic name). In addition, the user can also define other configurable options such as the termination condition. Then, the *FLEX-consumer* establishes the connection with the broker and, depending on the configuration used, waits until a repetition or halt condition is triggered. *FLEX-consumers* can behave as long-lived connection applications, i.e., they can be configured without halting conditions, using the persistent option to maintain the execution of the processing algorithm indefinitely. It is worth mentioning that there are some constraints in the combination of some of these parameters. For instance, in the networking configuration, it is mandatory to have at least the IP address of the message broker and the topic (or queue in AMQP) to subscribe. In addition, it seems natural that the processing algorithm is executed at least once. Therefore, if no repetition parameters or the persistent option are specified, a halting condition must be specified. This way, the *FLEX-consumer* could execute the algorithm once and finish. Furthermore, if there is a repetition argument, there is no need to define a halting condition of the *FLEX-consumer*, although they can also be combined.

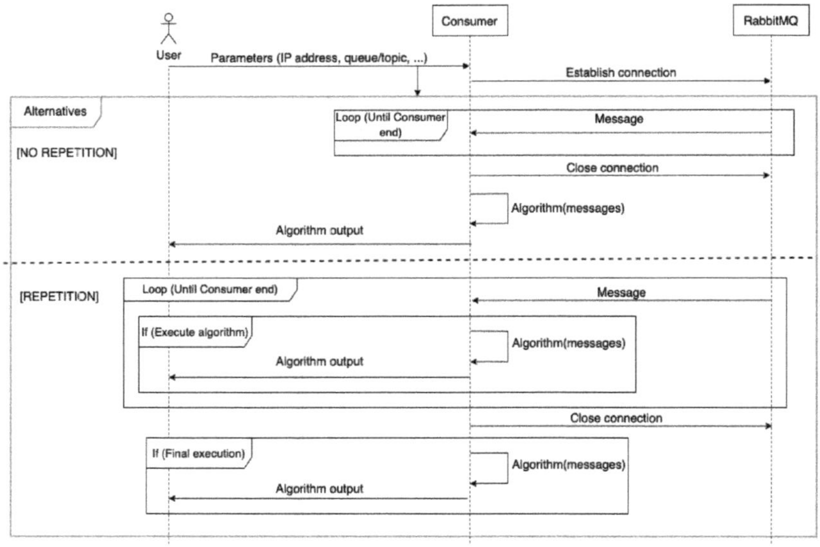


**Figure 10.** Message exchange between the *FLEX-consumer*, its user, and the message broker.

## 5. Illustrative Examples

In this section, we present two examples in which FLEXTORY can help boost the development of a *FLEX-consumer*. Both examples arise from the needs of real research projects in which the authors currently participate. FLEXTORY and all the material required to replicate these examples is published in a gitLab repository (https://gitlab.com/morse-uma/formal-methods/flextory/, accessed on 28 March 2024).

### 5.1. *Learning from Observations*

In the last few years, there has been rising interest in the so-called *digital twins*, that is, system models that can be enriched when new systems' behaviors are observed. These models can be used to make decisions or predict failures. In order to construct a digital twin, a lot of information has to be collected and concurrently processed using a learning algorithm. The LearnFDT project aims to automatically generate *formal digital twins*, i.e., models of systems described with a formal language, using Automata Learning techniques. In this example, we use FLEXTORY to generate a *FLEX-consumer* that constructs such formal digital twins.

In particular, the system to be learned is DASH [12], a protocol for adaptive video transmission. Thus, two entities are involved in DASH: a streaming video server and a client application. To learn the behavior of DASH, the *FLEX-consumer* application implements an algorithm based on Automata Learning techniques with passive learning [30]. The purpose of Automata Learning techniques is to build formal models that simulate the behavior of the systems under learning (SULs). The *passive* learning approach uses the *observed* behavior (execution traces) of the SUL to build the formal models.

Figure 11 shows a general overview of the case study. The objective is to generate a *FLEX-consumer* that is able to construct a digital twin of a DASH remote server. The setup for generating the digital twin consists of a publisher, a message broker, and the *FLEX-consumer*. The publisher sniffs traffic exchanged between the DASH server (available online [31]) and some clients during the execution of several video streaming sessions. Then, these traffic captures are packed in a message in JSON format and transmitted to the message broker. Since the publisher is beyond the scope of this work, we use a dummy publisher that reads the traces from a file and sends them to the broker. In this example, the message broker is a RabbitMQ instance that uses MQTT and has a topic "dash" where all DASH traces will be stored.

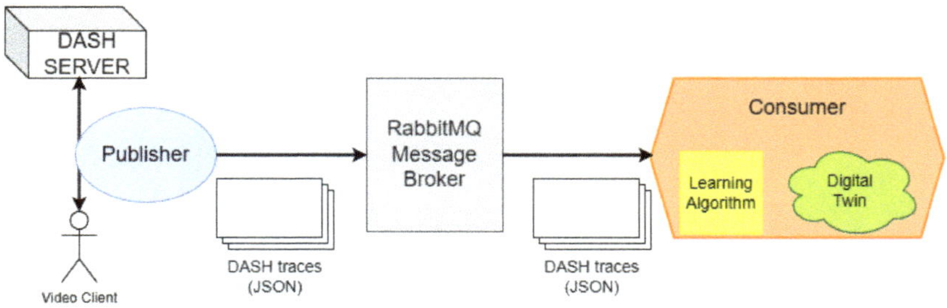


**Figure 11.** Deployment of the DASH case study with a DASH server acting as the publisher, a RabbitMQ message broker, and a *FLEX-consumer* instance integrating a learning algorithm.

The *FLEX-consumer* receives network traces and executes the learning algorithm in order to incrementally produce a model of the DASH protocol. In this case study, we have the role of developers (the FLEXTORY users) and also users of the *FLEX-consumer*. First, as developers, we provide FLEXTORY with the JSON schema defining the traces format (see Appendix A for JSON schema definition). Then, we provide the *Algorithm* class that launches the learning algorithm and feeds it with incoming traces. The algorithm constructs a model of the system in an incremental manner, extending the learned model when new behaviors are read. The *Algorithm* class also deploys a web server that allows us to inspect the model under construction. We have integrated a *learning automata* algorithm, in which the system models are described as *timed automata* with one timer. The details of the learning algorithm are beyond the scope of this paper, but, in general, we can integrate any learning algorithm by injecting it as a dependency. Finally, we select MQTT as the communicating protocol in order to communicate with the RabbitMQ broker.

Once the *FLEX-consumer* is built, it is invoked with the following configuration: "java -jar Dash.jar -ip <ip address of the message broker> -pers -t dash"; that is, the *FLEX-consumer* is configured to establish a connection with the message broker, subscribe to the topic "dash", and execute the processing algorithm in a persistent way. Then, the *FLEX-consumer* will start the connection and wait for new data. As mentioned before, the processing algorithm deploys a web server to check the progress of the Automata Learning algorithm. Figure 12 shows the timed automata learned during the *FLEX-consumer* execution (left) and the final automata produced after processing 92 traces.

### 5.2. Validating Data Format

The EPICENTRE project [32] proposes a 5G distributed experimentation platform. The platform, whose architecture is beyond the scope of this paper, includes a RabbitMQ broker with multiple queues. The first queue is used to inject the results of the experiments. These data are processed by a *consumer* application (called *Validator* in the project) that collects messages with a correct data format and injects them into a second queue in order to be processed by different analytics modules, which can also be considered as *consumer* applications. The broker communicates with all these entities using the MQTT protocol. In this project, most of these *consumers* (the *Validator* and the analytic modules) have been developed in Python. Anyhow, the programming language of the *consumer* is transparent to the broker message.

In this example, we use FLEXTORY to generate a Java *Validator* so that it subscribes to the first queue and collects the correct messages. We have limited this example to the *Validator*, since it is the module developed by our research group. However, the rest of the other analytic modules used in the 5G-EPICENTRE project could also be generated using FLEXTORY.

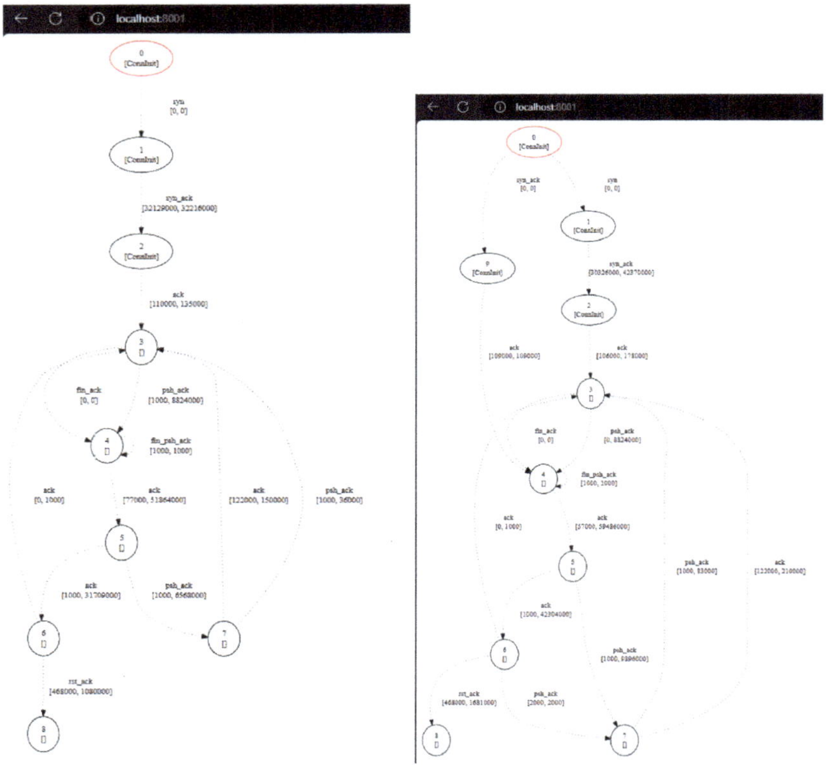

**Figure 12.** Intermediate automaton (**left**) and final automaton after learning 92 traces (**right**).

The development of the *Validator* follows the workflow of FLEXTORY. In the first step, we upload the JSON schema and in the second step, we provide the *Algorithm* class. Both the definition of the JSON schema and the implementation of the *Algorithm* class are included in Appendix A. In this case, the *Validator* just logs in a file whether the messages are either correct or not. Since the *Algorithm* class does not include third-party libraries, we can directly move to the fourth step, in which we select the MQTT protocol. The last step is the compilation of the *FLEX-consumer* that finishes without reporting errors.

Finally, we have executed the generated *FLEX-consumer* to collect real data from the EPICENTRE platform. In particular, we configure the *FLEX-consumer* to check the format of each message when it is received and to stop (terminate execution) when 100 messages have been processed. The results have been quite satisfactory; the *Validator* created is similar to the original one, and the time required to produce it is minimal once the validation algorithm is coded in Java.

## 6. Evaluation: User Study

We have conducted a user study with ten participants recruited from post-graduate students with different knowledge levels in IoT technologies and communication protocols (see Figure 13). In the study, each participant is assigned an exercise that consists of (1) creating a *FLEX-consumer* using FLEXTORY and (2) deploying and using it in a real environment. The *FLEX-consumer* has to process a sequence of messages coming from a message broker with information of different types of boats and use a machine learning algorithm to classify them. Finally, participants are requested to run the *FLEX-consumer* with different configurations. To reduce the time required to carry out the exercise, we have deployed a message broker that will interact with the resulting *FLEX-consumer*. Thus, the participants have to focus only on creating the consumer and using it. In order to successfully complete the implementation of the *FLEX-consumer*, they can follow the tutorial of FLEXTORY (https://gitlab.com/morse-uma/formal-methods/flextory/, accessed on 28 March 2024). In case a participant is not familiar with some of the technology (e.g., specifying a JSON schema), we provide some backup material.

We have collected the users' opinions and suggestions using an online questionnaire (https://forms.gle/DwBEA7mUw1jkw5Zv9, accessed on 28 March 2024). In general, users with a higher knowledge in IoT give a very positive feedback of FLEXTORY, whereas users with less experience do not have a clear picture of the utility of FLEXTORY and *FLEX-consumers*. The conclusions of the user study are summarized as follows:

- All participants have been able to complete the practical task correctly. On average, the task was completed in less than 1 h. A total of 90% of participants think that FLEXTORY is user-friendly and eases the task of implementing IoT consumers.
- As a possible improvement, some participants have suggested easing the installation process of FLEXTORY by automatically installing the Java Development Kit (JDK) if it is not present on the machine. We believe that manually installing the JDK is not a big deal and allows more flexibility to decide which distribution to use.
- We explicitly asked about the most confusing step when using FLEXTORY. As shown in Figure 14, there is not a consensus: 50% of the participants have not faced any issues using FLEXTORY whereas 20% found some difficulties with the definition of the JSON schema.
- With respect to *FLEX-consumers*, 100% of users believe that they are easy to configure and use. However, 30% of them are not sure if they are useful. We believe that this may be related to the case study proposed in the exercise that could not be relevant or attractive enough. One participant suggested to generate *FLEX-consumers* to collect network data and perform a characterization of its behavior.
- Finally, we asked participants suggestions. Among others, they proposed to improve the documentation of FLEXTORY and *FLEX-consumers*. In addition, they recommended to run FLEXTORY as a web service in such a way that no installation is required. Finally, they suggested the integration of *FLEX-consumers* with other message brokers such as Kafka. Since Kafka uses its own communication protocol, we think that this proposal implies to design the templates for a new *FLEX-consumer* that implements the protocol.

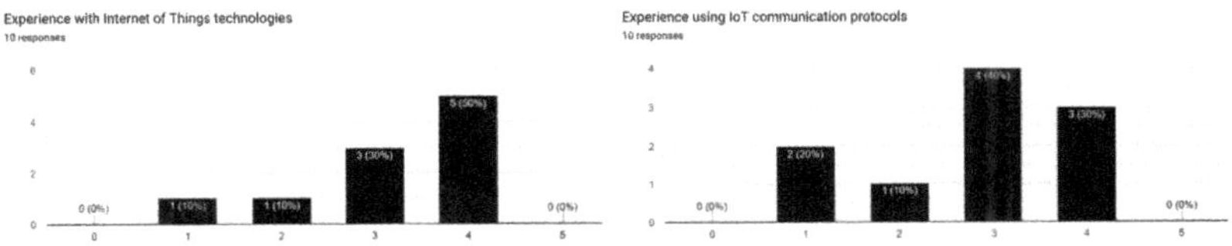


**Figure 13.** Participants' experience in IoT technologies (**left**) and IoT communication protocols (**right**).

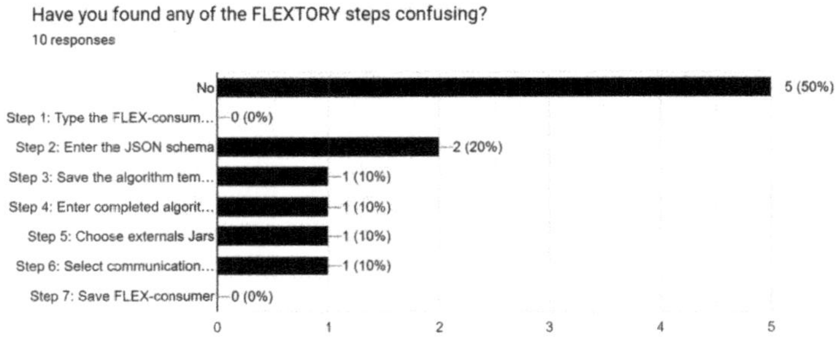


**Figure 14.** Users' opinions on FLEXTORY's most complex steps.

## 7. Discussion

In this section, we discuss the pros and cons of FLEXTORY and the *FLEX-consumers*. First, we would like to highlight that the target users of FLEXTORY are developers without extensive background in IoT communication protocols who want to process data. In this case, FLEXTORY is a valuable tool to produce, in *five steps,* a fully operational IoT consumer (*FLEX-consumer*) which is able to process data coming from a remote message broker. Since the IoT ecosystem is very diverse and changing, the *FLEX-consumer* execution mode can be re-configured without having to re-run FLEXTORY. Thus, a consumer can be used in different scenarios. Clearly, this greatly simplifies the development cycle, and the results of the user study (see Section 6) indicate that, in general, potential users of FLEXTORY are satisfied with it.

However, an expert user could find FLEXTORY a bit limited. The main weakness of the generated *FLEX-consumers* is, presumably, that it can only receive data from one message broker and from one topic (or queue). For instance, in [33], the authors present an IoT Edge-Cloud hybrid architecture in which consumers have to dynamically connect to different message brokers according to different conditions. In order to adapt *FLEX-consumers* to new IoT architectures or even to a changing environment, we can adopt solutions from the SPL community, such as [5,6], that require different models (e.g., variability and goal models) to generate IoT applications. It should be noted that these models can be very dependent on the case study and therefore the FLEXTORYuser would need some knowledge of the case study and modeling techniques.

Another weak point could be the limited set of execution modes of the *FLEX-consumers*. Based on the literature, FLEXTORY covers the most common execution modes, such as daemon mode and end processing after a time deadline. In addition, it is possible to set the frequency of execution of the processing algorithm according to the elapsed time or the number of messages received.

To finalize, the *FLEX-consumers* only support messages in JSON format. Although this format it is very flexible, thanks to the definition of the JSON schema, most message brokers support other formats. For example, RabbitMQ currently supports XML, Thrift, and MessagePack. Despite these limitations, FLEXTORY could be useful for fast prototyping *FLEX-consumers*.

Concerning the IoT communication protocols integrated in *FLEX-consumers*, MQTT was a clear option for many reasons. In general, determining the most proper IoT communication protocol to use in a particular application is an important engineering problem, as many factors have to be considered. Since IoT devices have limited hardware features, a wide variety of IoT communication protocols have been developed to overcome distinct application problems while aiming for low latency, maximum throughput, and low energy consumption. Different studies [34,35] have compared the characteristics and capabilities of the most popular protocols, such as MQTT, HTTP, COAP, AMQP, and XMPP. There is usually a consensus about MQTT being the most suitable option for the majority of the IoT case studies. Even if it is not the option with the best performance, the strong points of the MQTT protocol are its lightweight design and the fact that a message can be sent to multiple subscribers with maximum performance, thanks the publisher/subscriber model. Moreover, MQTT is often preferred when a secure communication environment is needed. In the design phase of FLEXTORY, our intention was to choose multiple communication protocols with features similar to MQTT's. We selected AMQP because it can behave in a manner quite similar to MQTT, enabling the same features in both implementations. In further FLEXTORY versions, we will study how to implement other protocols without changing the similar use of the *FLEX-consumers*.

## 8. Conclusions and Future Work

In this paper, we have presented FLEXTORY, a software factory tool whose objective is to simplify the challenging and time-consuming task of implementing IoT data *consumer* applications. Although there exist some frameworks that ease the implementation process, they do not address the heterogeneous case studies of the IoT domain and, in many cases, developers are forced to build ad hoc applications. For this reason, we propose FLEXTORY to guide a developer in the process of generating *consumer* applications that are characterized by connecting to a *message broker* and process received data. Thanks to FLEXTORY developers do not have to worry about implementation details such as the communication logic with the message broker, the integration of the processing algorithm in the *FLEX-consumer*, or the management of the incoming data. In addition, FLEXTORY produces configurable *FLEX-consumer* applications with a flexible behavior, in the sense that a *FLEX-consumer* user can define different conditions to trigger the processing algorithm or to finish the execution.

The current version of FLEXTORY produces Java *FLEX-consumers* that support MQTT or AMQP communication protocols. This paper presents the requirements and architectures of both the software factory FLEXTORY and the resulting *FLEX-consumers*.

To show the versatility of FLEXTORY and the *FLEX-consumers* generated, we have presented a running example to describe the methodology (Section 4.1) and a more complex case study in Section 5. In addition to these examples, FLEXTORY has also been used in other domains to show the wide variety of message formats and processing algorithms that can be included in the *FLEX-consumers*. In the field of computational phylogenetics, we have generated a *FLEX-consumer* that processes phylogenetic trees using the Sankoff [36,37] algorithm. Another case study is a *FLEX-consumer* that checks if data have been properly encoded using different cryptography algorithms. Finally, FLEXTORY has been used to build a *FLEX-consumer* application that collects the result messages from different experiments and transforms them into a specific data format, which can be found at gitLab.

To assess the user's opinion about FLEXTORY, we have carried out a user study. The overall feedback is very positive, and the participants have made some suggestions that we plan to address.

As future work, we plan to extend FLEXTORY in different ways to generate more flexible *FLEX-consumers* that can be used in different contexts. For instance, we have observed the importance of enabling subscription to multiple topics or queues. In addition, we would like to produce *FLEX-consumers* in Python, since it is a user-friendly language for non-software experts and has plenty of support to develop data analytics tools. Moreover, we would like to distribute FLEXTORY as a web application or even as an Integrated Development Environment (IDE) plugin.

We also aim to check the compatibility of the current *FLEX-consumers* with other message brokers. Although AMQP and MQTT are standardized protocols, there are different versions, and some compatibility issues may occur depending on the version used in the *FLEX-consumer* application and the message broker.

In recent years, new IoT architectures have arisen from the evolution of wireless and mobile networks and require new features in the IoT applications. We plan to study in depth these new architectures in order to generate *FLEX-consumers* suitable for these dynamic scenarios.

**Author Contributions:** Implementation and Evaluation, R.L.-G.; Research, R.L.-G., L.P. and M.-d.-M.G.; Funding acquisition, L.P. and M.-d.-M.G.; Writing, R.L.-G., L.P. and M.-d.-M.G. All authors have read and agreed to the published version of the manuscript.

**Funding:** This research was funded by the UNICO I+D Advanced 5G and 6G program (Spanish Ministry of Economy and Digital Transformation) grant number TSI-063000-2021-11 (5G+TACTILE) and the State Plan for Scientific, Technical and Innovation Research 2021-2023 (Spanish Ministry of Science, Innovation and Universities) grant number PID2022-142181OB-I00 (LearnFDT).

**Data Availability Statement:** Data is contained within the article.

**Conflicts of Interest:** The authors declare no conflicts of interest.

## Appendix A

This appendix includes supplementary material to clarify the implementation of FLEXTORY and the examples presented in Sections 4.1 and 5.

Figure A1 shows the class diagram of the current implementation of FLEXTORY, which is related to the components diagram shown in Figure 6. Classes with suffixes `Panel` and `Frame` are part of the View component. The `ConsumerTemplates` and `Compiler` classes are part of the Model component. The former class includes the code of a generic MQTT and AMQP client in String format. The AMQP client's template code uses an external library provided by RabbitMQ [38], while the MQTT client relies on the Eclipse Paho library [39]. The `Compiler` class includes methods to generate and compile all the Java code of the *FLEX-consumer* and package the result in JAR format. For instance, to convert the JSON schema into serializable classes, FLEXTORY uses the Jackson library (https://github.com/FasterXML/jackson, accessed on 28 March 2024). Finally, the `Controller` class includes all the events handlers in order to properly update the View and the Model.

Listings A1 and A2 show part of the code of the AMQP and MQTT *FLEX-consumers*. We would like to clarify that the main differences between them only concern the communication module. As mentioned above, the AMQP template (Listing A1) uses the RabbitMQ library that supports AMQP communications. In Figure A1, the method `connect` (lines 22–53) is in charge of establishing the connection with the broker to create a channel to receive the messages of a specific queue. In addition, we have to define a callback (`deliverCallback` lines 34–51) that will be executed when a new message arrives. Basically, depending on the configuration, this callback will launch the processing algorithm in a worker thread or will delete the queue, close the connection with the broker, and end the *FLEX-consumer* execution. The method `end` (lines 55–64) implements this functionality.

**Listing A1.** Communication module of an AMQP *FLEX-consumer*.

```
   public class ConsumerNameAMQP {
2    // Configuration attributes
     private Timer timerWaitTime, timerTimeRep, timerPersistent;
4    private int numMessages, numMessagescont, maxMessages, maxMessagescont;
     private boolean finalexec;
6
     // Connection attributes
8    private String host, queueName, username, password,consTag;
     private int port;
10   private Connection connection;
     private Channel channel;
12   //Queue to store messages
     private LinkedBlockingQueue<Schema> lbq;
14
     public ConsumerNameAMQP(String configuration[]){
16     //Initialize attributes, queue and timers tasks
```

```
    //..
    //Establish connection
    connect();
  }

  private void connect() throws Exception{
    ConnectionFactory factory = new ConnectionFactory();
    factory.setHost(host);
    if (port != -1) factory.setPort(port);
    if (!username.equals("")) factory.setUsername(username);
    if (!password.equals("")) factory.setPassword(password);

    connection = factory.newConnection();
    channel = connection.createChannel();
    channel.queueDeclare(queueName, false, false, false, null);
    channel.basicQos(2);
    //Definition of callback
    DeliverCallback deliverCallback = (consumerTag, delivery) -> {
      if (timerWaitTime != null)
        reset();
        try { processMessage(new String(delivery.getBody(), "UTF-8"));
          if (maxMessages != 0){
            maxMessagescont++;
            if (maxMessages == maxMessagescont)  end();
          }
          if (numMessages != 0){
            numMessagescont++;
            if (numMessages == numMessagescont) {
              doWork();
              numMessagescont = 0;
            }
          }
        }
        catch (Exception e) {e.printStackTrace(); }
    };
    consTag = channel.basicConsume(queueName, true, deliverCallback,
        consumerTag -> {});
  }

  private void end() throws Exception {
    channel.basicCancel(consTag);
    channel.queueDelete(queueName);
    channel.close();
    connection.close();
    if (timerWaitTime != null) timerWaitTime.cancel();
    if (timerTimeRep != null) timerTimeRep.cancel();
    if (finalexec) doWork();
    System.exit(0);
  }
}
```

In the MQTT client shown in Figure A1, we use the Eclipse Paho library. In this case, the `connect` method (lines 16–27) creates a session with the broker and subscribes to a specific topic. When a new message arrives, it is managed by the `messageArrived` method (lines 29–54). Observe that its implementation is similar to `deliveryCallback` of the AMQP template. Finally, the method `end` (lines 56–63) closes the session with the broker and ends the execution of the client.

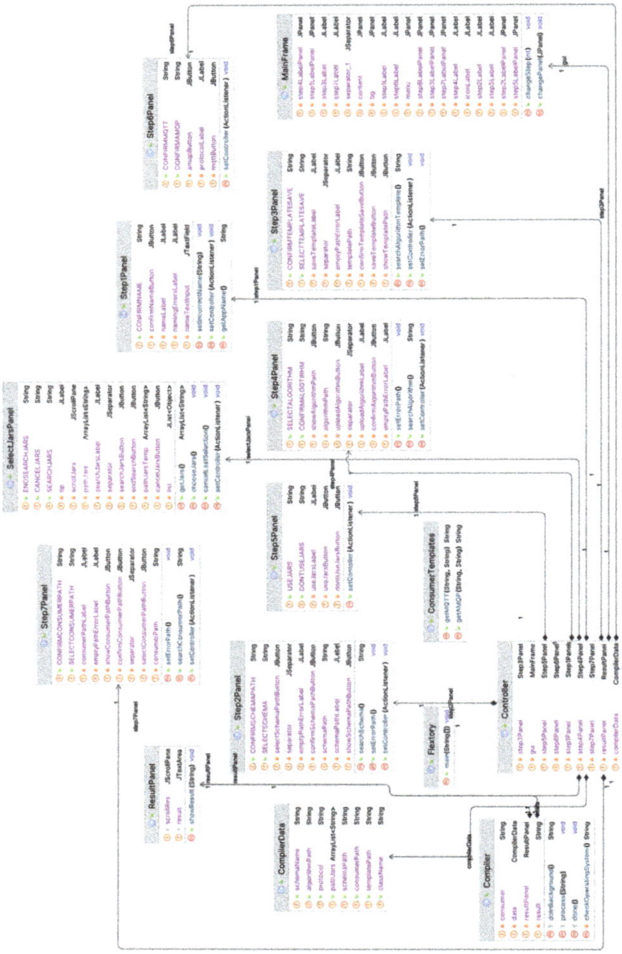

**Figure A1.** FLEXTORY class diagram.

**Listing A2.** Communication module of an MQTT *FLEX-consumer*.

```
public class ConsumerNameMQTT {
  // Configuration attributes (same as AMQP) ...
  // Connection attributes
  private String brokerUrl, host, topic, username, password, clientId;
  int port;
  MqttClient client;
  //Queue to store messages
  private LinkedBlockingQueue<Schema> lbq;

  public ConsumerNameMQTT(String configuration[]){
    //Initialize attributes, queue and timers tasks ...
    //Establish connection
    connect();
  }

  private void connect() throws Exception{
    client = new MqttClient(brokerUrl, clientId, null);
    MqttConnectOptions connOpts = new MqttConnectOptions();
    connOpts.setCleanSession(true);
    connOpts.setAutomaticReconnect(true);
```

```
    connOpts.setConnectionTimeout(0);
    if (!username.equals("")) connOpts.setUserName(username);
    if (!password.equals("")) connOpts.setPassword(password.toCharArray());
    client.connect(connOpts);
    client.setCallback(this);
    client.subscribe(topic);
  }

  public void messageArrived(String topic, MqttMessage message) {
    if (timerWaitTime != null) reset();
    try { processMessage(new String(message.getPayload(), "UTF-8"));
      if (maxMessages != 0) {
        maxMessagescont++;
        if (maxMessages == maxMessagescont) {
          timerEnd = new Timer();
          TimerTask task_end = new TimerTask() {
            @Override
            public void run() {
              try { end();}
              catch (Exception e) {e.printStackTrace(); }
            }
          };
          timerEnd.schedule(task_end, 0);
        }
      }
      if (numMessages != 0) {
        numMessagescont++;
        if (numMessages == numMessagescont) {
          doWork();
          numMessagescont = 0;
        }
      }
    } catch (Exception e) { e.printStackTrace(); }
  }

  private void end() throws Exception {
    client.disconnect();
    client.close();
    if (timerWaitTime != null) timerWaitTime.cancel();
    if (timerTimeRep != null) timerTimeRep.cancel();
    if (finalexec)  doWork();
    System.exit(0);
  }
}
```

We continue with the definition of the JSON schema of the examples presented in Sections 4.1 and 5. JavaScript Object Notation (JSON) is a lightweight data-interchange format commonly used for communication in the IoT domain. A JSON schema (https://json-schema.org/, accessed on 28 March 2024) provides a formal description of the expected format of JSON messages, including the data type of each field, any constraints on the data, and the relationships between different parts of the message. Using a JSON schema allows for validation of incoming and outgoing messages.

Listing A3 shows the JSON schemas used in the DASH example (left) and the Validator example (right). Both schemas include the fields `title`, `type`, `properties`, and `additionalProperties`. However, in each example, the messages include a different set of properties. For example, the DASH example includes information of the TCP header, such as the flags activated (`flags`) or the sequence number (`seqN`). In the Validator example, the properties include information of the experiments such as the experiment id or the scenario, among others.

**Listing A3.** JSON schemas of the DASH example (**left**) and the Validator example (**right**).

```
{
  "title": "Dash tcp format",
  "type":"object",
  "properties":{
    "timestamp":{"type":"string"},
    "flags":{"type":"integer"},
    "source":{"type":"integer"},
    "target":{"type":"integer"},
    "seqN":{"type":"integer"}
  },
  "additionalProperties":{
    "type":"string"
  }
}
```

```
{
  "title": "Epicentre exp. data",
  "type":"object",
  "properties":{
    "category":{"type":"string"},
    "testbed_id":{"type":"integer"},
    "scenario_id":{"type":"integer"},
    "use_case_id":{"type":"integer"},
    "experiment_id":{"type":"integer"},
    "netapp_id":{"type":"string"},
    "data":{
      "type":"array",
      "items":{
        "type":"object",
        "javaName":"Data",
        "properties":{
          "type":{"type":"string"},
          "timestamp":{"type":"number"},
          "origin":{"type":"string"},
          "unit":{"type":"string"}
        }
      }
    }
  },
  "additionalProperties": false
}
```

To conclude, we present the implementation of the *Algorithm* class used in the Validator case study (Listing A4). The *Algorithm* class implements the `Runnable` interface so that the processing algorithm can be executed in a different thread. The processing algorithm is coded in the run method and can include references to external libraries. In the example, it only logs in a file that the messages are either correct or not. Observe that the run method processes messages from the class attribute `data` (line 14), which consist of a queue where the communication module stores all incoming messages.

**Listing A4.** Implementation of the Algorithm class of the Validator consumer.

```
   public class Algorithm implements Runnable {
2    LinkedBlockingQueue<EpiSchema> data;
     private final static String log_filename = "output.txt";
4    public Algorithm(LinkedBlockingQueue<EpiSchema> data){
       this.data = data;
6    }

8    @Override
     public void run() {
10     ArrayList<String> category = new
           ArrayList<String>(Arrays.asList("5g_network", "nfv_mano",
           "vnf_chain", "experiment"));
       ArrayList<String> origin = new ArrayList<String>(Arrays.asList("UE",
           "RAN", "5GC", "EPC", "main data server", "edge"));
12     try {
         PrintWriter outputFile = new PrintWriter(new
             FileWriter(log_filename, true));
14       for(EpiSchema message : data) {
           if (!category.contains(message.getCategory())) {
16           outputFile.println("***Invalid or missing category***");
             outputFile.println(message.toString());
18         }
           if (message.getTestbedId() == null)
20           outputFile.println("***Missing testbed ID***");

22         for (Data d : message.getData()) {
             if (d.getType() == null || d.getTimestamp() == null ||
                 !origin.contains(d.getOrigin())) {
```

```
                outputFile.println("***Missing one or more required data
                    arguments**");
                break;
            }
        }
        /*Print rest of message content to outputFile*/
        }
        outputFile.close();
    } catch (Exception e) { e.printStackTrace();}
  }
}
```

## References


1. Hassan, R.; Qamar, F.; Hasan, M.K.; Aman, A.H.M.; Ahmed, A.S. Internet of Things and Its Applications: A Comprehensive Survey. *Symmetry* **2020**, *12*, 1674. [CrossRef]
2. Advanced Message Queuing Protocol. Available online: https://www.amqp.org/ (accessed on 28 March 2024).
3. MQTT: The Standard for IoT Messaging. Available online: https://mqtt.org/mqtt-specification/ (accessed on 28 March 2024).
4. Pohl, K.; Böckle, G.; Van Der Linden, F. *Software Product Line Engineering: Foundations, Principles, and Techniques*; Springer: Berlin/Heidelberg, Germany, 2005; pp. 1–467. [CrossRef]
5. Butting, A.; Kirchhof, J.C.; Kleiss, A.; Michael, J.; Orlov, R.; Rumpe, B. Model-Driven IoT App Stores: Deploying Customizable Software Products to Heterogeneous Devices. In Proceedings of the 21st ACM SIGPLAN International Conference on Generative Programming: Concepts and Experiences (GPCE 2022), New York, NY, USA, 6–7 December 2022; Association for Computing Machinery: New York, NY, USA, 2022; pp. 108–121. [CrossRef]
6. Ayala, I.; Amor, M.; Horcas, J.M.; Fuentes, L. A goal-driven software product line approach for evolving multi-agent systems in the Internet of Things. *Knowl. Based Syst.* **2019**, *184*, 104883. [CrossRef]
7. Cusumano, M.A. The Software Factory: A Historical Interpretation. *IEEE Softw.* **1989**, *6*, 23–30. [CrossRef]
8. Greenfield, J.; Short, K. Software factories: assembling applications with patterns, models, frameworks and tools. In Proceedings of the Companion of the 18th Annual ACM SIGPLAN Conference on Object-Oriented Programming, Systems, Languages, and Applications (OOPSLA'03), Anaheim, CA, USA, 26–30 October 2003; Association for Computing Machinery: New York, NY, USA, 2003; pp. 16–27. [CrossRef]
9. Benaddi, L.; Ouaddi, C.; Jakimi, A.; Ouchao, B. Towards A Software Factory for Developing the Chatbots in Smart Tourism Mobile Applications. *Procedia Comput. Sci.* **2024**, *231*, 275–280. [CrossRef]
10. Beneke, T. A Perfect Match: Java and the Internet of Things. 2014. Available online: https://www.oracle.com/technical-resources/articles/java/java-maker-iot.html (accessed on 28 March 2024).
11. Angluin, D. Learning regular sets from queries and counterexamples. *Inf. Comput.* **1987**, *75*, 87–106. [CrossRef]
12. International Organization for Standardization (ISO), I.E.C.I. Dynamic Adaptive Streaming over HTTP (DASH). Parts 1–9. Available online: https://standards.iso.org/ittf/PubliclyAvailableStandards/index.html (accessed on 28 March 2024).
13. MQTT Version 3.1.1. Available online: http://docs.oasis-open.org/mqtt/mqtt/v3.1.1/mqtt-v3.1.1.html (accessed on 28 March 2024).
14. MQTT Version 5.0. Available online: http://docs.oasis-open.org/mqtt/mqtt/v5.0/os/mqtt-v5.0-os.html (accessed on 28 March 2024).
15. OASIS Advanced Message Queuing Protocol (AMQP). Available online: https://docs.oasis-open.org/amqp/core/v1.0/os/amqp-core-overview-v1.0-os.html (accessed on 28 March 2024).
16. The Constrained Application Protocol (CoAP). Available online: https://datatracker.ietf.org/doc/html/rfc7252 (accessed on 28 March 2024).
17. Extensible Messaging and Presence Protocol (XMPP): Address Format. Available online: https://datatracker.ietf.org/doc/rfc7622/ (accessed on 28 March 2024).
18. RabbitMQ. Available online: https://www.rabbitmq.com/ (accessed on 28 March 2024).
19. Azure IoT Hub. Available online: https://azure.microsoft.com/products/iot-hub (accessed on 28 March 2024).
20. Apache Kafka. Available online: https://kafka.apache.org/ (accessed on 28 March 2024).
21. Eclipse Mosquitto. Available online: https://mosquitto.org/ (accessed on 28 March 2024).
22. AWS IoT Core. Available online: https://aws.amazon.com/iot-core/ (accessed on 28 March 2024).
23. de Souza Cimino, L.; de Resende, J.E.E.; Silva, L.H.M.; Rocha, S.Q.S.; de Oliveira Correia, M.; Monteiro, G.S.; de Souza Fernandes, G.N.; da Silva Moreira, R.; de Silva, J.G.; Santos, M.I.B.; et al. A middleware solution for integrating and exploring IoT and HPC capabilities. *Softw. Pract. Exp.* **2019**, *49*, 584–616. [CrossRef]
24. Cherrier, S.; Ghamri-Doudane, Y.M.; Lohier, S.; Roussel, G. D-LITe: Distributed logic for internet of things sErvices. In Proceedings of the 2011 IEEE International Conferences on Internet of Things and Cyber, Physical and Social Computing, iThings/CPSCom 2011, Dalian, China, 19–22 October 2011; pp. 16–24. [CrossRef]
25. Chauhan, S.; Patel, P.; Sureka, A.; Delicato, F.C.; Chaudhary, S. Demonstration Abstract: IoTSuite—A Framework to Design, Implement, and Deploy IoT Applications. In Proceedings of the 2016 15th ACM/IEEE International Conference on Information Processing in Sensor Networks (IPSN), Vienna, Austria, 11–14 April 2016. [CrossRef]

26. Akasiadis, C.; Pitsilis, V.; Spyropoulos, C.D. A Multi-Protocol IoT Platform Based on Open-Source Frameworks. *Sensors* **2019**, *19*, 4217. [CrossRef] [PubMed]
27. Nguyen, X.T.; Tran, H.T.; Baraki, H.; Geihs, K. FRASAD: A framework for model-driven IoT Application Development. In Proceedings of the IEEE World Forum on Internet of Things, WF-IoT 2015, Reston, VA, USA, 12–14 December 2015; pp. 387–392. [CrossRef]
28. Henelius, A.; Torniainen, J. MIDAS: Open-source framework for distributed online analysis of data streams. *SoftwareX* **2018**, *7*, 156–161. [CrossRef]
29. Iris Data Set. Available online: https://archive.ics.uci.edu/ml/datasets/iris (accessed on 28 March 2024).
30. Aichernig, B.K.; Muškardin, E.; Pferscher, A. Active vs. Passive: A Comparison of Automata Learning Paradigms for Network Protocols. *Electron. Proc. Theor. Comput. Sci.* **2022**, *371*, 1–19. [CrossRef]
31. DASH, HLS or PROGRESSIVE Stream Test. Available online: https://bitmovin.com/demos/stream-test?format=dash&manifest=https%3A%2F%2Fcdn.bitmovin.com%2Fcontent%2Fassets%2Fart-of-motion-dash-hls-progressive%2Fmpds%2Ff08e80da-bf1d-4e3d-8899-f0f6155f6efa.mpd (accessed on 28 March 2024).
32. Arampatzis, D.; Apostolakis, K.C.; Margetis, G.; Stephanidis, C.; Atxutegi, E.; Amor, M.; Di Pietro, N.; Henriques, J.; Cordeiro, L.; Carapinha, J.; et al. Unification architecture of cross-site 5G testbed resources for PPDR verticals. In Proceedings of the 2021 IEEE International Mediterranean Conference on Communications and Networking, MeditCom 2021, Athens, Greece, 7–10 September 2021; pp. 13–19. [CrossRef]
33. Pham, V.N.; Lee, G.W.; Nguyen, V.; Huh, E.N. Efficient Solution for Large-Scale IoT Applications with Proactive Edge-Cloud Publish/Subscribe Brokers Clustering. *Sensors* **2021**, *21*, 8232. [CrossRef] [PubMed]
34. Bayılmış, C.; Ebleme, M.A.; Ünal Çavuşoğlu.; Küçük, K.; Sevin, A. A survey on communication protocols and performance evaluations for Internet of Things. *Digit. Commun. Netw.* **2022**, *8*, 1094–1104. [CrossRef]
35. Wytrębowicz, J.; Cabaj, K.; Krawiec, J. Messaging Protocols for IoT Systems—A Pragmatic Comparison. *Sensors* **2021**, *21*, 6904. [CrossRef] [PubMed]
36. Sankoff, D. Minimal Mutation Trees of Sequences. *SIAM J. Appl. Math.* **1975**, *28*, 35–42. [CrossRef]
37. Sankoff, D.; Rousseau, P. Locating the vertices of a steiner tree in an arbitrary metric space. *Math. Program.* **1975**, *9*, 240–246. [CrossRef]
38. RabbitMQ Java Client Library. Available online: https://www.rabbitmq.com/java-client.html (accessed on 28 March 2024).
39. Eclipse Paho Java Client. Available online: https://github.com/eclipse/paho.mqtt.java (accessed on 28 March 2024).

sensors

MDPI

*Article*

# Dynamic-Distance-Based Thresholding for UAV-Based Face Verification Algorithms

**Julio Diez-Tomillo, Jose Maria Alcaraz-Calero * and Qi Wang**

School of Computing, Engineering and Physical Sciences (CEPS), University of the West of Scotland (UWS), Paisley PA1 2BE, UK; julio.diez-tomillo@uws.ac.uk (J.D.-T.); qi.wang@uws.ac.uk (Q.W.)

* Correspondence: jose.alcaraz-calero@uws.ac.uk

**Abstract:** Face verification, crucial for identity authentication and access control in our digital society, faces significant challenges when comparing images taken in diverse environments, which vary in terms of distance, angle, and lighting conditions. These disparities often lead to decreased accuracy due to significant resolution changes. This paper introduces an adaptive face verification solution tailored for diverse conditions, particularly focusing on Unmanned Aerial Vehicle (UAV)-based public safety applications. Our approach features an innovative adaptive verification threshold algorithm and an optimised operation pipeline, specifically designed to accommodate varying distances between the UAV and the human subject. The proposed solution is implemented based on a UAV platform and empirically compared with several state-of-the-art solutions. Empirical results have shown that an improvement of 15% in accuracy can be achieved.

**Keywords:** face verification; thresholds; siamese network; cosine distance; Euclidean distance

**Citation:** Diez-Tomillo, J.; Alcaraz-Calero, J.M.; Wang, Q. Dynamic-Distance-Based Thresholding for UAV-Based Face Verification Algorithms. *Sensors* **2023**, *23*, 9909. https://doi.org/10.3390/s23249909

Academic Editors: Behnam Mobaraki and Jose Turmo

Received: 13 November 2023
Revised: 5 December 2023
Accepted: 15 December 2023
Published: 18 December 2023



## 1. Introduction

Nowadays, face recognition and face verification are widely deployed in many areas of our lifes, for instance, at border control points between countries, for building access management in universities and companies, unlocking smartphones, service authentication/authorisation for individuals, and so on. Face recognition aims to identify all the concerned persons, and thus a database is usually needed to store all the people to be identified. In contrast, face verification seeks to identify only one person, so a database would not be needed as only one person would be being searched. The reasons for recognising or verifying a face are many, including customised services, targeted advertisement, health measures, or even security.

Moreover, unmanned aerial vehicles (UAVs) are increasingly deployed in multiple areas for Search and Rescue (SAR) operations [1], delivering goods [2], and many other purposes, as a cost-effective mobile platform. When UAVs meet face verification, the possibilities of potential use cases are immense in light of joining the capabilities of both technologies together in one platform. It is therefore promising to combine the two technologies and be able to recognise people from UAVs. Nevertheless, it is important to secure the UAV, as it will process sensitive privacy data, like the faces images for verification purposes [3].

Such a face verification-enabled UAV system allows for the recognition of a person from distances, without being intrusive to the user by removing the need for a camera close to the face, and opens up opportunities for new location or mobility-based services. Combining both technologies, face verification and UAVs could be highly useful for many use cases, lowering costs and extending the range. One main new use case for face verification from UAVs is the Drone Guard Angel for public safety escort services in the Arcadian-IoT European Project [4]. In this use case, a UAV comes to the customer's position on demand, recognises the person, verifies his/her face to authorise the services as subscribed, and accompanies the person home in a safe way, acting as an invigilator that can help in case of any malicious activity.

Currently, most face verification techniques are performed at close distances. For instance, when a phone is unlocked using face verification, it is usually positioned 30 cm from the face. At this distance, all the features of the face can be captured with a high resolution; therefore, face verification can be performed without significant difficulties using state-of-the-art technologies. However, if face verification is performed from a UAV, it is not possible to position it at 30 cm from the face due to safety reasons. Thus, the UAV has to be at long distances, which vary depending on its size. It is noted that most face verification algorithms are optimised for close distances. Moreover, face verification is usually effective at a fixed distance or within a small range, whereas, for a UAV-based use case, the distance between a face and the UAV is significantly higher; thus, the size of the face in pixels varies and the low pixel resolution of a face introduces significant difficulties in face verification, imposing the challenge of developing new techniques to improve the accuracy of these algorithms when face verification is carried out from longer distances.

In addition, for services like the Drone Guard Angel, the operational environments pose further challenges in terms of poor lighting conditions due to outdoor and late evening operations as well as the fast-changing positions and angles of the target face due to the movement of the UAV. For this reason, a new technique that is lighting- and position-adaptive is proposed in this manuscript to enable the successful deployment of the new services such as Drone Guard Angel.

Consequently, the main contribution of this work is a novel dynamic size-adaptive threshold technique for face verification from UAVs, where the face has to be verified from long distances and in a highly adaptive manner in response to the challenging operational environments where lighting conditions and the positioning of the face change rapidly. To this end, a smart pipeline has been developed for introducing dynamic thresholds.

Furthermore, in order to compare the proposed technique with the existing solution, an empirical analysis of different algorithms is performed, including four state-of-the-art face verification algorithms (ArcFace [5], VGG-Face [6], SFace [7], and Dlib [8]). To conduct a more complete analysis, two different similarity metrics for distance calculation are explored: cosine and Euclidean distances. A study of the suitability of each metric for the different face verification algorithms is conducted too.

The main contributions herein can then be summarised as follows:

- A novel adaptive threshold algorithm for enhancing different state-of-the-art face verification and recognition algorithms to suit UAV-based applications.
- A new pipeline with novel capabilities to integrate the proposed dynamic threshold algorithm.
- An empirical evaluation of the effectiveness of the proposed adaptive threshold technique when dealing with UAV-based face verification use cases using two different similarity metrics.

The rest of the paper is structured as follows: in Section 2, the literature in this research area is reviewed, especially the representative state-of-the-art face verification algorithms and metrics focused for enhancement in this paper. Section 3 describes the proposed design of the face verification pipeline. In Section 4, the algorithm to calculate the proposed thresholds, together with the dataset used, is explained. In Section 5, the implementation of the experiments is presented. In Section 6, the accuracy of the algorithms of the proposed dynamic-distance-based thresholds is compared with the state-of-the-art algorithms and the suitability of each metric for the algorithms is analysed. Finally, in Section 7 conclusions are drawn.

## 2. Related Work

Most of the state-of-the-art studies focus on face recognition rather than on face verification, whilst the latter is what this paper will emphasise. The primary distinction between them lies in the outcome: a face recognition algorithm provides the identity of the recognised person, whereas a face verification algorithm determines whether the person being assessed matches the intended individual or not.

Furthermore, the inference time of face recognition algorithms is always greater than that of face verification ones. This is due to the need to compare one face with all the present faces in the database to achieve face recognition. On the contrary, in face verification it is only necessary to compare one face to another to decide if they are or are not the same person.

Additionally, the decision could be made differently between face recognition and face verification. For the former, the decision can be performed through a K Nearest Neighbour (KNN) algorithm. It stores all the available classes and classifies the data based on similarity. The decision it makes relates to which person in the database is most similar to the one being compared to. On the other hand, face verification is based on a global threshold to decide if it is the same person or not [9]. For this reason, choosing an optimal threshold is critical. If it is not defined properly, many false positives or negatives are possible, leading to erroneous verification. Other thresholding techniques are based on having client-specific thresholds. This paper is based on applying different thresholds depending on the person to be identified [10].

Figure 1 shows the most common approach used by other research works to verify a person: a Siamese neural network [11]. In other words, two identical neural networks with the same network model/structure and the same weights. Two images are also needed. The first one is a picture of the face of the person already verified. The second one is a picture of the person to be verified. These pictures go through the Siamese neural network, obtaining as output two different vectors that can be compared as they are the result of the same neural network. Then, the distance between them is calculated. If the distance is above the threshold, the decision is that they are not the same person. If it is below the threshold, it is concluded that they are the same person.

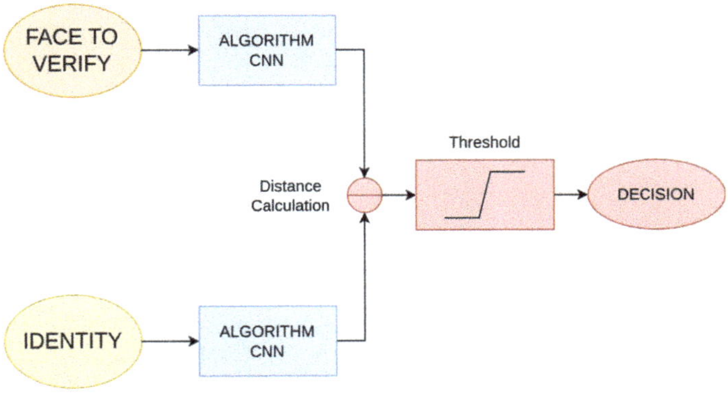


**Figure 1.** Simplified diagram of the face verification process.

Currently, there is a lack of literature on face verification from UAVs. A few studies can be found about detecting and finding people from UAVs [12], for instance, to conduct SAR operations [1]. There are studies in which face verification from UAVs is performed, although only from very short distances. Moreover, they focused on traditional face verification algorithms with low accuracy but high speed, such as LBP (Local Binary Pattern) [13]. Other papers develop algorithms for face verification; however, they do not use real images from UAVs.

For example, [14] uses a Haar Cascade for detecting faces, and an LBPH (LBP Histogram) Face Recognizer. A complete system for the acquisition of images from a UAV and their subsequent physical recognition is proposed. The results do not provide any real images from the UAV, so it has not been possible to test it in a real system. While the LBPH algorithm is easy to implement, it is not as accurate as more current face verification algorithms such as those selected in our contribution.

Another similar work is [15], which also employed an LBPH algorithm along with OpenCV to perform face recognition from a UAV. It used a Raspberry Pi and a Raspberry Pi Camera. The results present some images of face recognition while controlling the UAV. Although all the images are from very close distances, we are trying to perform face recognition from long distances, up to 30 m.

Several studies have explored both the influence of the distance from the UAV to the person and the angle between them. Certain findings suggest that the distance and the depressing angles are crucial constraints in face verification systems, particularly when the UAV is positioned at high altitudes [16]. Another study analysed the effects of the horizontal and vertical angles between the face and the camera, utilising a face recognition pipeline based on OpenCV and Dlib libraries [17]. Further research has developed an innovative deep learning-based face detection and recognition system for drones. The system was analysed by varying the distance and height of the drone to the person, demonstrating good accuracy at lower heights and close distances [18]. These existing studies did not address the adaptive dynamic thresholding technique proposed in this paper.

### 2.1. Empirical Comparison of Face Verification Algorithms

Table 1 shows an in-depth comparison among different face verification and recognition algorithms available in the literature. The verification performance and the different parameters of each algorithm have been obtained from its official paper. The table compares the input and output size of the network used by each algorithm, and the verification performance using two different datasets: Labeled Faces in the Wild (LFW) [19] and YouTube Faces (YTF) [20], as well as comparing the size of the pre-trained weights of the neural networks, and the number of images it has been trained with.

The human-level performance in face verification on LFW is 97.53% [21]. Four algorithms shown in the table do not surpass this value: Open-Face [22], Deep-Face [23], Fisher-Vector Faces [24], and TL Joint Bayesian [25].

VGG-Face [6], Deep-Face [23], and CosFace [26] are the algorithms with the heaviest weights, leading to slow performance when executed. On the other hand, DeepID2 [27] and Open-Face [22] have the lightest weights. Therefore, they will be fast but not as accurate as others. Furthermore, TL Joint Bayesian [25] and GaussianFace [28] algorithms do not have pre-trained weights available, and they have not been evaluated on the YTF dataset [20].

AdaFace [29], CurricularFace [30] and SFace [7] are new algorithms that achieve good accuracy in the LFW dataset (more than 99%). However, there is no information about the output size or the accuracy on the YTF dataset for the first two.

Four representative algorithms have then been selected to perform deeper, empirical analysis in this paper, including VGG-Face [6], ArcFace [5], SFace [7], and Dlib [8].

These four algorithms have been selected for several reasons. Firstly, they all achieve more than 98% accuracy in the LFW dataset; therefore, all four surpass the human-level performance of LFW (97.53%). Moreover, they have pre-trained weights available that can perform tests without training. ArcFace was selected because it provides the loss function with the best verification performance of the three algorithms compared (SphereFace [31], CosFace [26], and ArcFace [5]), whilst Face-Net [9] has been discarded because SFace [27], Dlib [8], and VGG-Face [6] have better verification performance. Furthermore, SFace explores a new method by training the model with synthetically generated data.

These four selected algorithms have the greatest potential to improve through training as they all have open-source code that can be modified according to our needs, and the datasets used are public so the results of these four algorithms can be replicated and used conveniently.

These face verification models are Convolutional Neural Networks (CNNs). They obtain an image as an input and have a vector as an output. They represent images of faces as vectors. The images captured with the UAV have to be resized in order to have the same size as the inputs for each algorithm. Then as output shapes, there is a vector with different sizes for each algorithm. These vectors will be the ones used to measure the distance between them to decide if the person is verified or not. The bigger the output shape, the greater the computational usage and processing time but the higher the accuracy.

**Table 1.** Comparison between different face verification state-of-the-art algorithms.

| Algorithm | Input Size | Output Size | Verification on LFW Dataset | Performance on YTF Dataset | Pre-Trained | Training | Other Information |
|---|---|---|---|---|---|---|---|
| **Human-beings** [21] | N/A | N/A | 97.53 | NG | N/A | N/A | Humans performance |
| **Open-Face** [22] | 96 × 96 × 3 | 128 | 92.92 | NG | 14 MB | 1 M | Open-source framework |
| **Deep-Face** [23] | 152 × 152 × 3 | 4096 | 97.35 | 91.40 | 551 MB | 4.4 M | Deep Convolutional Network |
| **Fisher-Vector Faces** [24] | 160 × 125 × 3 | 128 | 93.03 | NG | NG | NG | Fisher Vectors |
| **TL Joint Bayesian** [25] | Variable | NG | 96.33 | NG | NG | NG | Bayesian model |
| **VGG-Face** [6] | 224 × 224 × 3 | 2622 | 98.95 | 97.30 | 554 MB | 2.6 M | Convolutional Neural Network |
| **CosFace** [26] | 112 × 96 × 3 | NG | 99.73 | 97.60 | 214.3 MB | 5 M | Loss function |
| **DeepID2** [27] | 55 × 47 × 3 | 160 | 99.53 | 93.20 | 1.6 MB | 0.2 M | Deep Neural Network |
| **GaussianFace** [28] | 150 × 120 × 3 | NG | 98.52 | NG | NG | NG | Gaussian Model |
| **ArcFace** [5] | 112 × 112 × 3 | 512 | 99.83 | 98.02 | 131 MB | 5.8 M | Loss function |
| **Dlib** [8] | 150 × 150 × 3 | 128 | 99.38 | NG | 22 MB | 3 M | Machine Learning Toolkit |
| **SphereFace** [31] | 112 × 96 × 3 | 512 | 99.42 | 95.00 | 68.6 MB | 0.5 M | Loss function |
| **Face-Net** [9] | 160 × 160 × 3 | 128 | 98.87 | 95.12 | 88 MB | 200 M | Deep Convolutional Network |
| **AdaFace** [29] | 112 × 112 × 3 | NG | 99.82 | NG | 668 MB | 15.1 M | Loss function |
| **SFace** [7] | 112 × 112 × 3 | 128 | 99.13 | NG | 36.9 MB | 1.1 M | Synthetic dataset training |
| **CurricularFace** [30] | 112 × 112 × 3 | NG | 99.80 | NG | 249 MB | 6.3 M | Loss function |

NG = not given; N/A = not applicable.

#### 2.1.1. VGG-Face

This algorithm has been developed by the University of Oxford. Its architecture has 22 layers and 37 deep units. Also, the input is a face image of 224 × 224 pixels, which is the biggest input of the four algorithms analysed. Moreover, it also has the biggest output, a 2622 dimension vector. That is why, as will be seen below, this algorithm achieves the best accuracy but it is also the slowest [6].

#### 2.1.2. SFace

It has been trained using a synthetically generated face dataset to train the face verification model. To generate the data, a class-conditional generative adversarial network has been used. The generated dataset is composed of 634k synthetic images distributed on 10,575 classes. The face verification model uses ResNet-50 as the backbone and the loss function used is CosFace. The input size of the face is 112 × 112 pixels [7].

#### 2.1.3. ArcFace

It is an algorithm developed by the Imperial College London. It is based on MXNet [32] and Python, but in the framework used it is based on a re-implementation in Keras [33]. The model is based on ResNet34 Architecture. The algorithm has as input an image of 112 × 112 pixels and has as output a vector of 512 dimensions [5].

#### 2.1.4. Dlib

It is not an algorithm but a library written in C++ that also has a Python interface. This library contains machine learning algorithms like the one used for face recognition. It is based on a Resnet34 model but with some modifications consisting of 29 convolutional layers. Its input is a 150 × 150 face image and its output is a 128 dimensions vector [8].

### 2.2. Distance Calculation Methods

For the calculation of the distance between the two images (the one of the person to compare and the image to be compared), two different metrics will be used: Euclidean distance and cosine distance. For each one of them, a different dynamic threshold will be calculated.

## 3. Proposed Pipeline Architecture of Face Verification Algorithm

This research develops a new system with a size-adaptive dynamic threshold for enhancing face verification capabilities and then compares four state-of-the-art face verification algorithms. Figure 2 shows the proposed face verification system. The proposed system needs two inputs to perform face verification. The first one is a video received from a UAV, and the second one is the face of the person to verify. Each input will go through the same steps in parallel up to the distance calculation. The system could be divided into five different stages as shown in the figure:

1. **Face Detection:** This is the first stage in our pipeline. The images are processed with RetinaFace [34], which is a face detection algorithm with high performance for both high- and low-resolution faces. Notice that face detection has nothing to do with face identification or face verification. Face detection is the process of determining that there is a face in a set of pixels. Thus, the output of this algorithm is the number of faces in the image and the coordinates of each of them in the image. For research purposes, there will be only one face in each frame. This stage is the slowest of the pipeline as the detection lasts on average 170 ms, causing the maximum processing speed to be less than 6 fps (frames per second). This face-detection algorithm has been selected for our pipeline because of its considerable accuracy in detecting low-resolution faces.
2. **Preprocessing:** It is divided into two different components. First, the detected face is cropped from the frame using the coordinates provided by RetinaFace and the OpenCV library. Then, the cropped face is resized to the input size of the algorithm that is going to be used according to Table 1 using OpenCV. This step is really fast as it only takes 0.3 ms to complete. In most cases, the image ratio of the cropped face will not be the same as the expected size of the algorithms. Therefore, in order not to distort the image, black pixels are included to reach the expected size while maintaining the same image ratio. Further preprocessing is not executed in order to compare the performance of the face verification algorithms at different distances without any image enhancement methods.
3. **Siamese network:** This stage leverages a Siamese network, which contains two CNNs. They have the same architecture and the same weights. Therefore, for example, if the same two images are introduced in the Siamese network the same outputs would be obtained in both. It can be used with any face verification algorithm. In this research, the four selected ones are explored. The inputs are the two cropped faces that are going to be compared with each other. The size of both images is the same as the input required for the CNNs. The outputs are the features of each face converted into a vector. The length of each vector is different according to the definition of the algorithm, as can be seen in Table 1.
4. **Distance calculation:** In this stage of the pipeline, the distance between the features of the faces is calculated. Two different metrics are used for the distance calculation: Euclidean distance and cosine distance, as described in the previous section. Meanwhile, more metrics can be explored, for instance, Manhattan distance. The metric used depends on the face verification algorithm in the Siamese network. Depending on how it obtains the features, one metric can perform better than others in determining if the faces are the same or not. Therefore, each algorithm has a better performance with one of the metrics.

5. **Decision making:** This is the most critical stage where a decision is made based on a threshold. If the distance is above the threshold, the faces are from different people; otherwise, they are the same person. Thus, it is crucial to define an appropriate threshold to be able to have the true positives while minimizing the false negatives. Our main contribution at this stage is a size-adaptive dynamic threshold; depending on the size of the cropped face from the video, different thresholds are used. Just as the size of the image varies, so do the features obtained from the Siamese network. Thus, the distances between the faces will vary too. That is why a dynamic-distance-based threshold will improve the accuracy of the face verification algorithm, leading to more true positives and fewer false negatives. A different threshold will be used depending on the selection of the algorithm, the metric, and the distance between the face and the UAV (calculated in the last stage). Finally, the pipeline ends by showing the decision that is made: whether it is the same person or not. More details of the proposed threshold algorithm are presented in the next section.

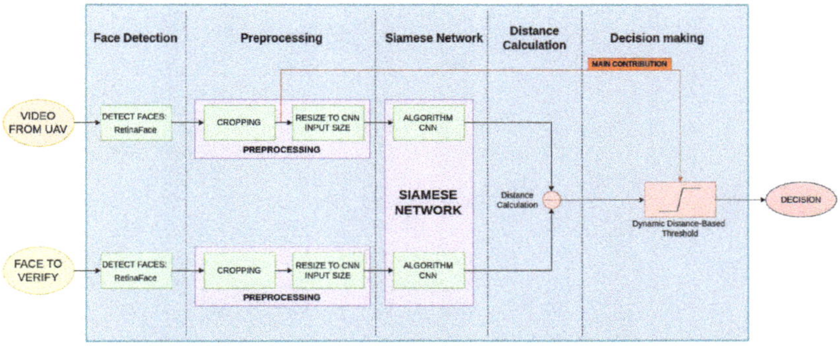


**Figure 2.** Block diagram of the proposed face verification pipeline composed of five stages: face detection, preprocessing, Siamese network, distance calculation, and decision making. There are two inputs: a video from a UAV and the face of the person to identify, and an output that is the decision.

## 4. Proposed Dynamic-Distance-Based Thresholding

The problems of the existing face verification algorithms and the distances at which they are least efficient have been identified and can be improved by defining appropriate thresholds for each distance.

In this section, a scale of sizes is defined and then the proposed distance-based thresholds are calculated for each range using a new algorithm. The calculations are data-driven as the thresholds are calculated based on the data obtained from a dataset.

### 4.1. Dataset Description

A new dataset of UAV-recorded human subjects has been created. This is due to the need for a dataset with a list of specific characteristics for our research and use case, and the fact of the lack of public availability of such an existing dataset.

This new dataset is composed of videos recorded from a UAV at different distances. The UAV used for the recording is a DJI Mini 2. The videos obtained have a resolution of 4K (3840 $\times$ 2160 pixels) and a frame rate of 30 fps.

For the purpose of this research, 20 volunteers were recorded individually in an open field with no one else in the video—one face per video. Each video lasted 30 s, during which a human subject was asked to make head movements to obtain different angles of the face and to stare at the camera for a few seconds. Per subject, eight videos were recorded with each subject from different distances (2, 5, 7, 10, 15, 20, 25, and 30 m). Figure 3 shows the distances arranged for the recordings. The distances are measured from the face of the subject to the UAV camera, and in all cases the UAV has 30 degrees of elevation above the face of the person. The dataset contains 60 images of each person per distance; as there are eight distances, the total number of images per person is 480 plus the identity image. Therefore, the total number of images in the dataset is 9620. The identity face is a close image taken from a smartphone with the volunteer facing the camera. The height of the identity face in pixels is approximately 800 px.

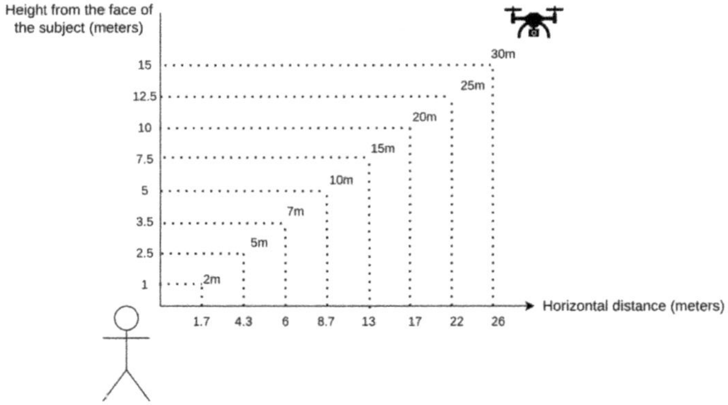


**Figure 3.** Schematic of the dataset recording distances.

The dataset has been divided arbitrarily into two subsets of the same size. One is going to be used for the calculation of the proposed thresholds in Section 4 and the other for the comparison between the accuracy using the proposed and the original thresholds in Section 6. A summary of the dataset can be seen in Table 2.

It is worth noting that GDPR (General Data Protection Regulation) compliance is a critical consideration in our data-gathering process, as we are committed to respecting individuals' privacy and adhering to the regulations. All the volunteers were asked to give informed consent to allow us to record their faces and to use them for the purpose of this research only.

The minimum distance of our videos is 2 m. This is a safe distance for the user with a small drone like the one used to record. At further distances, the volunteers felt more comfortable as they did not have a drone close to them. Therefore, to enhance the user experience, it is important to increase the accuracy of the face verification algorithms at long distances.

**Table 2.** Specifications of the novel created dataset.

| **Number of people** | 20 volunteers |
|---|---|
| **Distances recorded** | 2, 5, 7, 10, 15, 20, 25 and 30 m |
| **Videos resolution** | 4K (3840 × 2160 px) |
| **Videos frame rate** | 30 FPS |
| **Recording device** | DJI Mini 2 Drone |
| **Duration of each video** | 30 s |
| **Total number of images** | 9620 |
| **Age range** | 18–60 years |
| **Other characteristics** | Different ethnicities<br>Different genders<br>Different lighting conditions |
| **Dataset division** | 50% Tresholds Calculation<br>50% Testing |

Moreover, Table 3 shows the relation between the distance of the face from the UAV and the average size in pixels of the face recorded. The large difference in size depending on the distance can be observed. These face sizes have been acquired using a 4k camera. If a lower resolution camera is used, the size of the face will be smaller too at each distance as the size of the whole image is also smaller.

**Table 3.** Distance from the camera to the face versus the size of the detected face in pixels.

| Distance | Size Face |
|---|---|
| 2 m | 125 × 170 px |
| 5 m | 80 × 98 px |
| 7 m | 50 × 59 px |
| 10 m | 38 × 44 px |
| 15 m | 25 × 31 px |
| 20 m | 20 × 24 px |
| 25 m | 17 × 19 px |
| 30 m | 14 × 19 px |

Furthermore, Figure 4 shows the cropped faces from the eight different distances of the dataset. In Figure 4a, all features of the face can be seen without any problem. The person can easily be verified. On the other hand, in Figure 4h it is challenging to verify the person of the image due to the low resolution and its small size.

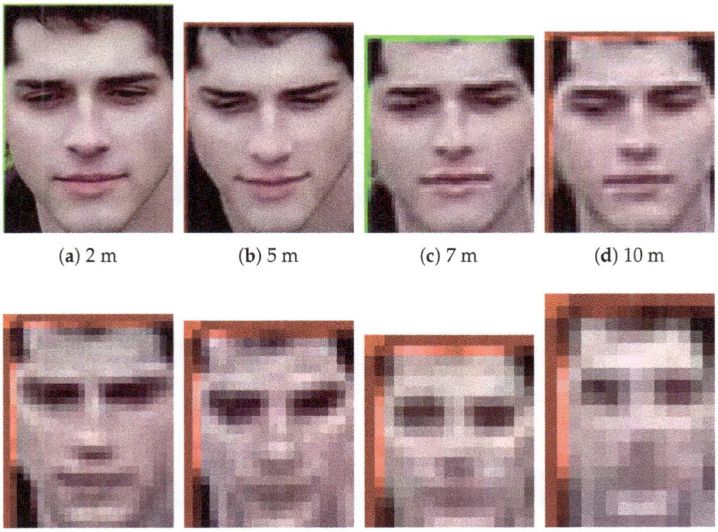

(**a**) 2 m (**b**) 5 m (**c**) 7 m (**d**) 10 m

(**e**) 15 m (**f**) 20 m (**g**) 25 m (**h**) 30 m

**Figure 4.** Cropped faces at the eight distances recorded.

### 4.2. Empirical Definition of Face Scale Based on Distance

To be able to calculate the new dynamic thresholds, a new scale for them has to be established. Two different approaches can be adopted to define the new scale. The first one uses the distances from the face to the UAV's onboard camera. Depending on the distance of the person, one threshold or another is applied. This approach has several problems. For example, it is hard to know exactly the distance from the face. The UAV should have a rangefinder mounted to be able to find the distance. However, not all UAVs have one mounted, thereby being unable to take this approach. Moreover, if the resolution of the camera varies, the size of the detected face in pixels will vary too. Therefore, we can have different face sizes for one distance, depending on the resolution of the camera. That is why using the distance as a scale is not efficient or useful.

The alternative and more advantageous approach is to directly utilise the size of the face for the scale. One threshold would be applied depending on the pixels of the face. Moreover, by taking this approach and knowing the resolution of the camera, the distance to the face could be calculated approximately if needed.

Once the approach is determined, the scale has to be created. As Table 3 shows, there are major changes in the width rather than the height of the image. Hence, the thresholds will be decided depending on the width of the face. Then, the scale is defined using four different ranges, as Figure 5 shows. It also shows which distances of our dataset each of the ranges cover. The principle of the proposed scale scheme is simple and intuitive, whilst the performance is effective. If the detected face width is less than 20 px (pixels), it will be considered that the face is very far and the corresponding threshold will be used. If the face width is between 20 and 32 px, the person is considered to be far and a second threshold will be selected instead. If it is between 32 and 75 px, it is a medium distance, and if the face is larger than 75 px, the person is very close and then a corresponding threshold will be applied. The regions have been defined based on empirical experimentation and because those pixel values correspond to the different distances shown in Figure 5.

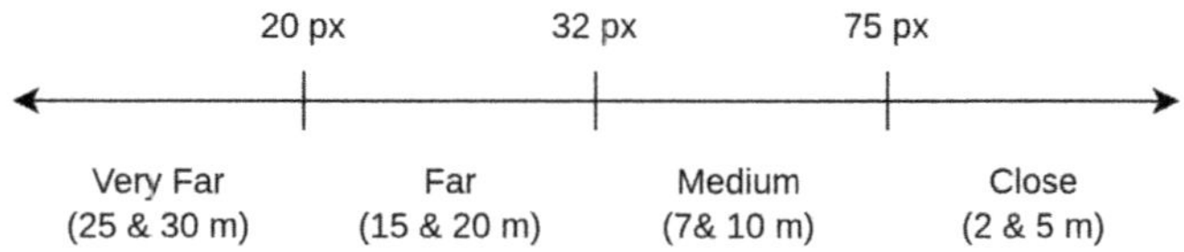


**Figure 5.** Scale of defined distances depending on the width of the cropped face.

By using this scale and changing the thresholds depending on the detected face width, it is expected to achieve an improvement in the accuracy of the face verification algorithms at all distances. Moreover, by adopting this approach we have made the system independent of the camera resolution because it only depends on the size in pixels of the face, not the distance to the person.

### *4.3. Methodology for Empirical Threshold Calculation*

Algorithm 1 shows the proposed algorithm to calculate the dynamic-distance-based thresholds for each distance, metric, and face verification algorithm.

The verification set is composed of a close image of the face of each person appearing in the dataset named as 'face identity'. Then, the positive pairs are the frames from the dataset associated with a specific identity and the same 'face identity' of that person. In contrast, the negative pairs include the same 'face identity', along with all the remaining frames that are not associated. The similarity index is defined as the distance between two faces. Therefore, each negative or positive pair will have an associated similarity index. A hyperparameter is used to iterate along the threshold candidates in the loop. Its value depends on the metric or the face verification algorithm used.

In the algorithm, first, two ratios are calculated, one for positive pairs and another for negative ones. As the number of samples in both pairs are different, this study establishes this ratio to provide the same importance to both pairs. Otherwise, the pairs with more samples (in this case the negative pairs) will have more importance when calculating the thresholds. Afterwards, a for loop is defined from the minimum ($E^+_{min}$) to the maximum similarity index ($E^+_{max}$) of all positive pairs using as step the hyperparameter.

**Algorithm 1** Proposed optimal threshold calculation algorithm.

Let us define the verification set ($V$) as one sample ($V_i$) of each one of the y face identifies to be verified ($V_i$) and a dataset ($D$), composed of x samples ($S_j$) of the same face identities ($F_i$).

$V = \sum_{i=0}^{y} V(i),\ D = \sum_{i=0}^{y} \sum_{j=0}^{x} F_i(S_j)$

Let us create now the **Positive Pairs of identity** $I$, defined as:

$P_I^+ = (V_i, \sum_{j=0}^{x} F_I(S_j))$

And the **Negative Pairs of identity** $I$, defined as:

$P_I^- = (V_i, \sum_{i=0}^{y} \sum_{j=0}^{x} if(i \neq I)[F_I(S_j)], otherwise\,0)$

Let us define the **similarity index of a pair** as: $E(x,y)$. And let us define the **minimum similarity index of all the positive pairs** as :

$E_{min}^+ = min(\sum_{i=0}^{y} E(P_i^+))$

Analogously, let us define the **maximum similarity index of all the positive pairs** as:

$E_{max}^+ = max(\sum_{i=0}^{y} E(P_i^+))$

Let us define the **hyperparameter** as the step used to iterate the threshold candidates in the *for* loop. The value depends on the algorithm and the metric (the cosine or Euclidean distance) used.

And let us apply the proposed dynamic threshold calculation as follows:

**Input:** $P_I^+, P_I^-, E_{min}^+, E_{max}^+, hyperparameter$
**Output:** $Threshold,\ MaxAccuracy$

Initialize $threshold = 0, MaxAccuracy = 0$

Calculate Ratio of all Positive Pairs:

$R(P^+) = \frac{\sum_{i=0}^{y} Size(P_i^+)}{\sum_{i=0}^{y} Size(P_i^+) + \sum_{i=0}^{y} Size(P_i^-)}$

Calculate Ratio of all Negative Pairs:

$R(P^-) = \frac{\sum_{i=0}^{y} Size(P_i^-)}{\sum_{i=0}^{y} Size(P_i^+) + \sum_{i=0}^{y} Size(P_i^-)}$

**for** $candidate = E_{min}^+; candidate < E_{max}^+; candidate = candidate + hyperparameter$ **do**

$TP(TruePositives) = count(\sum_{i=0}^{y} if(E(P_i^+) < candidate))\,1,\ otherwise\,0$

$TN(TrueNegatives) = count(\sum_{i=0}^{y} if(E(P_i^-) < candidate))\,1,\ otherwise\,0$

$A = 100 \times \frac{TP \times R(P^+) + TN \times R(P^-)}{\sum_{i=0}^{y} Size(P_i^+) \times R(P^+) + \sum_{i=0}^{y} Size(P_i^-) \times R(P^-)}$

**if** $A > MaxAccuracy$ **then**
$MaxAccuracy = A$
$Threshold = candidate$
**end if**
**end for**

The threshold candidate value is the one to be iterated. In the loop, the numbers of True Positives ($TP$) and True Negatives ($TN$) are calculated first. $TP$ is calculated as the number of positive pairs whose similarity index is less than the threshold candidate. TN are the result of the number of negative pairs whose similarity index is less than the threshold candidate. Then, the accuracy is calculated using $TP$, $TN$, and the ratios previously calculated. The formula for the accuracy is as follows:

$$Accuracy = \frac{TP + TN}{TP + TN + FP + FN}$$

where $FP$ means False Positives (negative pairs similarity indexes are higher than the threshold candidate) and $FN$ means False Negatives (positive pairs similarity indexes are higher than the threshold candidate). The sum of $TP$ and $FN$ corresponds to all the positive pairs, and the sum of $TN$ and $FP$ corresponds to all the negative pairs. Therefore, the accuracy is calculated by dividing the sum of $TP$ and TN by the sum of all positive and negative pairs, applying the ratios calculated.

Finally, if the accuracy is the maximum value calculated so far, the threshold candidate is the chosen threshold for this moment. When the loop finishes, the threshold candidate that achieves the highest accuracy will be the chosen one.

### 4.4. *Application of the Proposed Methodology for Close (5 m) and Far (15 m) Distances*

The thresholds will be calculated in order to maximise the accuracy in each range of the scale, as seen in the proposed algorithm (Algorithm 1). This section further explains how the algorithm works to maximise the accuracy.

Figure 6 shows the graphics for the four face verification algorithms while using a cosine distance at 5 m. Figure 7 is an analogous figure for 15 m. In each of these figures, two graphs are shown per the algorithms analysed. The first graph shows the distribution of the similarity indexes. It is composed of two plots: one for the positive pairs and another one for the negative pairs. The second graph shows the value of the accuracy as a function of the threshold. Additionally, this second graph shows the two plots depicting the original threshold and the proposed one.

Regarding the distributions of the similarity indexes, they have been obtained by executing the face verification pipeline on the created dataset using the four state-of-the-art algorithms analysed: ArcFace, SFace, Dlib, and VGG-Face. By using our own dataset, we have obtained well-distributed distances both for positive and negative pairs. The positive pair distances have been obtained by executing the pipeline using the videos of one person and comparing them with their photo. The negative pairs have been obtained in the opposite way, using the videos of one person and comparing them with the photos of the other people from the dataset, as explained in the proposed algorithm. The further apart the two different plots are, the greater the accuracy that will be achieved by returning more true positives whilst minimising the false ones. If they are too overlapped, it will be difficult to verify a person and many mistakes will occur. Nevertheless, by choosing an adequate threshold the accuracy can be maximised.

Let us focus, for instance, on ArcFace at 5 m using cosine distance Figure 6a,e. In Figure 6a, it can be seen that both plots are well separated, and thus high accuracy should be achieved. The positive pairs have lower values than the negative pairs. This is because the positive pairs are more similar than the negative ones, as they should be. In Figure 6e, the accuracy across the thresholds can be seen. Below 0.2, the accuracy is 50% because all the pairs would be identified as false due to none of the similarity indexes being below the threshold. Then, we have our proposed threshold in green and the original one in orange. As shown, there is an improvement in the accuracy of choosing the proposed one.

It is worth noting that the thresholds chosen do not match the maximum of the accuracy plots. This is because what has been maximised is a range of our scale and not every distance of the UAV from the face. Therefore, what is optimised is a range of distances rather than every fixed distance.

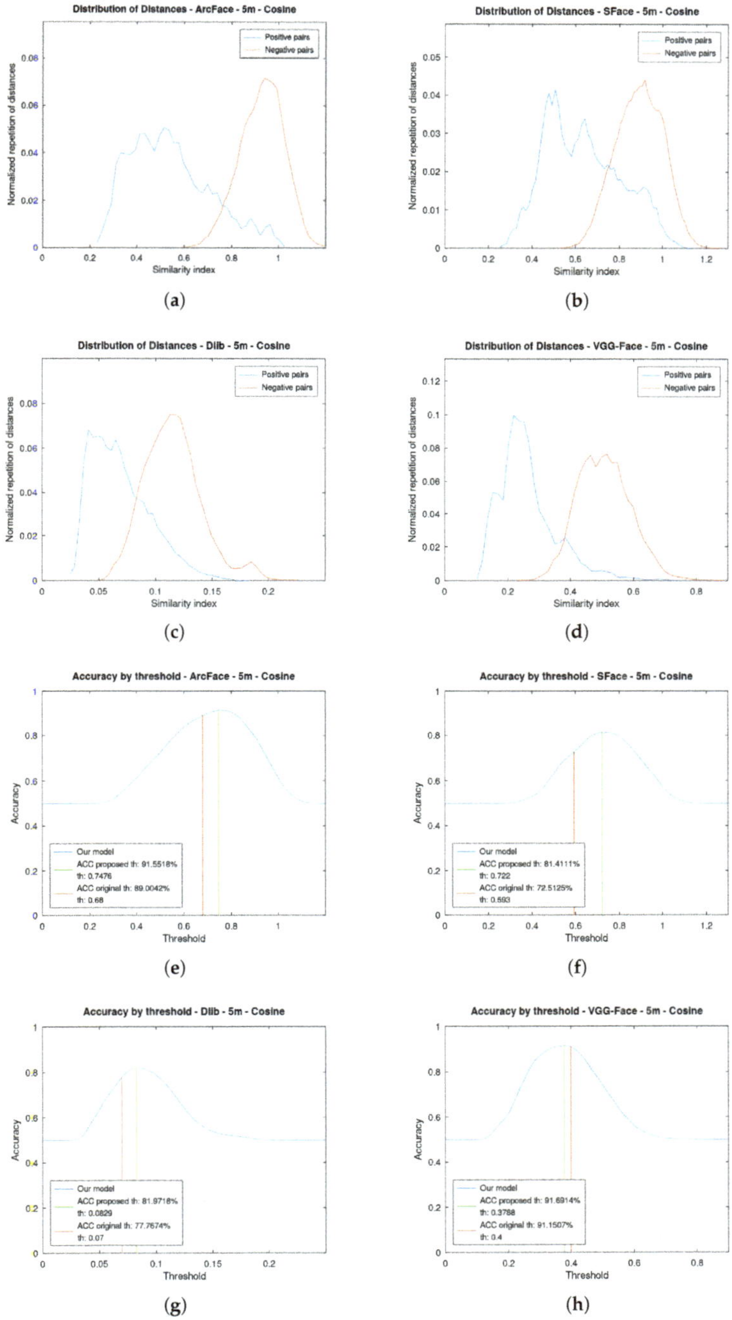


**Figure 6.** Distribution of the similarity indexes and accuracy by threshold at **5 m** of the four algorithms using **cosine distance** as metric. (**a**) Distribution of Similarity indexes of ArcFace. (**b**) Distribution of Similarity indexes of SFace. (**c**) Distribution of Similarity indexes of Dlib. (**d**) Distribution of Similarity indexes of VGG-Face. (**e**) Accuracy by threshold of ArcFace. (**f**) Accuracy by threshold of SFace. (**g**) Accuracy by threshold of Dlib. (**h**) Accuracy by threshold of VGG-Face.

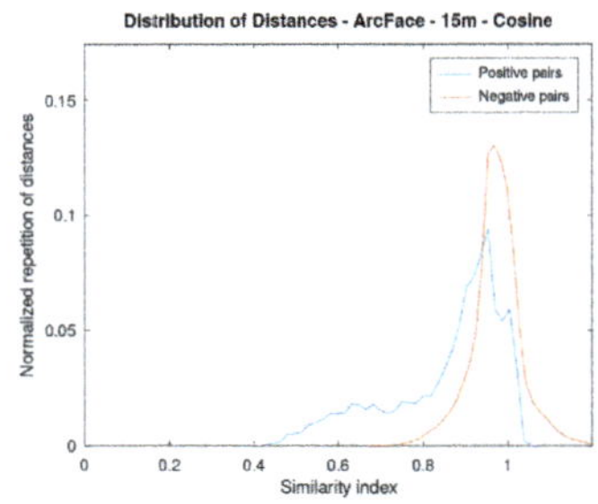


(**a**)

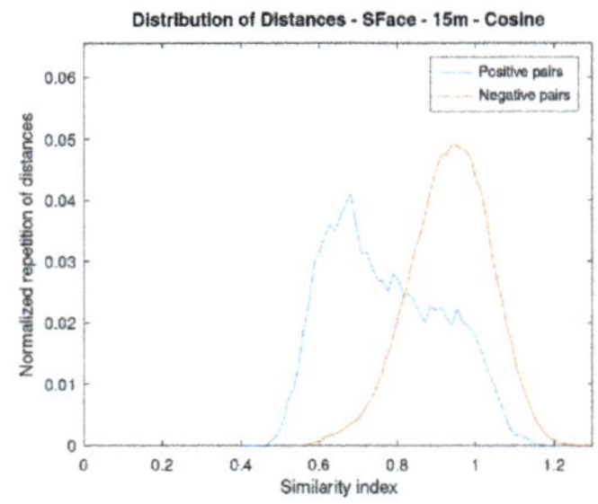


(**b**)

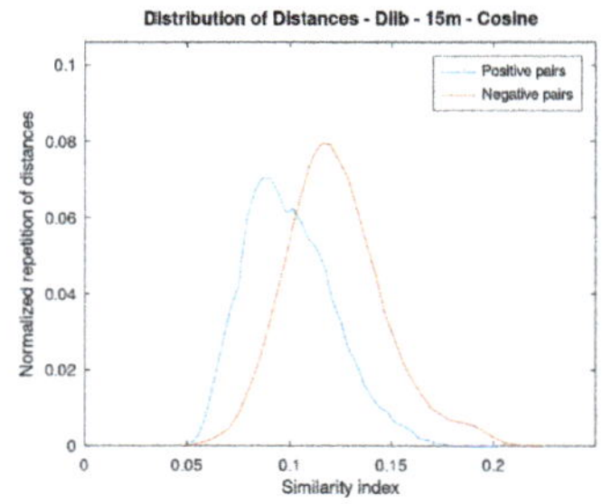


(**c**)

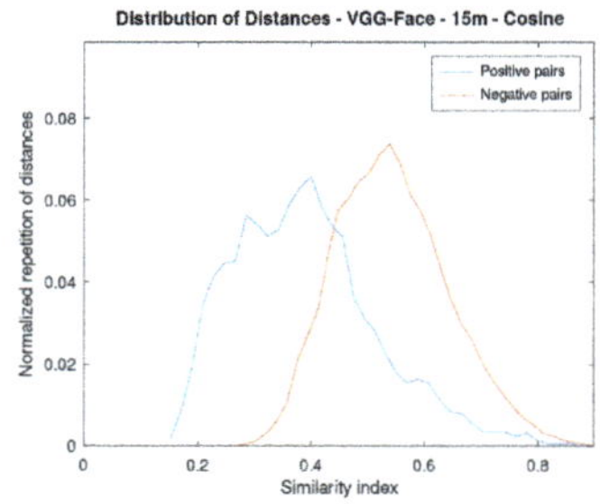


(**d**)

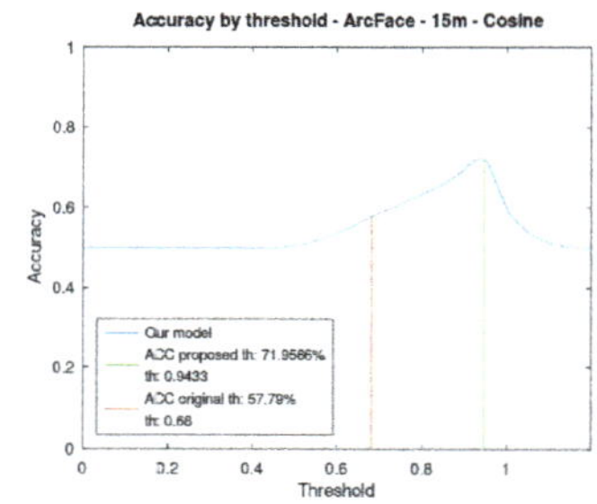


(**e**)

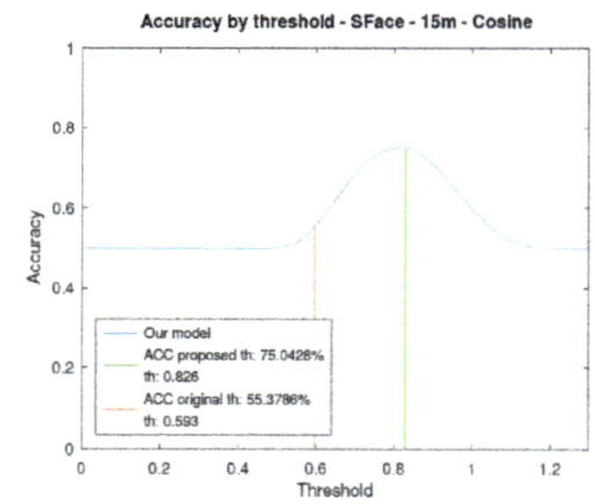


(**f**)

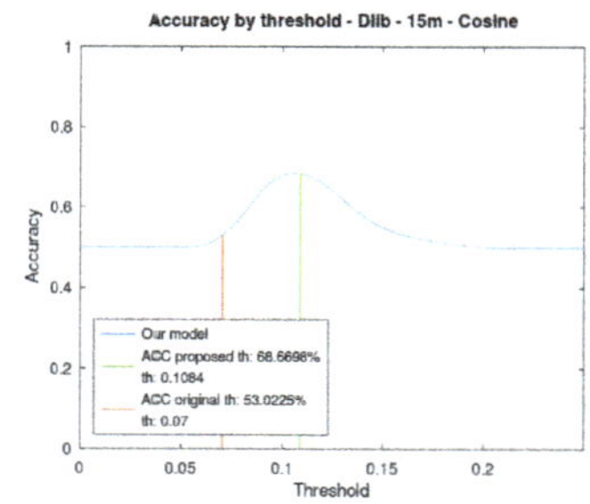


(**g**)

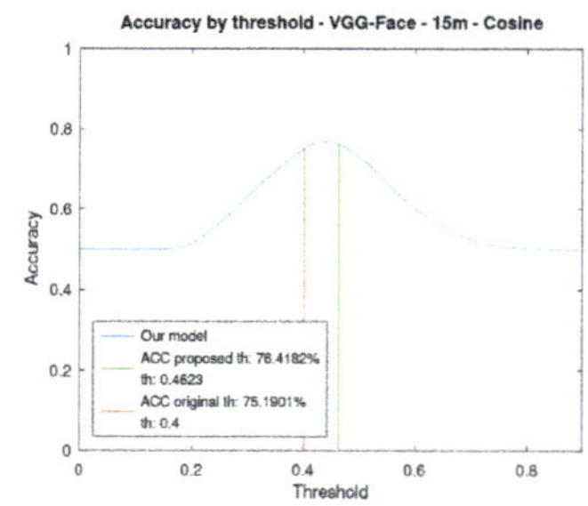


(**h**)

**Figure 7.** Distribution of similarity indexes and accuracy by threshold at **15 m** of the four algorithms using **cosine distance** as metric. (**a**) Distribution of Similarity indexes of ArcFace. (**b**) Distribution of Similarity indexes of SFace. (**c**) Distribution of Similarity indexes of Dlib. (**d**) Distribution of Similarity indexes of VGG-Face. (**e**) Accuracy by threshold of ArcFace. (**f**) Accuracy by threshold of SFace (**g**) Accuracy by threshold of Dlib. (**h**) Accuracy by threshold of VGG-Face.

Notice how at 15 m, for instance, Figure 7a, the plots of the distribution of the positive and negative pairs are more overlapped. This is due to the fact that the faces are increasingly different as their resolution is lowered. Therefore, it is more difficult to choose an optimal threshold. However, it can be seen in the distribution of positive pairs that there is a tail to the left, which facilitates the presence of a substantial number of true positives, even in cases of plot overlap.

All of the algorithms approximately maintain their distribution, but the positive pairs have moved to the right. This supports the necessity of varying the thresholds depending on the distance to the face.

Figures 8 and 9 show analogous results but using Euclidean distance as metric. As can be seen, the plots are further apart and do not overlap much. By using Euclidean distance, the value of the thresholds is greater than in cosine because the similarity indexes also have greater values. The cosine distance range is limited between 0 and 2, while the Euclidean distance range is unlimited.

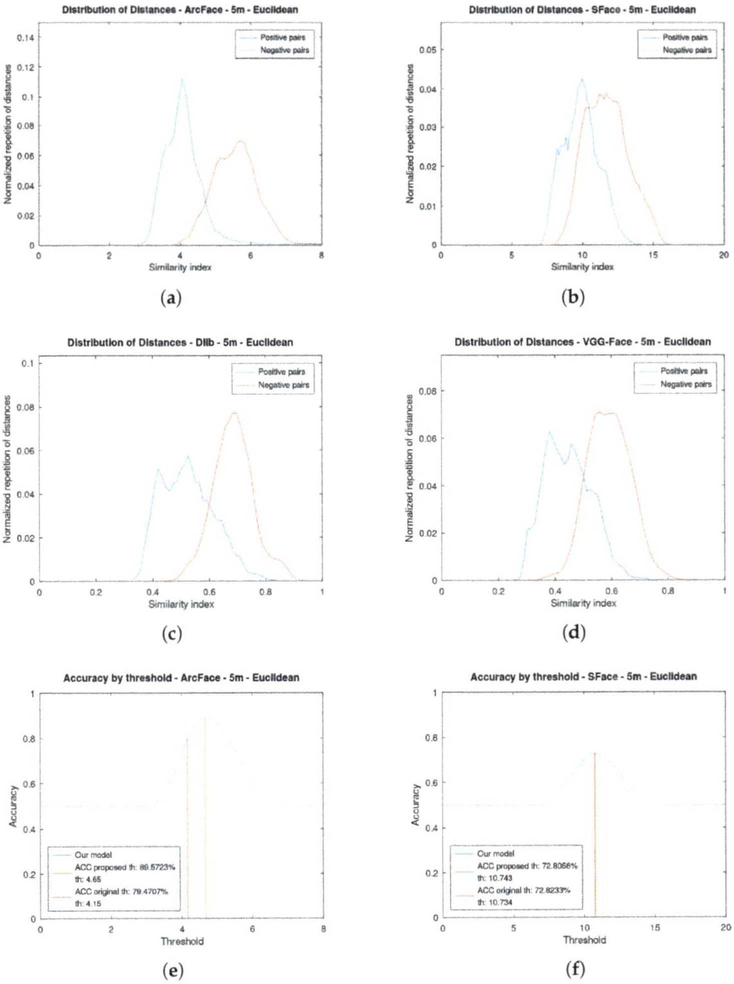


(a) (b) (c) (d) (e) (f)

**Figure 8.** *Cont.*

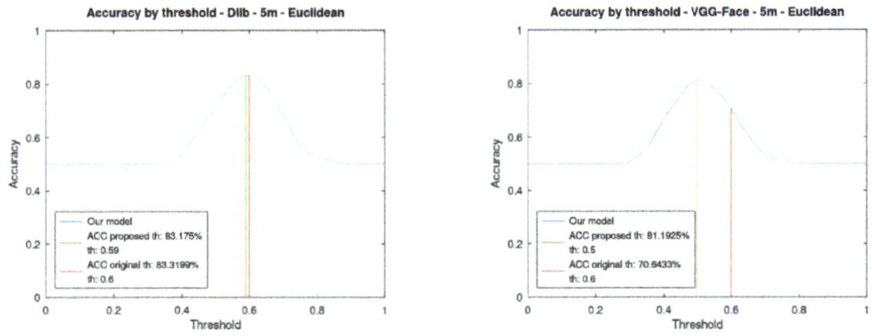


(g) (h)

**Figure 8.** Distribution of similarity indexes and accuracy by threshold at **5 m** of the four algorithms using **Euclidean distance** as metric. (**a**) Distribution of Similarity indexes of ArcFace. (**b**) Distribution of Similarity indexes of SFace. (**c**) Distribution of Similarity indexes of Dlib. (**d**) Distribution of Similarity indexes of VGG-Face. (**e**) Accuracy by threshold of ArcFace. (**f**) Accuracy by threshold of SFace. (**g**) Accuracy by threshold of Dlib. (**h**) Accuracy by threshold of VGG-Face.

In all figures at 15 m (Figure 9a–d), it can be seen that the positive pairs plots are shifted to the right: the two pairs of faces are less alike. This is due to the fact that the resolution is lower as the face is further from the drone, so the two faces are going to be more different.

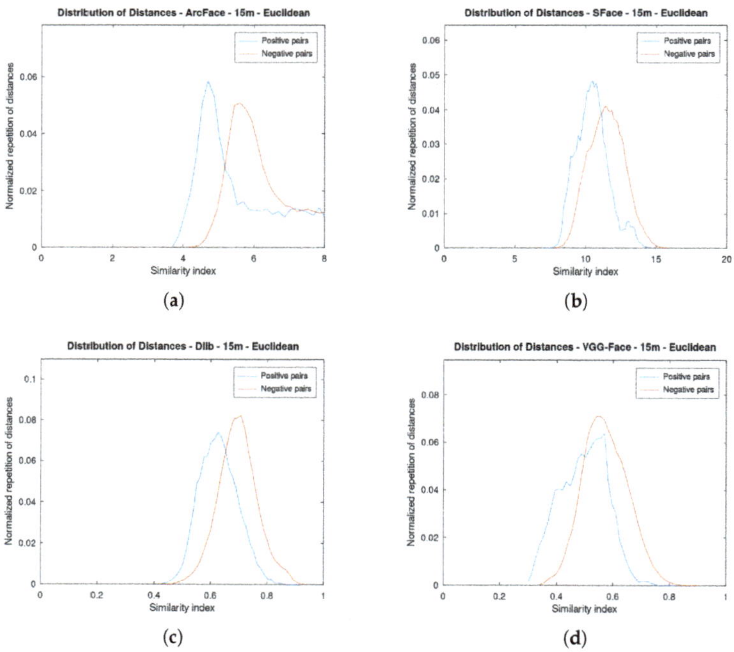


**Figure 9.** *Cont.*

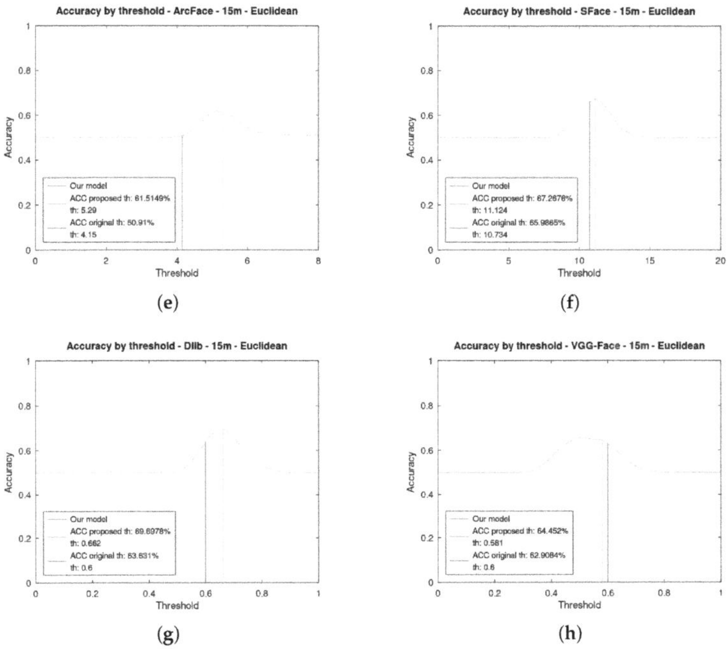


**Figure 9.** Distribution of the similarity indexes and accuracy by threshold at **15 m** of the four algorithms using **Euclidean distance** as metric. (**a**) Distribution of Similarity indexes of ArcFace. (**b**) Distribution of Similarity indexes of SFace. (**c**) Distribution of Similarity indexes of Dlib. (**d**) Distribution of Similarity indexes of VGG-Face. (**e**) Accuracy by threshold of ArcFace. (**f**) Accuracy by threshold of SFace. (**g**) Accuracy by threshold of Dlib. (**h**) Accuracy by threshold of VGG-Face.

The use of our dynamic thresholds will not significantly increase the computational requirements of a face verification system. Instead of having one fixed threshold for all the distances, in the proposed scheme a threshold will be dynamically selected based on the distance to the person. Our technique will only add 1 us to the execution time of the pipeline. As the selection is not complex, the computational requirements or the power consumption will not vary greatly. Furthermore, our system will still work in different operational conditions. For instance, in crowded environments, the threshold will be adjusted for the size of each of the faces detected. Moreover, in adverse weather conditions, our proposed technique will also work as far as the face detection algorithm is able to detect faces correctly under such circumstances. Finally, it is worth highlighting that the empirical work of this paper is based on real implementation of the proposed system.

### *4.5. Recommended Thresholds for the Selected Face Verification Algorithms*

Tables 4 and 5 show the proposed dynamic-distance-based thresholds calculated for each range of the scale for both cosine and Euclidean distance, respectively. They are the result of applying the proposed algorithm (Algorithm 1) to every range, face verification algorithm, and metric selected.

**Table 4.** Calculated dynamic thresholds for cosine distance as metric.

| Algorithm | Very Far | Far | Medium | Close |
|---|---|---|---|---|
| ArcFace | 0.9522 | 0.9433 | 0.7958 | 0.7476 |
| SFace | 0.9 | 0.826 | 0.772 | 0.722 |
| Dlib | 0.1245 | 0.1084 | 0.0882 | 0.0829 |
| VGGFace | 0.5241 | 0.4623 | 0.3924 | 0.3788 |

**Table 5.** Calculated dynamic thresholds for Euclidean distance as metric.

| Algorithm | Very Far | Far | Medium | Close |
|---|---|---|---|---|
| ArcFace | 9.65 | 5.29 | 4.91 | 4.65 |
| SFace | 10.916 | 11.124 | 10.659 | 10.743 |
| Dlib | 0.721 | 0.662 | 0.603 | 0.59 |
| VGGFace | 0.62 | 0.581 | 0.528 | 0.5 |

## 5. Implementation

As mentioned before, four different face verification algorithms are explored: ArcFace, SFace, Dlib, and VGG-Face. For the empirical calculation of the threshold, a common framework where all of them are integrated is created in the same machine in this research. The implemented framework is referred to as DeepFace and includes four different face verification algorithms with their pre-trained weights and two different distance calculation metrics [35,36].

It also implements several face detection algorithms such as RetinaFace [34], MTCNN [37], Dlib [8], or Face-SSD [38]. As mentioned, RetinaFace is the face detection algorithm used in this research. It has been chosen due to its great accuracy at all distances in our dataset. It was able to detect faces even at 30 m with a small reduced number of false positives. As this paper is focused on the face verification algorithms, there is no analysis of the face detection algorithms, and, therefore, the one with the best performance on our dataset was the one chosen.

Deepface is run on Python version 3.8. The library is mainly powered by TensorFlow [39] and Keras [33]. For the preprocessing and postprocessing of the videos, OpenCV is used. The framework has been executed on the same machine using a Focal Ubuntu version 20.04.3. Pycharm is the IDE (Integrated Development Environment) used to write and execute the code. The threshold calculations were performed using GNU Octave.

The proposed pipeline has been implemented using the mentioned framework. The proposed algorithm and associated techniques have been integrated to the four state-of-the-art face verification algorithms on the same testbed for evaluation and comparison purposes

## 6. Experimental Results

### 6.1. Testbed Description

All experiments were carried out on a single NVIDIA GeForce GTX TITAN X (Nvidia Corporation, Santa Clara, CA, USA) with 12 GB of onboard memory. The experiments were repeated five times, and the average of the results of the five runs was then calculated to obtain the final results. The UAV used to obtain the videos is a Mini 2 model created by DJI. The videos were recorded with 4K resolution (3840 $\times$ 2160 px) at 30fps.

### 6.2. Empirical Validation of the Improvement Achieved by the Dynamic Thresholding Technique Proposed for UAV-Based Face Verification

The accuracy is compared between the original thresholds and the proposed ones. Table 6 shows the accuracy using the proposed thresholds and with the cosine distance as metric, and Table 7 shows the accuracy using Euclidean distance.

Figure 10 illustrates the accuracy of the four face verification algorithms depending on the distance using the cosine distance as the metric. It shows the accuracy both using the proposed thresholds and the original ones. It is important to mention that our proposed technique improves all the analyzed face verification algorithms, and thus it is not tailored for a concrete algorithm. For instance, in VGG-Face (Figure 10d) the best improvement is achieved at long distances, whilst at close distances, the accuracy is not greatly increased as it already had good results. Both ArcFace and Dlib have a greater improvement at medium and far distances. SFace (Figure 10b) has improved the accuracy in all instances, with more than a 25% of improvement at some distances.

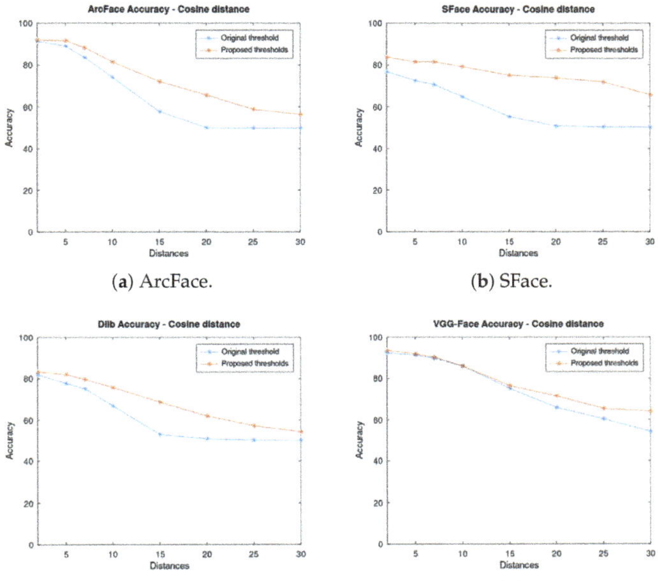


(a) ArcFace. (b) SFace. (c) Dlib. (d) VGG-Face.

**Figure 10.** Accuracy of the algorithms at fixed distances using the original and the proposed thresholds and cosine distance as metric.

**Table 6.** Accuracy of the algorithms at different distances using our dynamic thresholds—Cosine Distance.

| Algorithm | 2 m | 5 m | 7 m | 10 m | 15 m | 20 m | 25 m | 30 m |
|---|---|---|---|---|---|---|---|---|
| **ArcFace** | 91.91 | 91.55 | 88.19 | 81.41 | 71.96 | 65.59 | 58.82 | 56.39 |
| **SFace** | 83.74 | 81.41 | 81.48 | 79.13 | 75.04 | 73.76 | 71.81 | 65.70 |
| **Dlib** | 83.26 | 81.97 | 79.58 | 75.68 | 68.67 | 61.95 | 57.21 | 54.22 |
| **VGG-Face** | 93.39 | 91.69 | 90.09 | 85.87 | 76.42 | 71.49 | 65.43 | 64.10 |

**Table 7.** Accuracy of the algorithms at different distances using our dynamic thresholds—Euclidean distance.

| Algorithm | 2 m | 5 m | 7 m | 10 m | 15 m | 20 m | 25 m | 30 m |
|---|---|---|---|---|---|---|---|---|
| **ArcFace** | 91.69 | 89.57 | 85.91 | 77.17 | 61.52 | 52.28 | 50.58 | 50.41 |
| **SFace** | 76.70 | 72.81 | 73.51 | 70.41 | 67.27 | 62.23 | 60.64 | 56.80 |
| **Dlib** | 84.19 | 83.18 | 81.07 | 77.55 | 69.70 | 62.47 | 58.07 | 56.42 |
| **VGG-Face** | 82.47 | 81.19 | 77.36 | 75.28 | 64.45 | 62.83 | 58.84 | 57.17 |

Then, Figure 11 shows the accuracy of each algorithm for both thresholds using Euclidean distance as a metric. In contrast to cosine, if using the VGG-Face algorithm, a better improvement is achieved at close distances (Figure 11d), but still, the accuracy is lower than that of using cosine. In the ArcFace algorithm, similar results are obtained, and there are greater improvements at close and medium distances (Figure 11a), whilst at long distances, the accuracy does not improve much. One conclusion that can be drawn from this is that Euclidean distance is not optimal for these algorithms as the accuracy does not improve at long distances, which is where it is needed most for a UAV use case. Dlib (Figure 11c) improves at far distances, while at close distances the same results are obtained.

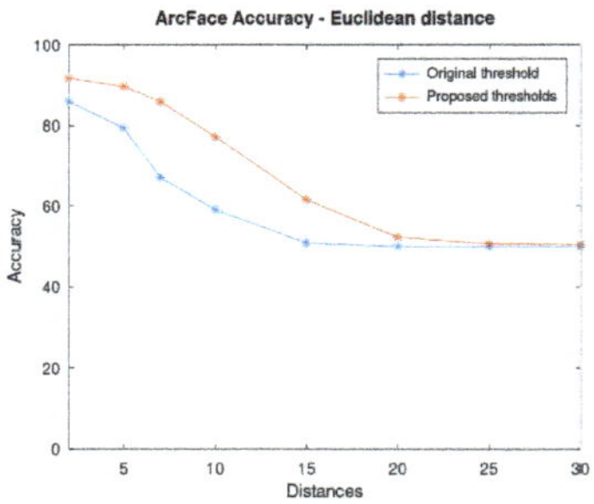


(**a**) ArcFace.

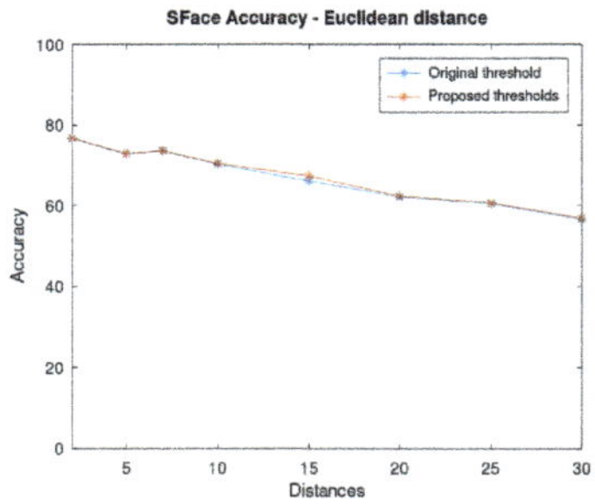


(**b**) SFace.

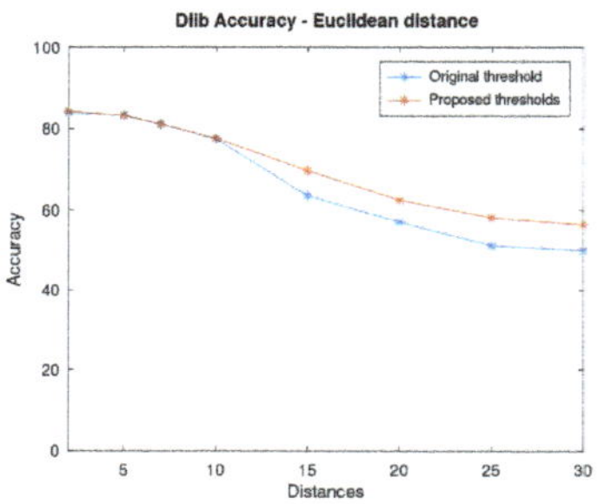


(**c**) Dlib.

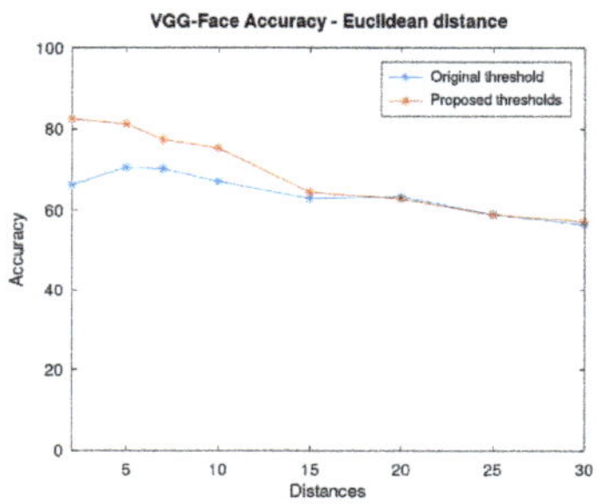


(**d**) VGG-Face.

**Figure 11.** Accuracy of the algorithms at fixed distances using the original and the proposed thresholds and Euclidean distance as metrics.

SFace (Figure 11b) has minimum improvement while using Euclidean distance as a metric. It can be seen that SFace does not have good accuracy while using this metric. The proposed thresholds are all very close to the original one; therefore, the improvement is minimal. This algorithm could not be improved using our method of Euclidean distance.

On the other hand, as seen previously, while using cosine distance (Figure 10b) there is a high improvement in accuracy. Therefore, it can be concluded that SFace performs far better while using cosine distance than while using Euclidean distance.

As can be seen, there is a huge improvement in accuracy when the new thresholds are used. This improvement is more significant at long distances where the accuracy was very low with the original thresholds as they were optimised for close distances. For instance, at 20 m with ArcFace and cosine distance, an improvement of 15% in accuracy has been achieved.

To the best of our knowledge, existing face verification studies simply apply a static fixed threshold to the distance between the embeddings as their thresholding technique. In these results, we have shown that when executing face verification algorithms from drones, the accuracy can be significantly improved by applying a dynamic threshold depending on the distance instead of using a fixed threshold.

Our system has some limitations that are the same as the ones of the face verification algorithms being used. If the algorithm cannot perform correctly at a specific distance, our dynamic thresholds will not be able to increase its accuracy significantly.

### 6.3. Empirical Validation of the Best Similarity Index Based on the Selected Algorithm

Table 8 shows, with each metric, at which distances the greatest improvements are achieved. Also, the final column shows which metric is better to use with each algorithm. Dlib is the only one that achieves a greater performance using Euclidean distance at all distances.

It is worth mentioning that the proposed technique is able to enhance the accuracy of all the proposed face verification algorithms using any of the similarity metrics analysed. Thus, it has been proven that the methodology and approach provided in this contribution is robust enough to be considered in this kind of AI pipeline.

**Table 8.** Distances at which the greatest improvements are achieved with each metric and what is the recommended metric to use with each algorithm.

| Algorithm | Improvement with Cosine Distance | Improvement with Euclidean Distance | Metric Recommended |
|---|---|---|---|
| **ArcFace** | All distances | Close Distances | **Cosine Distance** |
| **SFace** | All distances | None distance | **Cosine Distance** |
| **Dlib** | All distances | Far distances | **Euclidean Distance** |
| **VGG-Face** | Far distances | Close distances | **Cosine Distance—Close Distances**<br>**Euclidean Distance—Far Distances** |

### 6.4. Analysis of Inference Times

The speed of the pipeline is assessed by measuring the inference time. It is defined as the time spent since the frame enters the processing pipeline until the decision is received, and it is measured empirically to compare the speed of the algorithms. Figure 12 shows the cumulative average inference time per frame. A great difference between the face verification algorithms can be appreciated.

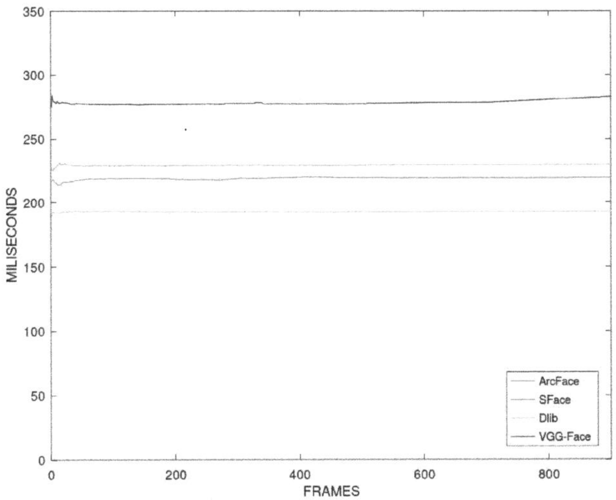


**Figure 12.** Cumulative average of the inference time in milliseconds per frame for each algorithm.

VGG-Face is by far the slowest algorithm. When used, the inference time takes approximately 275 ms, and we can only achieve a maximum processing speed of 4 fps. On the other hand, Dlib is the fastest algorithm. The pipeline processes each frame in approximately 180 ms. Furthermore, it is noted that RetinaFace takes approximately 165 ms, which means that Dlib can obtain the features in less than 15 ms.

The second fastest algorithm is SFace, with approximately 220 ms per frame. Finally, ArcFace is the second slowest algorithm, taking approximately 230 ms per frame to obtain the result of the pipeline.

Our distance-based thresholds do not add any significant delay to the system as one threshold or another will be selected based on the size of the face. Our technique will only add 1 us to the execution time of the pipeline. The calculation of the thresholds using the proposed algorithm will be made before the system is deployed. Therefore, the threshold selection will be almost instant, and real-time processing could be achieved by improving the face detection and verification algorithms.

## 7. Concluding Remarks

This paper presents a novel technique to perform face verification by modifying the thresholds depending on the distance of the UAV from the face. Moreover, an empirical study of four state-of-the-art face verification algorithms has been performed including a comparison with the proposed technique.

By adding the dynamic size-adaptive thresholds to the face verification pipeline, we have significantly improved the accuracy of these face verification algorithms. Furthermore, two metrics have been employed to conduct the comparison (cosine and Euclidean distance). Another analysis has been performed to conclude which of the metrics better suits each algorithm in order to achieve higher accuracy for face verification.

Empirical results have shown that a high improvement in the accuracy of the face verification algorithms at different distances has been achieved. Using the ArcFace algorithm at 20 m with cosine distance as a metric, there is an improvement of 15% in the accuracy. The SFace algorithm has been improved by more than 20% at some distances. At 30 m and using cosine distance, the VGG-Face algorithm has improved its accuracy by 10%.

Our proposed thresholding algorithm has improved all four face verification algorithms used. This proves that by using a dynamic threshold, the accuracy of the face verification algorithms can be improved, leading to fewer false positives and maximising the true ones. Furthermore, our technique is not limited to UAV-based public safety use cases and can be applied to any application domain where face verification is required, such as intruder detection in industrial premises.

For future work, an enhanced dataset can be created using more people at more distances and angles from the UAV to the face. The lighting is important too because in the dataset for this research, the data were obtained in various light conditions. For future research, more lighting conditions can be considered to have a more complex and complete dataset, for example, at night or low visibility.

Moreover, the thresholds have been calculated to maximise the accuracy, but other metrics can be maximised. For example, the same calculations can be made but maximising the F1-Score or defining a maximum acceptable value of false positives. The maximised metric would vary depending on the use case.

Some potential future research directions involve integrating other techniques to improve the accuracy of the algorithms depending on the environment. For instance, light or image enhancement methods can be added to increase the brightness or resolution of the images. Furthermore, the backbones of the algorithms can also be modified to increase the accuracy by adding or modifying layers while trying not to reduce their speed.

**Author Contributions:** Conceptualization, J.M.A.-C. and Q.W.; methodology, J.D.-T.; software, J.D.-T.; validation, J.D.-T., J.M.A.-C. and Q.W.; formal analysis, J.D.-T.; investigation, J.D.-T.; resources, J.D.-T.; data curation, J.D.-T.; writing—original draft preparation, J.D.-T.; writing—review and editing, J.D.-T.; visualization, J.D.-T.; supervision, J.M.A.-C. and Q.W.; project administration, J.M.A.-C. and Q.W.; funding acquisition, J.M.A.-C. and Q.W. All authors have read and agreed to the published version of the manuscript.

**Funding:** This work was in part funded by the EU Horizon 2020 ARCADIAN-IoT project ("Autonomous Trust, Security and Privacy Management Framework for IoT") under grant number H2020-SU-DS-2020/101020259.

**Informed Consent Statement:** Informed consent was obtained from all subjects involved in the study.

**Data Availability Statement:** The novel UAV-recorded dataset used in the current study is not publicly available due to the General Data Protection Regulation (GDPR) requirements as it contains personally identifiable information.

**Acknowledgments:** The authors would like to thank all the partners in the project for their support.

**Conflicts of Interest:** The authors declare no conflict of interest.

## References

1. Martinez-Alpiste, I.; Golcarenarenji, G.; Wang, Q.; Alcaraz-Calero, J.M. Search and rescue operation using UAVs: A case study. *Expert Syst. Appl.* **2021**, *178*, 114937. [CrossRef]
2. Zakaria, A.H.; Mustafah, Y.M.; Hatta, M.M.M.; Azlan, M.N.N. Development of load carrying and releasing system of hexacopter. In Proceedings of the 2015 10th Asian Control Conference (ASCC), Kota Kinabalu, Malaysia, 31 May–3 June 2015; pp. 1–6. [CrossRef]
3. Derhab, A.; Cheikhrouhou, O.; Allouch, A.; Koubaa, A.; Qureshi, B.; Ferrag, M.A.; Maglaras, L.; Khan, F.A. Internet of drones security: Taxonomies, open issues, and future directions. *Veh. Commun.* **2023**, *39*, 100552. [CrossRef]
4. EU H2020 Project ARCADIAN-IoT. Autonomous Trust, Security and Privacy Management Framework for IoT. Available online: https://www.arcadian-iot.eu/ (accessed on 10 November 2023 ).
5. Deng, J.; Guo, J.; Xue, N.; Zafeiriou, S. Arcface: Additive angular margin loss for deep face recognition. In Proceedings of the IEEE/CVF Conference on Computer Vision and Pattern Recognition, Long Beach, CA, USA, 15–19 June 2019; pp. 4690–4699.
6. Parkhi, O.M.; Vedaldi, A.; Zisserman, A. Deep Face Recognition. In Proceedings of the British Machine Vision Conference (BMVC), Swansea, UK, 7–10 September 2015; Xie, X., Tam, G.K.L., Eds.; BMVA Press: Durham, UK, 2015; pp. 41.1–41.12. [CrossRef]
7. Boutros, F.; Huber, M.; Siebke, P.; Rieber, T.; Damer, N. Sface: Privacy-friendly and accurate face recognition using synthetic data. In Proceedings of the 2022 IEEE International Joint Conference on Biometrics (IJCB), Abu Dhabi, United Arab Emirates, 10–13 October 2022; IEEE: Piscataway, NJ, USA, 2022; pp. 1–11.
8. King, D.E. Dlib-ml: A Machine Learning Toolkit. *J. Mach. Learn. Res.* **2009**, *10*, 1755–1758.
9. Schroff, F.; Kalenichenko, D.; Philbin, J. Facenet: A unified embedding for face recognition and clustering. In Proceedings of the IEEE Conference on Computer Vision and Pattern Recognition, Boston, MA, USA, 7–12 June 2015; pp. 815–823.
10. Matas, J.; Hamouz, M.; Jonsson, K.; Kittler, J.; Li, Y.; Kotropoulos, C.; Tefas, A.; Pitas, I.; Tan, T.; Yan, H.; et al. Comparison of face verification results on the XM2VTFS database. In Proceedings of the 15th International Conference on Pattern Recognition, ICPR-2000, Barcelona, Spain, 3–7 September 2000; Volume 4, pp. 858–863.
11. Chicco, D. Siamese neural networks: An overview. In *Artificial Neural Networks*; Springer: Berlin/Heidelberg, Germany, 2021; pp. 73–94.
12. Golcarenarenji, G.; Martinez-Alpiste, I.; Wang, Q.; Alcaraz-Calero, J.M. Efficient Real-Time Human Detection Using Unmanned Aerial Vehicles Optical Imagery. *Int. J. Remote. Sens.* **2021**, *42*, 2440–2462. [CrossRef]
13. Rahim, M.A.; Azam, M.S.; Hossain, N.; Islam, M.R. Face recognition using local binary patterns (LBP). *Glob. J. Comput. Sci. Technol.* **2013**, *13*, 1–18.
14. Shanthi, K.G.; Sivalakshmi, P. Smart drone with real time face recognition. *Mater. Today Proc.* **2021**, *80*, 3212–3215. [CrossRef]
15. Saha, A.; Kumar, A.; Sahu, A.K. Face Recognition Drone. In Proceedings of the 2018 3rd International Conference for Convergence in Technology (I2CT), Pune, India, 6–8 April 2018; pp. 1–5. [CrossRef]
16. Hsu, H.J.; Chen, K.T. Face recognition on drones: Issues and limitations. In Proceedings of the First Workshop on Micro Aerial Vehicle Networks, Systems, and Applications for Civilian Use, Florence, Italy, 18 May 2015; pp. 39–44.
17. Jurevičius, R.; Goranin, N.; Janulevičius, J.; Nugaras, J.; Suzdalev, I.; Lapusinskij, A. Method for real time face recognition application in unmanned aerial vehicles. *Aviation* **2019**, *23*, 65–70. [CrossRef]
18. Rostami, M.; Farajollahi, A.; Parvin, H. Deep learning-based face detection and recognition on drones. *J. Ambient. Intell. Humaniz. Comput.* **2022**, 1–15 . [CrossRef]
19. Huang, G.B.; Ramesh, M.; Berg, T.; Learned-Miller, E. *Labeled Faces in the Wild: A Database for Studying Face Recognition in Unconstrained Environments*; Technical Report 07-49; University of Massachusetts: Amherst, MA, USA, 2007.
20. Wolf, L.; Hassner, T.; Maoz, I. Face recognition in unconstrained videos with matched background similarity. In Proceedings of the Conference on Computer Vision and Pattern Recognition, Colorado Springs, CO, USA, 20–25 June 2011; Volume 2011, pp. 529–534. [CrossRef]
21. Kumar, N.; Berg, A.C.; Belhumeur, P.N.; Nayar, S.K. Attribute and simile classifiers for face verification. In Proceedings of the 2009 IEEE 12th International Conference on Computer Vision, Kyoto, Japan, 29 September–2 October 2009; pp. 365–372. [CrossRef]

22. Baltrušaitis, T.; Robinson, P.; Morency, L.P. OpenFace: An open source facial behavior analysis toolkit. In Proceedings of the 2016 IEEE Winter Conference on Applications of Computer Vision (WACV), Lake Placid, NY, USA, 7–10 March 2016; pp. 1–10. [CrossRef]
23. Taigman, Y.; Yang, M.; Ranzato, M.; Wolf, L. DeepFace: Closing the Gap to Human-Level Performance in Face Verification. In Proceedings of the 2014 IEEE Conference on Computer Vision and Pattern Recognition, Columbus, OH, USA, 23–28 June 2014; pp. 1701–1708. [CrossRef]
24. Simonyan, K.; Parkhi, O.M.; Vedaldi, A.; Zisserman, A. Fisher vector faces in the wild. *BMVC* **2013**, *2*, 4.
25. Cao, X.; Wipf, D.; Wen, F.; Duan, G.; Sun, J. A practical transfer learning algorithm for face verification. In Proceedings of the IEEE International Conference on Computer Vision, Sydney, Australia, 1–8 December 2013; pp. 3208–3215.
26. Wang, H.; Wang, Y.; Zhou, Z.; Ji, X.; Gong, D.; Zhou, J.; Li, Z.; Liu, W. CosFace: Large Margin Cosine Loss for Deep Face Recognition. In Proceedings of the IEEE Conference on Computer Vision and Pattern Recognition, Salt Lake City, UT, USA, 18–23 June 2018. [CrossRef]
27. Sun, Y.; Wang, X.; Tang, X. Deep learning face representation from predicting 10,000 classes. In Proceedings of the IEEE Conference on Computer Vision and Pattern Recognition, Columbus, OH, USA, 23–28 June 2014; pp. 1891–1898.
28. Lu, C.; Tang, X. Surpassing human-level face verification performance on LFW with GaussianFace. In Proceedings of the Twenty-Ninth AAAI Conference on Artificial Intelligence, Austin, TX, USA, 25–30 January 2015.
29. Kim, M.; Jain, A.K.; Liu, X. AdaFace: Quality Adaptive Margin for Face Recognition. In Proceedings of the IEEE/CVF Conference on Computer Vision and Pattern Recognition, New Orleans, LA, USA, 18–24 June 2022.
30. Huang, Y.; Wang, Y.; Tai, Y.; Liu, X.; Shen, P.; Li, S.; Li, J.; Huang, F. Curricularface: Adaptive curriculum learning loss for deep face recognition. In Proceedings of the IEEE/CVF Conference on Computer Vision and Pattern Recognition, Seattle, WA, USA, 13–19 June 2020; pp. 5901–5910.
31. Liu, W.; Wen, Y.; Yu, Z.; Li, M.; Raj, B.; Song, L. SphereFace: Deep Hypersphere Embedding for Face Recognition. In Proceedings of the IEEE Conference on Computer Vision and Pattern Recognition, Honolulu, HI, USA, 21–26 July 2017. [CrossRef]
32. Chen, T.; Li, M.; Li, Y.; Lin, M.; Wang, N.; Wang, M.; Xiao, T.; Xu, B.; Zhang, C.; Zhang, Z. MXNet: A Flexible and Efficient Machine Learning Library for Heterogeneous Distributed Systems. *arXiv* **2015**, arXiv:1512.01274. https://doi.org/10.48550/ARXIV.1512.01274.
33. Keras. 2015. Available online: https://keras.io (accessed on 11 November 2023).
34. Deng, J.; Guo, J.; Ververas, E.; Kotsia, I.; Zafeiriou, S. Retinaface: Single-shot multi-level face localisation in the wild. In Proceedings of the IEEE/CVF Conference on Computer Vision and Pattern Recognition, Seattle, WA, USA, 13–19 June 2020; pp. 5203–5212.
35. Serengil, S.I.; Ozpinar, A. LightFace: A Hybrid Deep Face Recognition Framework. In Proceedings of the 2020 Innovations in Intelligent Systems and Applications Conference (ASYU), Istanbul, Turkey, 15–17 October 2020; pp. 23–27. [CrossRef]
36. Serengil, S.I.; Ozpinar, A. HyperExtended LightFace: A Facial Attribute Analysis Framework. In Proceedings of the 2021 International Conference on Engineering and Emerging Technologies (ICEET), Istanbul, Turkey, 27–28 October 2021; pp. 1–4. [CrossRef]
37. Zhang, K.; Zhang, Z.; Li, Z.; Qiao, Y. Joint Face Detection and Alignment Using Multitask Cascaded Convolutional Networks. *IEEE Signal Process. Lett.* **2016**, *23*, 1499–1503. [CrossRef]
38. Jang, Y.; Gunes, H.; Patras, I. Registration-free Face-SSD: Single shot analysis of smiles, facial attributes, and affect in the wild. *Comput. Vis. Image Underst.* **2019**, *182*, 17–29. . [CrossRef]
39. Abadi, M.; Agarwal, A.; Barham, P.; Brevdo, E.; Chen, Z.; Citro, C.; Corrado, G.S.; Davis, A.; Dean, J.; Devin, M.; et al. TensorFlow: Large-Scale Machine Learning on Heterogeneous Systems. 2015. Available online: https://www.tensorflow.org (accessed on 10 November 2023).

*sensors*

MDPI

*Article*

# Autonomous Internet of Things (IoT) Data Reduction Based on Adaptive Threshold

**Handuo Zhang [1], Jun Na [2] and Bin Zhang [2,*]**

[1] School of Computer Science and Engineering, Northeastern University, Shenyang 110167, China; 1910630@stu.neu.edu.cn
[2] Software College, Northeastern University, Shenyang 110167, China; najun@mail.neu.edu.cn
* Correspondence: zhangbin@mail.neu.edu.cn

**Abstract:** With the development of intelligent IoT applications, vast amounts of data are generated by various volume sensors. These sensor data need to be reduced at the sensor and then reconstructed later to save bandwidth and energy. As the reduced data increase, the reconstructed data become less accurate. Usually, the trade-off between reduction rate and reconstruction accuracy is controlled by the reduction threshold, which is calculated by experiments based on historical data. Considering the dynamic nature of IoT, a fixed threshold cannot balance the reduction rate with the reconstruction accuracy adaptively. Aiming to dynamically balance the reduction rate with the reconstruction accuracy, an autonomous IoT data reduction method based on an adaptive threshold is proposed. During data reduction, concept drift detection is performed to capture IoT dynamic changes and trigger threshold adjustment. During data reconstruction, a data trend is added to improve reconstruction accuracy. The effectiveness of the proposed method is demonstrated by comparing the proposed method with the basic Kalman filtering algorithm, LMS algorithm, and PIP algorithm on stationary and nonstationary datasets. Compared with not applying the adaptive threshold, on average, there is an 11.7% improvement in accuracy for the same reduction rate or a 17.3% improvement in reduction rate for the same accuracy.

**Keywords:** data reduction; Internet of Things; Kalman filtering; concept drift detection

**Citation:** Zhang, H.; Na, J.; Zhang, B. Autonomous Internet of Things (IoT) Data Reduction Based on Adaptive Threshold. *Sensors* **2023**, *23*, 9427. https://doi.org/10.3390/s23239427

Academic Editor: Claudia Campolo

Received: 23 October 2023
Revised: 18 November 2023
Accepted: 21 November 2023
Published: 26 November 2023



## 1. Introduction

The rapid advancement in mobile smart hardware has enabled the creation of intelligent IoT applications, which generate a vast amount of sensor data [1]. The amount and geographic distribution of these data make them distinctive. Processing such data requires sending the data to a remote processor, such as a sink node, an edge device, or a cloud center, as sensor devices frequently lack the computing and storage power to do so [2]. These sensor data are often reduced at the sensor and then reconstructed at the data processor to save bandwidth and communication costs.

Research on data collection and reduction in wireless sensor networks (WSNs) aims to reduce IoT nodes' energy consumption by reducing data transmission volume [3]. Data compression, data prediction, and data aggregation are the three main types of data reduction algorithms [4]. Data prediction is a more popular and effective strategy because it may achieve a substantial data reduction ratio in contrast with other strategies [5]. Data prediction executes by building a data prediction model with the same parameters at both the sensor and the remote processor. The data predicted by the sensor and the remote processor are identical at one time. Therefore, the sensor only needs to determine if the predicted data are accurate or not before uploading the data. No data transmission is necessary if the difference between the predicted value and the collected value is smaller than the reduction threshold. If not, the remote processor receives the data sensor gathered, and the prediction model is updated [4]. As the reduced data increase, the reconstructed data become less accurate. Usually, the trade-off between reduction rate and reconstruction accuracy is controlled by the reduction threshold, which is calculated by experiments based on historical data.

Considering the dynamic nature of IoT, a fixed threshold can rarely maintain the optimal balance, resulting in a lower reduction rate and reconstruction accuracy. If the fixed threshold is smaller than expected, less data are reduced and data reconstruction becomes easier. While the reconstruction accuracy is superior, the data reduction rate is compromised, leading to unacceptable energy and bandwidth consumption. On the other hand, if the fixed threshold is excessive, too much data are reduced, and it is difficult to reconstruct accurate data at a remote processor. Experiments in the literature [6] also showed that as the reduction parameter increased, the degree of simplification of a reduced object also increased. Thus, we consider dynamically adjusting the threshold to increase the reduction effect in a dynamic IoT. When IoT data change frequently, the threshold can be lower to upload additional data for reconstruction. As the data stabilize, the threshold can gradually increase to reduce more unessential data.

Aiming to dynamically balance the reduction rate with the reconstruction accuracy, an autonomous IoT data reduction method based on an adaptive threshold is proposed. The proposed method consists of a data reduction phase and a data reconstruction phase. During the reduction phase, concept drift detection is performed to capture IoT data changes and trigger threshold adjustment. The threshold is adjusted to be lower if concept drift occurs, and higher otherwise. During reconstruction, data trends are introduced to improve reconstruction accuracy. When concept drift detection identifies data changes, a data trend is introduced to replace a fixed linear rate from a Kalman filter for higher data reconstruction accuracy. To verify the applicability of the proposed method, experiments are executed in seven properties on three datasets, including stationary and nonstationary types. Then, a comparative analysis with the basic Kalman filtering algorithm [7], LMS filter algorithm [8], and critical+PIP algorithm [9] is conducted. Moreover, our main contribution is as follows.

- First, to the best of our knowledge, this is the first scheme to incorporate an adaptive reduction threshold into a data reduction algorithm based on Kalman filtering, which enables autonomous IoT data reduction without the need for cloud.
- Second, aiming to execute reduction threshold adjustment dynamically, a concept drift detection to capture IoT changes is introduced.
- Third, we add a data trend in the data reconstruction stage to further improve data reconstruction accuracy.

The rest of the paper is organized as follows. In Section 2, we analyze the lack of autonomy from traditional data reduction and several IoT data reduction algorithms, then give essential background knowledge of Kalman filtering and data reduction based on prediction. The autonomous IoT data reduction algorithm is presented in Section 3 in the order of two steps. Section 4 consists of experimental evaluations of data reduction rate and data reconstruction accuracy on stationary and nonstationary datasets. Finally, Section 5 concludes the paper and provides insights on autonomous edge data reduction.

## 2. Related Works

In related works, first, they introduce data compression and data aggregation methods in wireless sensor networks and why they are less autonomous and suitable for a dynamic IoT. Second, data reduction methods in IoT are introduced, especially data prediction that takes into account a dynamic IoT. After that, we summarize the basic process and mathematical basis of using a Kalman filter for data reduction. This process and mathematical notation will continue to be used for the rest of this article.

### 2.1. Data Compression

Data compression techniques [10,11], also known as compressive sampling or compressive sensing, are based on the inherent sparsity properties of natural signals and reduce the original signal according to the Shannon–Nyquist theorem. Data compression can significantly reduce the energy consumption used for data acquisition in IoT nodes. For example, Chang et al. [12] applied the mean difference (MD) for filtering data noise and redundant values in the proposed an AIoT architecture. Gilles et al. [13] used a compressed sensing approach based on a sub-Nyquist scheme, known as a modulated wideband converter, to solve wideband spectrum sensing. Aniol et al. [14] proposed an algorithm based on linear prediction that can perform both the lossless and near-lossless compression of RF signals. The proposed algorithm is coupled with two signal detection methods to determine the presence of relevant signals and apply varying loss levels as needed. In data compression, the amount of reduced data depends on the compression algorithm, and thus, the reduction rate can rarely be adjusted autonomously according to the dynamic IoT. Meanwhile, the real-time compression and decompression also put pressure on the storage and computing capabilities of IoT devices.

### 2.2. Data Aggregation

Data aggregation [15] is mainly used at the sink node to regulate sensor sampling frequency and thus optimize energy consumption. It works in two ways. First, it dynamically adjusts the sensor-sampling-frequency-based variance between sensor data at a given epoch, which reduces the energy consumption of the sensing unit by preventing the sensor from collecting redundant information. Second, it dynamically adjusts the rate at which features are computed from the original signal. Chen et al. [16] proposed to extract the data features based on fast Fourier transform (FFT) and apply K-means to generate a set of patterns to represent the time-series data in the application of reducing real-time bridge vibration data. Wang et al. [17] proposed an energy-efficient load balancing tree-based data aggregation scheme (LB-TBDAS) for grid-based WSNs. In the scheme, the sensing area is partitioned into many cells of a grid, and the treelike path is established by using the minimum spanning tree algorithm. Zhang et al. [18] proposed a lightweight and privacy-friendly data aggregation scheme against abnormal data, in which the valid data can correctly be aggregated, but abnormal data will be filtered out during the aggregation process. Data aggregation emphasizes the task allocation of data reduction and reconstruction at the physical level. Then, additional data reduction algorithms are required at each node. In this paper, the sensor performs data reduction, and the remote processor performs reconstruction.

### 2.3. IoT Data Reduction

To reduce data transferred to an edge node, Bhargava et al. [19] came up with the idea of only storing values that cannot be predicted accurately based on history. According to an analysis of geographical restrictions from Cao et al. [20], only data along the trajectory for local services should be collected. Wang et al. [21] built an RNN by edge-cloud cooperating for performing data prediction on the edge node and selecting necessary data for updating the data prediction model to upload. The edge data were divided into known situations and unknown scenarios by Zhang et al. [22] for learning model updates. Only the recognized unknown situations are sent to the cloud, and other redundant data are discarded.

An in-networking approach is proposed in [7] based on data prediction. The proposed approach consists of data filtering and data fusion layers. The data filtering layer aims to minimize the number of transmissions. At the same time, the data fusion layer fuses the data based on the minimum squared error criterion. Its Kalman filter double-layer architecture is used in this paper as the base model and comparison method. The least mean square (LMS) algorithm is proposed in [8]. The algorithm is based on two decoupled LMS windowed filters combined convexly with different sizes. It estimates future readings at both the sink and sensor nodes. Data transmission occurs if the current reading deviates significantly from a predefined threshold.

Given the dynamic nature of IoT, several existing data prediction approaches focus on dynamic edge resources and sensor hardware for data reduction. Fuzzy redundancy elimination for data deduplication (FREDD) [23] finds that traditional data reduction overlooks the context and dynamics of the network, meanwhile relying on a fixed threshold to execute data reduction. Simple natural language rules represent domain knowledge and expert preferences regarding data duplication boundaries. It is adapted for multiple scenarios, considering both static and mobile devices, with different configurations of hard-separated and soft-separated zones and sensor coverage areas. Data redundancy management for leaf-edges (DRMF) [24] allows for identifying and removing data redundancies in connected environments at the device level. DRMF considers static and mobile edge devices and provides two temporal and spatiotemporal redundancy detection algorithms. Once redundancies are identified, DRMF performs data deduplication, considering the dynamic requirements of data consumers and device resources. Meanwhile, data inaccuracies and unreliability due to sensor dynamics are usually ignored [5]. Thus, data reduction and faulty data detection are proposed while enhancing data reliability.

The following is a summary of how our approach differs from other data reduction techniques. First, existing data reduction techniques rarely consider the dynamic balance between reduction rate and reconstruction accuracy in a dynamic IoT. Second, we adaptively adjusted thresholds for autonomous data reduction using concept drift detection [25]. Finally, current Kalman-filter-based data prediction techniques assume that the IoT data vary linearly [26] due to low computing capabilities of sensors. They do not use an adaptive data trend [27] to forecast future data.

### 2.4. Kalman Filtering Basics

To introduce data reduction based on Kalman filtering, we give a brief review of Kalman filtering [26], which contains two steps, named the prediction step and the correction step. The prediction step can be described as

$$x_k = A_k x_{k-1} + B_k u_k \tag{1}$$

$$P_k = A_k P_{k-1} A_k^T + Q_k, \tag{2}$$

where $x_k$ is the estimate of the state at time step $k$, $A_k$ is the state transition matrix, $B_k$ is the control input matrix, $u_k$ is the control input, $P_k$ is the estimate of the covariance matrix of the state estimate, and $Q_k$ is the process noise covariance matrix.

The correction step can be described as

$$y_k = z_k - H_k x_k \tag{3}$$

$$S_k = H_k P_k H_k^T + R_k \tag{4}$$

$$K_k = P_k H_k^T S_k^{-1} \tag{5}$$

$$x_k = x_k + K_k y_k \tag{6}$$

$$P_k = (I - K_k H_k) P_k, \tag{7}$$

where $y_k$ is the innovation, $z_k$ is the measurement, $H_k$ is the measurement matrix, $S_k$ is the covariance of the innovation, $K_k$ is the Kalman gain, and $R_k$ is the measurement noise covariance matrix.

In basic data prediction based on Kalman filtering methods, $z_k$ denotes the real-time data collected by the sensor, and $x_k$ the data predicted by the Kalman filter. Data prediction executes by building a data prediction model with the same Kalman filter parameters at both the sensor and the remote processor. The data predicted by the sensor and the remote processor are identical at one time. Therefore, the sensor only needs to determine if the predicted data $x_k$ are accurate or not before uploading the data. No data transmission is necessary if the difference between the predicted value $x_k$ and the collected value $z_k$ is smaller than the reduction threshold.

$$e_k = |z_k - x_k| \quad (8)$$

When $e_k$ calculated by Equation (8) is less than $e_{max}$, the error is accepted, and data do not need to be uploaded. The parameter $e_{max}$ determines the accuracy tolerance and reconstruction accuracy. Thus, the value of $e_{max}$ is crucial in balancing the reduction rate with reconstruction accuracy.

## 3. Proposed Adaptive Reduction Threshold Data Reduction Method

Aiming to dynamically balance the reduction rate with reconstruction accuracy, we propose an autonomous IoT data reduction method based on an adaptive threshold. The proposed method consists of five modules: sensor data acquisition, concept drift detection, threshold adaptive adjustment, data reduction, and data reconstruction. As shown in Figure 1, the modules are divided into two main components: the sensor and the remote processor. The sensor is responsible for data acquisition and reduction, while the remote processor is for data reconstruction.

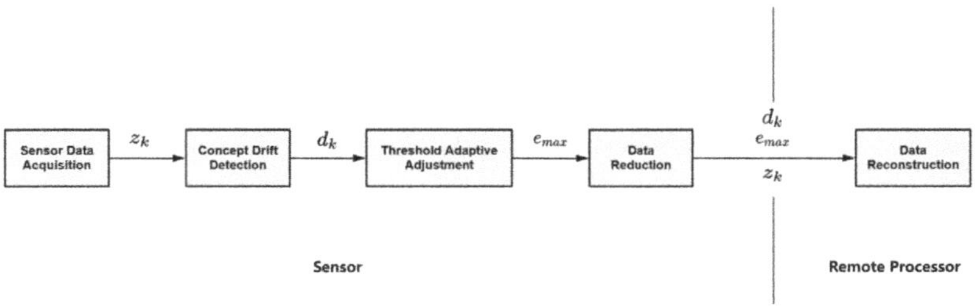


**Figure 1.** Proposed adaptive reduction threshold data reduction method.

Sensor data are transmitted to the concept drift detection after a sensor data acquisition module. The concept drift detection module is responsible for detecting IoT data changes. If concept drift is found, the adaptive threshold adjustment module lowers the threshold $e_{max}$. In other cases, $e_{max}$ rises and transmits to the data reduction module. The basic Kalman filter was used to execute the reduction in the data reduction module. Next, it was chosen whether to transmit the real data $z_k$ to the remote processor based on the comparison with the threshold. If $x_k$ is similar to $z_k$, there is no need to transmit $z_k$ to a remote processor, and a remote processor uses $x_k$ predicted locally in the same parameters with a sensor. Otherwise, $z_k$ should transmit to a remote processor and be assigned to $z_k$ for accurate later prediction. This assignment is an update to the remote processor's Kalman filter, which failed to predict at time k and needs to be updated for later predictions. Without $z_k$, the data reconstruction module forecasts $x_k$ based on the data trend $d_k$ and Kalman filter. Each algorithm is analyzed in subsections next in this paper.

### 3.1. *Adaptive Adjustment for Reduction Threshold Based on Concept Drift Detection*

As mentioned above, adaptive threshold adjustment based on concept drift detection is vital for balance reduction rate and reconstruction accuracy. The detected concept drift indicates a change in the data pattern in a given time window, necessitating a lower data reduction rate to capture more data. Without drift, the data reduction rate gradually increases to filter out irrelevant data.

The Kalman filter assumes that observed data vary linearly [26] and that the linear change rate is constant. Since IoT is dynamic, the linear rate may change sometimes. The linear rate will likely change when the absolute value of a cumulative increment over a time window is abnormal. Thus, the cumulative sum (CUSUM) algorithm [28] is employed to detect concept drift. CUSUM is a statistical control method that detects small shifts in the mean value of a process by monitoring it over time. The CUSUM algorithm accumulates and amplifies persistent biases, thus allowing earlier detection of concept drift, such as linear rate changes. Furthermore, we demonstrate that the CUSUM algorithm can be integrated with other concept drift detection methods by merely swapping out the drift detection module with a different algorithm.

The algorithm works as follows. To address detected concept drift, if the current value of $e_{max}$ exceeds the established $error_{min}$, $e_{max}$ decreases to lower the reduction rate and enhance the reconstruction accuracy. Without concept drift, $e_{max}$ increases for a higher reduction rate. Adjustments of $e_{max}$ are subject to the constraint that they must remain within the specified $error_{max}$ and $error_{min}$. When the values of $error_{max}$ or $error_{min}$ are large, a higher data reduction rate is chosen at the expense of a lesser level of reconstruction accuracy, which is suited for sensors with limited processing power. When $error_{max}$ or $error_{min}$ is small, a higher reconstruction accuracy can be guaranteed instead of pursuing a higher data reduction rate. More complex and intelligent decision-making behaviors can be performed based on more accurate data. The settings of $error_{max}$ and $error_{min}$ need to be analyzed and set after particular experiments on different datasets. In the experimental section of this paper, the data reduction rate and reconstruction accuracy are compared and analyzed in detail for different threshold values. Meanwhile, the step size of each threshold change depends on the experimental setup and preferences for how fast or slow the concept drift needs to be adapted.

### 3.2. *Autonomous Data Reduction Algorithm Based on Adaptive Reduction Threshold*

Next, we describe how to execute autonomous data reduction with an adaptive threshold. In addition, a mechanism for calculating and uploading data trend is shown. When the data are initialized, $z_1$ is uploaded and stored into the cachedval. With using historical data, cache the actual value before calculating $d_k$. To determine whether concept drift has taken place and to establish the new threshold, Algorithm 1 is performed. If the threshold value has changed, it suggests there may have been a change in the linear rate, in which case, $d_k$ should be uploaded instead of $H_k$ to forecast future data. Uploading $d_k$ is not necessary in any other case. The estimated value $x_k$ is then calculated using the Kalman filter, and the gap between the estimated value and the actual value is compared with $e_{max}$. The real value $z_k$ should be submitted when the difference exceeds $e_{max}$.

$$d_k = \begin{cases} z_k - z_{k-1}, & \text{k} = 2 \\ \alpha(z_k - z_{k-1}) + (1-\alpha)d_{k-1}, & \text{k} > 2 \end{cases} \tag{9}$$

**Algorithm 1** Adaptive Adjustment Algorithm for Reduction Threshold

**Input:** current threshold $e_{max}$, threshold minimum $error_{min}$, threshold maxmum $error_{max}$

```
while True do
    Call the CUSUM algorithm to determine if concept drift has occurred
    if Concept drift occurs then
        if e_max > error_min then
            Lower the current threshold e_max
            if e_max < error_min then
                e_max = error_min
    else
        if c < error_max then
            Raise the current threshold e_max
            if e_max > error_max then
                e_max = error_max
```

In Equation (9), $d_k$ represents the data trend at k and is smoothed with a weight $\alpha$, which lies in the range [0, 1]. A value of $\alpha$ close to 1 prioritizes the most recent trend.

Both the true data value and data trend $d_k$ are transmitted to the remote processor for data reconstruction, as shown in Algorithm 2.

**Algorithm 2** Data Reduction Algorithm Based on Adaptive Reduction Threshold

**Input:** current threshold $e_{max}$, sensor reading $z_k$, data trends $d_{k-1}$, data cache cachedval

```
while True do
    if k = 1 then
        Insert z_k into cachedval
        Send z_k to remote processor
    else
        Insert z_k into cachedval
        Calculate d_k
        Call Algorithm1 to calculate adaptive reduction threshold
        if e_max changes then
            send d_k to edge server
        Call the Kalman filter to calculate estimated value x_k
        e_k = z_k − x_k
        if | e_k | > e_max then
            send z_k to remote processor
```

Concept drift indicates the possibility of linear rate change, invalidating the original Kalman filter assumption of a constant linear rate. As a result, $d_k$ should be submitted for prediction instead of $H_k$. The observation does not match the sensor's predicted value when $e_k$ is larger than the threshold, and it is also challenging to reconstruct. However, these may be anomaly data or a measurement error rather than a concept drift or linear rate change. In this scenario, $H_k$ remains valid to forecast future data.

### 3.3. Data Reconstruction Algorithm Based on Data Trend

After autonomous data reduction, the remote processor does not receive the data uploaded from the sensor in every time window. When the data processor receives $z_k$, there is no need for data reconstruction. Nevertheless, when the remote processor fails to receive sensor data, data reconstruction is performed using a Kalman filter assisted by the data trend $d_k$. The data reconstruction procedure is detailed in Algorithm 3. The Kalman filter assumes that the observed data vary linearly. Since IoT is dynamic, nonlinear changes could occur sometimes. Nonlinear Kalman filters, however, are challenging to implement in IoT due to limited computing and storage capacity. As a result, when concept drift detection identifies data changes, we use a data trend to replace the fixed Hk and forecast future value.

**Algorithm 3** Data Reconstruction Algorithm Based on Kalman Filtering and Data Trend

**Input:** sensor reading $z_k$, data trends $d_k$, data cache *cachedval*

```
while True do
    if sensor reading z_k is not None then
        Insert z_k into cachedval
        Calculate d_k
    else
        Call the Kalman filter to calculate estimated value x_k
        if | e_k | > e_max then
            z_k = z_{k-1} + d_k
        else
            z_k = x_k
```

Upon receiving of the data $z_k$ from the sensor, the Kalman filter at the remote processor undergoes a data reconstruction phase. Data trends are stored to facilitate data reconstruction in subsequent cycles. Reversely, the Kalman filter is utilized to predict $z_k$ based on $x_k$. First, the difference between the Kalman filter's predicted value $x_k$ and the reconstructed data is calculated. If this difference exceeds a specified threshold $e_{max}$, the trend of the data $d_k$ is utilized for reconstruction. Otherwise, the result of the Kalman filter is employed as the reconstruction outcome.

## 4. Experiments

### 4.1. Datasets and Experiment Setting

For the experiments, three datasets were selected for analysis. The first dataset, Intel Lab data (Bodik P, Hong W, Guestrin C. Intel Lab data. http://db.csail.mit.edu/labdata/labdata.html, 2004), comprises information on data collected from 54 sensors deployed at Intel Lab from 28 February 2004, to 5 April 2004. Data were collected at a frequency of 30 seconds per sample, and temperature, humidity, light, and voltage properties were included. Experimental comparisons were performed using 6000 temperature, humidity, and light sensor data from this dataset. The second dataset, the Individual Household Electric Power Consumption dataset (Lichman M, UCI Machine Learning Repository. University of California, Irvine, School of Information and Computer Sciences, 2013), encompasses 2,075,259 measurements collected from December 2006, to November 2010 in residences in Sceaux, France. The data were acquired at 60 seconds per sample frequency and included attributes such as voltage, current, and power. Experimental comparisons were performed using 6000 voltage, current, and power sensor data from this dataset. The third dataset is the Dodgers Loop Sensor dataset (Lichman M, UCI Machine Learning Repository. University of California, Irvine, School of Information and Computer Sciences, 2013), which contains data collected from 10 April 2005, to 1 October 2005, on the Glendale ramp of the Los Angeles 101 North Freeway. Experimental comparisons were performed using 6000 data points within this dataset.

Upon conducting ADF root mean square tests on the above properties, we found $p$-values of 0.937 and 0.9024 for Intel Lab data, and 0.7437 and 0.7598 for current and power in the Household Power Consumption data. These values were significantly higher than 0.05, leading to the acceptance of the null hypothesis $H_0$ and indicating that the data exhibited stationary patterns. In contrast, ADF test results for the illumination attribute in Intel Lab data, voltage attribute in Household Power consumption data, and vehicle count attribute in Dodge Loop Sensor Data reveal $p$-values of $6.54 \times 10^{-16}$, $1.31 \times 10^{-12}$ and 0.0, respectively. These values were close to 0, leading to the rejection of $H_0$ and suggesting that these data exhibited nonstationary patterns, as shown in Table 1.

**Table 1.** Datasets.

| Dataset | Attribute | $p$-Value | Decision | Stationary |
|---|---|---|---|---|
| Intel Lab Data | Temperature | 0.937 | Retain $H_0$ | Stationary |
| | Humidity | 0.9024 | Retain $H_0$ | Stationary |
| | Light | $6.54 \times 10^{-16}$ | Reject $H_0$ | Nonstationary |
| Power Consumption | Voltage | $1.31 \times 10^{-12}$ | Reject $H_0$ | Nonstationary |
| | Current | 0.7437 | Retain $H_0$ | Stationary |
| | Power | 0.7598 | Retain $H_0$ | Stationary |
| Dodgers Loop Sensor | Count | 0 | Reject $H_0$ | Nonstationary |

This paper compared two aspects to evaluate the effectiveness of the proposed method in data reduction: data reduction rate (DRR) and data reconstruction accuracy (DRA). The definition of the data reduction rate is shown in Equation (10), where DRR represents the data rate, AD represents the total amount of data, and RD represents the total amount of remaining data after reduction. The data reconstruction accuracy is inspired by the Jaccard similarity between reconstructed and original data of the same length.

Let $T_1 = [z_1, z_2, \ldots, z_n]$ be the actual collected data, and $T_2 = [r_1, r_2, \ldots, r_n]$ be the reconstructed data. The Jaccard similarity between $T_1$ and $T_2$ is calculated using Equation (11), where DRA represents the data reconstruction accuracy, and n represents the number of reconstructed data. Meanwhile, we calculate the transmission of $d_k$ when computing the transmission of our method.

$$DRR = \frac{AD - RD}{AD \times 100} \tag{10}$$

$$DRA = \frac{\sum_{i=1}^{n} \min(z_i, r_i)}{\sum_{i=1}^{n} \max(z_i, r_i) \times 100} \tag{11}$$

### 4.2. Experiments on Adaptive Reduction Threshold $e_{max}$

To ensure that the threshold varies within a specific range, the algorithm is executed on the same dataset, the range of threshold variation is calculated and shown in Tables 2 and 3. The calculated range of threshold variation is also used as a statistical value in subsequent data reduction comparison experiments.

**Table 2.** Threshold setting for stationary datasets.

| | **Intel Lab Data** | | **Household Power Consumption** | |
|---|---|---|---|---|
| Threshold | Temperature | Humidity | Current | Power |
| Min(e_max) | 0.01 | 0.013 | 0.21 | 0.2 |
| Max(e_max) | 0.09 | 0.14 | 1.5 | 4 |

**Table 3.** Threshold setting for nonstationary datasets.

| | **Intel Lab Data** | **Household Power Consumption** | **Dodgers Loop Sensor** |
|---|---|---|---|
| Threshold | Light | Voltage | Count |
| Min(e_max) | 0.1 | 0.21 | 1 |
| Max(e_max) | 0.9 | 1.5 | 7 |

Experiments were conducted using the proposed data reduction method on the temperature and humidity attributes of the Intel Lab data dataset, as well as the current and power characteristics of the Household Power Consumption dataset with stationary-type variations. The threshold range for the temperature attribute was set between 0.01 and 0.1 °C, with an average adaptive threshold of 0.0598 °C, a median threshold of 0.07 °C, a mode threshold of 0.09 °C, and a Pearson correlation coefficient of −0.432 between the threshold variation process and the temperature attribute. The threshold range for the humidity attribute was set to 0.01–0.14%, with an average adaptive threshold of 0.0654%, a median threshold of 0.06%, a mode threshold of 0.01%, and a Pearson correlation coefficient of 0.4882 between the threshold variation process and the humidity attribute. For the current attribute of the Household Power Consumption dataset, the threshold range was set between 0.2 A and 4 A, with an average adaptive threshold of 2.45 A, a median threshold of 3.4 A, a mode threshold of 4 A, and a Pearson correlation coefficient of −0.548 between the threshold variation process and the current attribute. The threshold range for the power attribute was set between 0.25 and 7.2 W, with an average adaptive threshold of 2.26 W, a median threshold of 0.85 W, a mode threshold of 0.25 W, and a Pearson correlation coefficient of 0.5003 between the threshold variation process and the humidity attribute. The threshold variation for all attributes showed a moderate correlation with the Data, demonstrating the effectiveness of the proposed dynamic threshold adjustment mechanism in the Data reduction mechanism. The adaptive adjustment mechanism based on concept drift detection can adjust the reduction rate as the data change pattern evolves.

The following two figures depict the threshold variation process of stationary data. Figure 2 corresponds to the Intel Lab dataset, where Figure 2a shows the temperature data change, Figure 2b shows the temperature threshold change, Figure 2c shows the humidity data change, and Figure 2d shows the humidity threshold change. Figure 3 corresponds to the Household Power Consumption dataset, where Figure 3a shows the current data change, Figure 3b shows the current threshold change, Figure 3c shows the power data change, and Figure 3d shows the power threshold change.

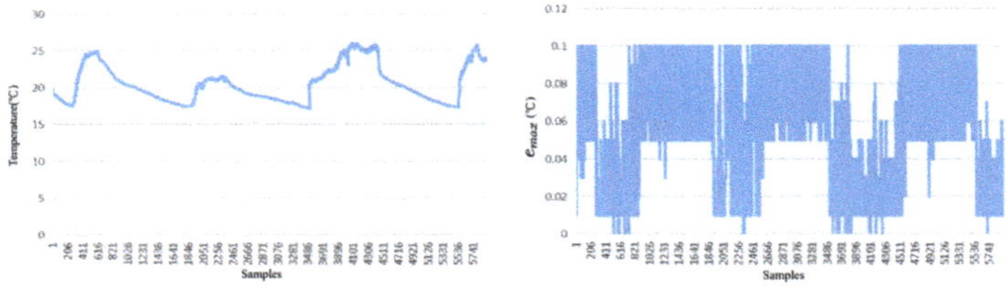


(**a**) temperature data (**b**) temperature threshold

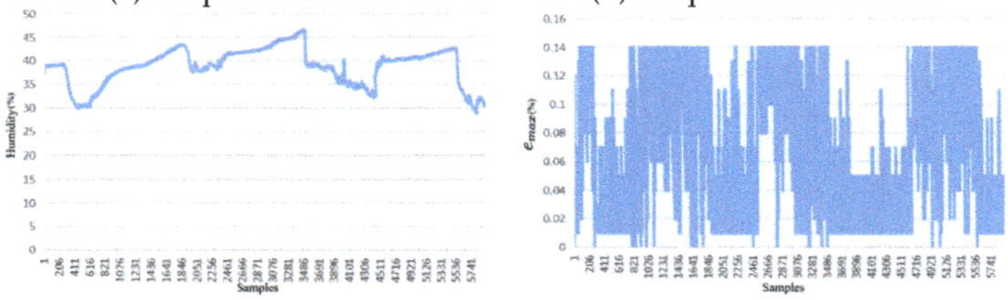


(**c**) humidity data (**d**) humidity threshold

**Figure 2.** Adaptive threshold chart of Intel Lab data.

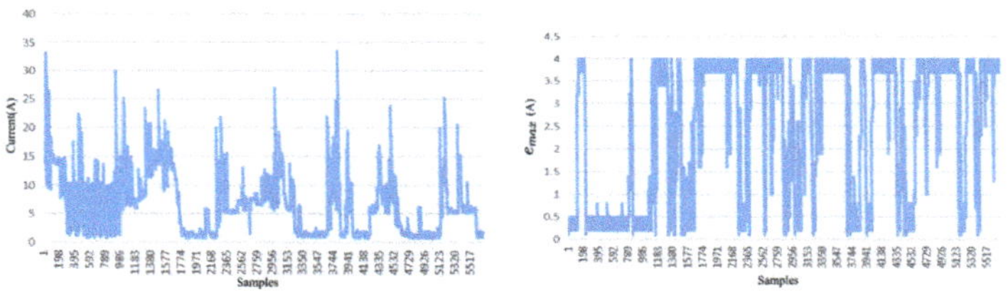


(**a**) current data (**b**) current threshold

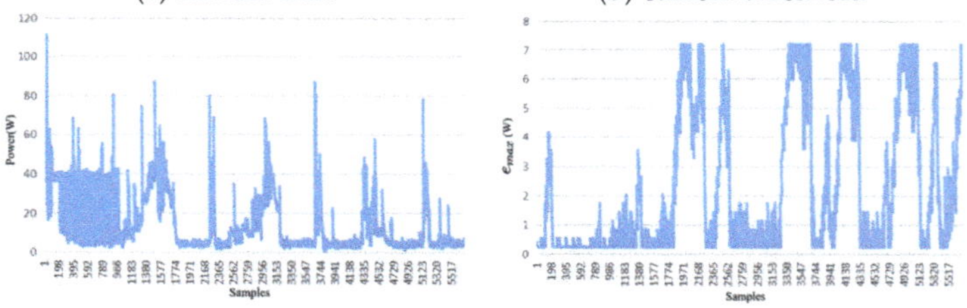


(**c**) power data (**d**) power threshold

**Figure 3.** Adaptive threshold chart of Household Power Consumption.

In this study, the proposed method for data reduction was applied to nonstationary data from the Intel Lab data for light intensity, the Household Power Consumption dataset for voltage, and the Dodgers Loop Sensor dataset for vehicle count. The threshold range for the light intensity attribute was set from 0.1 to 0.9 Lux, with an adaptive threshold average of 0.5139 Lux, a median threshold of 0.5 Lux, and a mode threshold of 0.4 Lux. The threshold change process showed a weak negative correlation with the current attribute, with a Pearson correlation coefficient of −0.398. For the voltage attribute, the threshold range was set from 0.21 to 1.5 V, with an adaptive threshold average of 1.02 V, a median threshold of 1 V, and a mode threshold of 1 V. The threshold change process showed a weak positive correlation with the humidity attribute, with a Pearson correlation coefficient of 0.344. For the Dodgers Loop Sensor, the vehicle count threshold range was set from 1 to 7, with an adaptive threshold average of 1.383, a median threshold of 1, and a mode threshold of 1. The threshold change process showed a weak negative correlation with the humidity attribute, with a Pearson correlation coefficient of −0.372. When dealing with nonstationary data, the threshold change process is weakly correlated with the data attributes. The data fluctuation is relatively large, resulting in significant differences between adjacent data points. Consequently, the Kalman filter model may fail to predict the next data value accurately, and the error threshold will continue to decrease. The error threshold is maintained at a relatively low level to ensure data accuracy while the data reduction rate is decreased.

The following images depict the threshold variation process for nonstationary data. Figure 4a shows the change in light data, Figure 4b represents the corresponding threshold variation, Figure 4c shows the variation in voltage intensity, and Figure 4d shows the voltage threshold variation. Figure 4e illustrates the variation in count data, while Figure 4f presents the corresponding threshold variation.

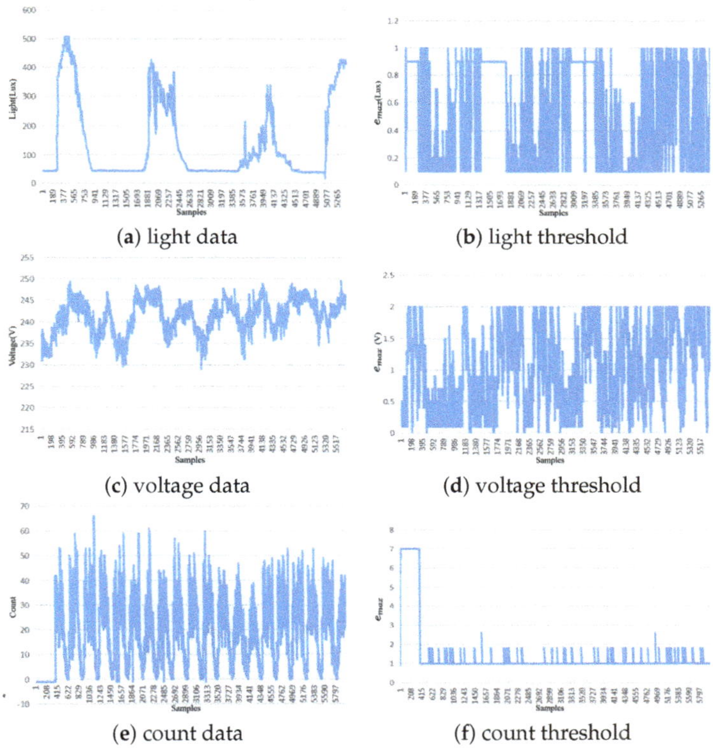

(**a**) light data (**b**) light threshold

(**c**) voltage data (**d**) voltage threshold

(**e**) count data (**f**) count threshold

**Figure 4.** Adaptive threshold chart of nonstationary datasets.

After $error_{max}$ and $error_{min}$ are set, we calculate a tenth of the difference between $error_{max}$ and $error_{min}$ as the step size. Each time $e_{max}$ increases or decreases, it changes by one step.

### *4.3. Experiments on Adaptive Reduction Rate and Reconstruction Accuracy*

In this section, the effectiveness of the proposed method is validated for both stationary and nonstationary datasets by comparing it with fixed threshold reduction methods, such as basic Kalman filter and LMS filter reduction methods, as well as the non-threshold reduction method and the critical+PIP reduction method. The proposed method adjusts the reduction rate dynamically based on the data change pattern, and the reduction threshold changes during the reduction process. Due to various external factors that affect the data, the change patterns of single-dimensional sensor data may differ at different stages, leading to differences in data reduction rate and reconstruction accuracy. Therefore, the minimum, maximum, mean, and mode of the threshold values that are adaptively adjusted by the proposed method in different datasets are taken as the fixed threshold values in traditional methods for comparison with basic Kalman filter and LMS filter data reduction methods. The critical+PIP algorithm is a non-threshold data reduction algorithm, and its efficiency is measured by comparing the data reduction rate and data reconstruction accuracy of the critical+PIP algorithm under both stationary and nonstationary datasets.

#### 4.3.1. Experiments on Stationary Attributes Compared with Fixed Threshold

By conducting experiments on temperature and humidity data from Intel Lab data, it is found that the proposed method has a higher data reduction rate than the basic Kalman and LMS filter data reduction methods with threshold values set by mean, median, and mode. As shown in Tables 4 and 5, when the threshold is set as the maximum value, the proposed method has only a slight reduction rate lower than that of Kalman and LMS filter. Moreover, the data reconstruction accuracy of the proposed method is higher than that of traditional Kalman and LMS filter data reduction methods with threshold values set by mean and median.

**Table 4.** Experimental results of temperature data.

| | Basic Kalman Filter | | LMS Filter | | Proposed Method | |
|---|---|---|---|---|---|---|
| | DRR | DRA | DRR | DRA | DRR | DRA |
| Mean (0.07) | 64.60% | 77.12% | 27.10% | 98.11% | 68.60% | 94.11% |
| Medium (0.05) | 59.80% | 82.77% | 26.30% | 99.29% | | |
| Mode (0.04) | 56.40% | 85.11% | 26.00% | 99.57% | | |
| Min (0.01) | 36.10% | 92.61% | 25.10% | 99.97% | | |
| Max (0.1) | 68.10% | 69.41% | 27.40% | 96.35% | | |

**Table 5.** Experimental results of humidity data.

| | Basic Kalman Filter | | LMS Filter | | Proposed Method | |
|---|---|---|---|---|---|---|
| | DRR | DRA | DRR | DRA | DRR | DRA |
| Mean (0.06) | 60.25% | 92.02% | 56.60% | 59.07% | 68.90% | 95.42% |
| Medium (0.06) | 60.25% | 92.10% | 56.60% | 63.66% | | |
| Mode (0.01) | 32.30% | 99.60% | 31.25% | 99.79% | | |
| Min (0.01) | 32.30% | 99.60% | 31.25% | 99.79% | | |
| Max (0.14) | 81.85% | 67.84% | 79.50% | 52.06% | | |

Through a comparative experiment on the current and power data in the Household Power Consumption dataset, it is found that for stationary datasets, the data reduction rate of the proposed method is higher than that of the basic Kalman filter with mean or median as the threshold. As shown in Tables 6 and 7, the data reconstruction accuracy is better than that of basic Kalman filtering and LMS filtering under different threshold values.

**Table 6.** Experimental resultsof current data.

| | Basic Kalman Filter | | LMS Filter | | Proposed Method | |
|---|---|---|---|---|---|---|
| | DRR | DRA | DRR | DRA | DRR | DRA |
| Mean (2.45) | 69.10% | 78.87% | 66.65% | 75.24% | 69.93% | 80.81% |
| Medium (3.4) | 93.90% | 63.23% | 75.70% | 63.95% | | |
| Mode (4) | 95.50% | 61.57% | 80.15% | 57.42% | | |
| Min (0.2) | 58.10% | 99.60% | 46.05% | 98.53% | | |
| Max (4) | 95.50% | 61.57% | 80.15% | 57.42% | | |

**Table 7.** Experimental results of power data.

| | Basic Kalman Filter | | LMS Filter | | Proposed Method | |
|---|---|---|---|---|---|---|
| | DRR | DRA | DRR | DRA | DRR | DRA |
| Mean (0.85) | 53.85% | 70.30% | 16.60% | 72.30% | 69.05% | 75.57% |
| Medium (2.26) | 67.70% | 62.10% | 36.95% | 61.76% | | |
| Mode (0.25) | 22.30% | 91.60% | 8.35% | 88.06% | | |
| Min (0.25) | 22.30% | 91.60% | 8.35% | 88.06% | | |
| Max (7.2) | 81.95% | 67.84% | 72.55% | 52.11% | | |

For stationary datasets, as the reduction threshold increases, the data reduction rate of the basic Kalman filter and LMS filter will continue to increase, but the data reconstruction accuracy will decrease. The data reduction rate and reconstruction accuracy of the traditional Kalman filter are both higher than those of the LMS filter. Compared with the proposed method, the data reduction algorithm is a dynamic mechanism for controlling the reduction rate, which can adjust the reduction rate dynamically according to the changing patterns of the data. The proposed method achieves higher data reconstruction accuracy when the data reduction rate is equal to that of the traditional Kalman filter.

#### 4.3.2. Experiments on Stationary Attributes Compared with Critical+PIP

The critical+PIP algorithm is a non-threshold-based data reduction algorithm. This article measures the efficiency of the proposed algorithm by comparing its data reduction rate and data reconstruction accuracy with those of another algorithm in the Intel Lab data dataset, specifically for the current and power attributes of Household Power Consumption and the temperature and humidity attributes.

The experimental results in Table 8 show that the data reconstruction accuracy of the critical+PIP algorithm is unstable. To observe the difference in data reconstruction accuracy between the two algorithms, the data reduction rate is controlled between 20% and 80%. For the Intel Lab data, the temperature data reconstruction accuracy of the critical+PIP algorithm decreased from 91.28% to 69.73%, and the humidity data reconstruction accuracy decreased from 68.91% to 52.60%. When processing current data, the data reconstruction accuracy of the critical+PIP algorithm decreased from 99.12% to 73.97%, and when processing power data, the data reconstruction accuracy decreased from 97.10% to 60.49%. The proposed method achieves higher and more stable data reconstruction accuracy with the same data reduction rate.

**Table 8.** Experimental results of stationary attributes compared with critical+PIP.

| | Intel Lab Data | | | | Household Power Consumption | | | |
|---|---|---|---|---|---|---|---|---|
| | Temperature | | Humidity | | Current | | Power | |
| DRR | C+PIP | Our | C+PIP | Our | C+PIP | Our | C+PIP | Our |
| 20% | 91.28% | 97.35% | 68.91% | 86.03% | 99.12% | 99.18% | 97.10% | 97.35% |
| 40% | 86.84% | 96.17% | 66.56% | 82.28% | 95.69% | 98.37% | 84.45% | 95.73% |
| 60% | 75.63% | 96.26% | 59.57% | 80.33% | 82.64% | 94.40% | 78.37% | 94.48% |
| 80% | 69.73% | 95.27% | 52.60% | 74.35% | 73.97% | 91.72% | 60.49% | 93.26% |

### 4.3.3. Experiments on Nonstationary Attributes Compared with Fixed Threshold

In the following, we will compare the fixed threshold methods in the nonstationary attributes. Analysis of the Intel Lab light data indicates that the data values remain mostly unchanged most of the time. As shown in Table 9, the proposed method exhibits a similar data reduction rate and data reconstruction accuracy to those of the other compared methods. We calculate the transmission of $d_k$ when computing the transmission of our method. Therefore, the effect may not be significant in light attribute in Intel Lab data, which may be caused by the fact that there are fewer nonlinear cases and $d_k$ does not need to be transmitted.

**Table 9.** Experimental results of light data.

| | Kalman Filter | | LMS Filter | | Proposed Method | |
|---|---|---|---|---|---|---|
| | DRR | DRA | DRR | DRA | DRR | DRA |
| Mean (0.51) | 83.15% | 93.89% | 82.65% | 92.68% | 82.30% | 93.11% |
| Medium (0.5) | 83.30% | 93.89% | 82.65% | 92.68% | | |
| Mode (0.9) | 84.45% | 93.94% | 82.65% | 92.68% | | |
| Min (0.1) | 82.75% | 93.89% | 82.65% | 92.68% | | |
| Max (1) | 85.65% | 93.52% | 82.65% | 92.68% | | |

In the case of voltage data shown in Table 10 and count data shown in Table 11, the proposed method achieves a data reduction rate that is 10% lower than that of the traditional Kalman filter with a mean value threshold for nonstationary datasets. However, the data reconstruction accuracy of the proposed method is better than that of the traditional Kalman filter and LMS filter under different threshold conditions.

**Table 10.** Experimental results of voltage data.

| | Kalman Filter | | LMS Filter | | Proposed Method | |
|---|---|---|---|---|---|---|
| | DRR | DRA | DRR | DRA | DRR | DRA |
| Mean (1.02) | 82.20% | 65.48% | 70.80% | 60.15% | 48.05% | 99.85% |
| Medium (1) | 82.00% | 65.48% | 70.70% | 60.15% | | |
| Mode (1) | 82.00% | 65.48% | 70.70% | 60.15% | | |
| Min (0.21) | 23.45% | 97.89% | 18.10% | 99.96% | | |
| Max (1.5) | 94.25% | 56.92% | 86.25% | 49.28% | | |

**Table 11.** Experimental results of count data.

| | Kalman Filter | | LMS Filter | | Proposed Method | |
|---|---|---|---|---|---|---|
| | **DRR** | **DRA** | **DRR** | **DRA** | **DRR** | **DRA** |
| Mean (2.13) | 55.75% | 64.25% | 44.35% | 65.69% | 45.55% | 81.46% |
| Medium (1) | 41.80% | 75.57% | 35.45% | 70.43% | | |
| Mode (1) | 41.80% | 75.57% | 35.45% | 70.43% | | |
| Min (1) | 41.80% | 75.57% | 35.45% | 70.43% | | |
| Max (7) | 90.30% | 47.96% | 35.45% | 44.98% | | |

When dealing with nonstationary datasets, the data reduction rate of the traditional Kalman filter and LMS filter will continue to increase with the increase in the reduction threshold. Still, the data reconstruction accuracy will be very low, and sudden changes in data anomalies cannot be observed in a timely manner. In the case of processing nonstationary datasets, the proposed method can automatically reduce the data reduction rate, maintain sensitivity, and sustain high data accuracy.

#### 4.3.4. Experiments on Nonstationary Attributes Compared with Critical+PIP

This subsection will measure the efficiency of the proposed algorithm by comparing the data reconstruction accuracy of critical+PIP data reduction methods with the same data reduction rate. As Shown in the Table 12, the data reconstruction accuracy of the critical+PIP algorithm is not stable. When dealing with light data, the data reduction rate is controlled from 50% to 80%, the data reconstruction accuracy of critical+PIP decreases from 88.48% to 77.70%, and the data reconstruction accuracy of the proposed method decreases from 96.69% to 93.05%. When processing voltage data, the data reduction rate is controlled to 20–80%, and the data reconstruction accuracy of critical+PIP data decreases from 92.36% to 61.73%, while the data reconstruction accuracy of the proposed method in this paper decreases from 97.35% to 80.36%. It can be seen that the proposed method in this paper has higher data reconstruction accuracy with the same data reduction rate, while the data reconstruction accuracy is more stable, and the data reduction effect is better. The data reconstruction accuracy of the critical+PIP algorithm decreases from 75% to 64.54%, and the data reconstruction accuracy of this paper decreases from 83.24% to 67.70%. This paper's data reconstruction accuracy is higher with the same data reduction rate. Meanwhile, the data reconstruction accuracy of this paper's method is more stable than that of the critical+PIP algorithm when facing a nonstationary dataset.

**Table 12.** Experimental results of nonstationary attributes compared with critical+PIP.

| Data Reconstruction Accuracy | | | | | | | | |
|---|---|---|---|---|---|---|---|---|
| Light | | | Power | | | Count | | |
| **DRR** | **C+PIP** | **Our** | **DRR** | **C+PIP** | **Our** | **DRR** | **C+PIP** | **Our** |
| 50% | 88.48% | 96.69% | 20% | 92.36% | 97.35% | 15% | 75.00% | 83.24% |
| 60% | 84.24% | 94.69% | 40% | 86.84% | 91.44% | 20% | 71.21% | 75.84% |
| 70% | 79.23% | 93.69% | 60% | 75.63% | 84.97% | 25% | 67.30% | 72.33% |
| 80% | 77.70% | 93.05% | 80% | 61.73% | 80.36% | 30% | 64.54% | 67.70% |

## 5. Conclusions

The large amount of data generated by the sensor needs to be reduced at the sensor and subsequently reconstructed to save bandwidth and energy. As the reduced data increase, the reconstructed data become less accurate. The trade-off between reduction rate and reconstruction accuracy is commonly controlled by the reduction threshold, which is calculated by experiments based on historical data. The motivation is that the basic assumption of the Kalman filter is to remove the influence of noise in the case of static linear rate, while a dynamic IoT may have special cases, such as linear rate change and concept drift. Using the original threshold significantly harms the reduction rate and reconstruction accuracy, and persists for long periods of time when concept drift occurs. In order to dynamically balance the reduction rate with the reconstruction accuracy, we propose an autonomous IoT data reduction method based on an adaptive threshold. During the data reduction phase, concept drift detection is performed to capture the IoT dynamic changes and trigger threshold adjustment. During the data reconstruction phase, a trend is added to the data to improve the reconstruction accuracy. The effectiveness of the proposed method is demonstrated by comparing the proposed method with the basic Kalman filtering algorithm, LMS algorithm, and PIP algorithm on stationary and nonstationary datasets. Compared with not applying the adaptive threshold, on average, we have an 11.7% improvement in accuracy for the same reduction rate or a 17.3% improvement in reduction rate for the same accuracy. The proposed approach focuses on addressing ongoing changes autonomously without cloud involvement, rather than short-term fluctuations, such as noise. Not limited to the IoT environment, the autonomous data reduction is also important to enable green and efficient data mining through energy and bandwidth saving.

**Author Contributions:** Conceptualization, H.Z., J.N. and B.Z.; methodology, H.Z. and J.N.; validation, H.Z.; investigation, J.N. and H.Z.; data curation, H.Z.; writing—original draft preparation, H.Z.; writing—review and editing, H.Z.; supervision, J.N. and B.Z.; funding acquisition, B.Z. All authors have read and agreed to the published version of the manuscript.

**Funding:** This research was funded by the Key Project of the National Natural Science Foundation of China: U1908212.

**Institutional Review Board Statement:** Not applicable.

**Informed Consent Statement:** Not applicable.

**Data Availability Statement:** The datasets can be obtained from http://db.csail.mit.edu/labdata/labdata.html and http://archive.ics.uci.ed accessed on 1 September 2023.

**Conflicts of Interest:** The authors declare no conflict of interest.

## References

1. Mahdavinejad, M.S.; Rezvan, M.; Barekatain, M.; Adibi, P.; Barnaghi, P.; Sheth, A.P. Machine learning for Internet of Things data analysis: A survey. *Digit. Commun. Netw.* **2018**, *4*, 161–175. [CrossRef]
2. Ahmed, E.; Yaqoob, I.; Hashem, I.A.T.; Khan, I.; Ahmed, A.I.A.; Imran, M.; Vasilakos, A.V. The role of big data analytics in Internet of Things. *Comput. Netw.* **2017**, *129*, 459–471. [CrossRef]
3. Dias, G.M.; Bellalta, B.; Oechsner, S. A survey about prediction-based data reduction in wireless sensor networks. *ACM Comput. Surv. (CSUR)* **2016**, *49*, 1–35. [CrossRef]
4. Sadri, A.A.; Rahmani, A.M.; Saberikamarposhti, M.; Hosseinzadeh, M. Data Reduction in Fog Computing and Internet of Things: A Systematic Literature Survey. *Internet Things* **2022**, *20*, 100629. [CrossRef]
5. Wang, H.; Yemeni, Z.; Ismael, W.M.; Hawbani, A.; Alsamhi, S.H. A reliable and energy efficient dual prediction data reduction approach for WSNs based on Kalman filter. *IET Commun.* **2021**, *15*, 2285–2299. [CrossRef]
6. Mujta, W.; Wlodarczyk-Sielicka, M.; Stateczny, A. Testing the Effect of Bathymetric Data Reduction on the Shape of the Digital Bottom Model. *Sensors* **2023**, *23*, 5445. [CrossRef] [PubMed]
7. Ismael, W.M.; Gao, M.; Al-Shargabi, A.A.; Zahary, A. An in-networking double-layered data reduction for internet of things (IoT). *Sensors* **2019**, *19*, 795. [CrossRef] [PubMed]
8. Tan, L.; Wu, M. Data reduction in wireless sensor networks: A hierarchical LMS prediction approach. *IEEE Sensors J.* **2015**, *16*, 1708–1715. [CrossRef]
9. Wong, S.L.; Ooi, B.Y.; Liew, S.Y. Data Reduction with Real-Time Critical Data Forwarding for Internet-of-Things. In Proceedings of the 2019 International Conference on Green and Human Information Technology (ICGHIT), Kuala Lumpur, Malaysia, 15–17 January 2019; pp. 1–6.
10. Luo, C.; Wu, F.; Sun, J.; Chen, C.W. Compressive data gathering for large-scale wireless sensor networks. In Proceedings of the 15th Annual International Conference on Mobile Computing and Networking, Beijing, China, 20–25 September 2009; pp. 145–156.
11. Srisooksai, T.; Keamarungsi, K.; Lamsrichan, P.; Araki, K. Practical data compression in wireless sensor networks: A survey. *J. Netw. Comput. Appl.* **2012**, *35*, 37–59. [CrossRef]

12. Chang, K.; Chiang, M. Design of Data Reduction Approach for AIoT on Embedded Edge Node. In Proceedings of the IEEE 8th Global Conference on Consumer Electronics, GCCE 2019, Osaka, Japan, 15–18 October 2019.
13. Burel, G.; Radoi, E.; Gautier, R.; Le Jeune, D. Wideband Spectrum Sensing Using Modulated Wideband Converter and Data Reduction Invariant Algorithms. *Sensors* **2023**, *23*, 2263. [CrossRef] [PubMed]
14. Martí, A.; Portell, J.; Riba, J.; Mas, O. Context-Aware Lossless and Lossy Compression of Radio Frequency Signals. *Sensors* **2023**, *23*, 3552. [CrossRef] [PubMed]
15. Maraiya, K.; Kant, K.; Gupta, N. Wireless sensor network: A review on data aggregation. *Int. J. Sci. Eng. Res.* **2011**, *2*, 1–6.
16. Chen, A.; Liu, F.; Wang, S. Data Reduction for real-time bridge vibration data on Edge. In Proceedings of the 2019 IEEE International Conference on Data Science and Advanced Analytics, DSAA 2019, Washington, DC, USA, 5–8 October 2019.
17. Wang, N.C.; Lee, C.Y.; Chen, Y.L.; Chen, C.M.; Chen, Z.Z. An Energy Efficient Load Balancing Tree-Based Data Aggregation Scheme for Grid-Based Wireless Sensor Networks. *Sensors* **2022**, *22*, 9303. [CrossRef] [PubMed]
18. Zhang, J.; Han, H. A lightweight and privacy-friendly data aggregation scheme against abnormal data. *Sensors* **2022**, *22*, 1452. [CrossRef] [PubMed]
19. Bhargava, K.; Ivanov, S.; Donnelly, W.; Kulatunga, C. Using Edge Analytics to Improve Data Collection in Precision Dairy Farming. In Proceedings of the 41st IEEE Conference on Local Computer Networks Workshops, LCN Workshops 2016, Dubai, United Arab Emirates, 7–10 November 2016.
20. Cao, X.; Madria, S. Efficient Geospatial Data Collection in IoT Networks for Mobile Edge Computing. In Proceedings of the 18th IEEE International Symposium on Network Computing and Applications, NCA 2019, Cambridge, MA, USA, 26–28 September 2019.
21. Wang, Z.; Wang, J. An IOT Data Collection Mechanism Based on Cloud-Edge Coordinated Deep Learning. In Proceedings of the Wireless Sensor Networks—13th China Conference, CWSN 2019, Chongqing, China, 12–14 October 2019.
22. Zhang, H.; Na, J.; Zhang, B. Scenario Adaptive Edge Data Reduction. In Proceedings of the 2021 IEEE International Conference on Edge Computing (EDGE), Chicago, IL, USA, 5–10 September 2021; pp. 9–16.
23. Yakhni, S.; Tekli, J.; Mansour, E.; Chbeir, R. Using fuzzy reasoning to improve redundancy elimination for data deduplication in connected environments. *Soft Comput.* **2023**, *27*, 12387–12418 [CrossRef]
24. Mansour, E.; Shahzad, F.; Tekli, J.; Chbeir, R. Data redundancy management for leaf-edges in connected environments. *Computing* **2022**, *104*, 1565–1588. [CrossRef]
25. Lu, J.; Liu, A.; Dong, F.; Gu, F.; Gama, J.; Zhang, G. Learning under concept drift: A review. *IEEE Trans. Knowl. Data Eng.* **2018**, *31*, 2346–2363. [CrossRef]
26. Khodarahmi, M.; Maihami, V. A review on Kalman filter models. *Arch. Comput. Methods Eng.* **2023**, *30*, 727–747. [CrossRef]
27. Gardner, E.S., Jr.; McKenzie, E. Forecasting trends in time series. *Manag. Sci.* **1985**, *31*, 1237–1246. [CrossRef]
28. Healy, J.D. A note on multivariate CUSUM procedures. *Technometrics* **1987**, *29*, 409–412. [CrossRef]

*sensors*

MDPI

*Article*

# A Vehicle-to-Grid System for Controlling Parameters of Microgrid System

**Jigar Sarda [1], Yashrajsinh Raj [1], Arpita Patel [2], Aasheesh Shukla [3], Satish Kachhatiya [1] and Mangal Sain [4,*]**

[1] M. & V. Patel Department of Electrical Engineering, Chandubhai S. Patel Institute of Technology, Charotar University of Science & Technology, Anand 388421, Gujarat, India; jigarsarda.ee@charusat.ac.in (J.S.); 19ee018@charusat.edu.in (Y.R.); 19ee006@charusat.edu.in (S.K.)

[2] V. T. Patel Department of Electronics & Communication Engineering, Chandubhai S. Patel Institute of Technology, Charotar University of Science & Technology, Anand 388421, Gujarat, India; arpitapatel.ec@charusat.ac.in

[3] Department of Electronics and Communication Engineering, GLA University, Mathura 281406, Uttar Pradesh, India; aasheesh.shukla@gla.ac.in

[4] Division of Computer & Information Engineering, Dongseo University, Jurye-ro, Sasang-gu, Busan 47011, Republic of Korea

* Correspondence: mangalsain1@gmail.com

**Abstract:** The power system for large-scale adoption of hybrid electric vehicles can benefit from a distributed reserve provided by the vehicle-to-grid (V2G) concept. This study suggests a V2G technology that can effectively control frequency on a microgrid throughout a 24-h cycle. When usage is at its lowest in the spring or fall, a microgrid is intended to be large enough to simulate a community of 2000 households. A 1:5 ratio of cars to households is realized by modelling 400 electric vehicles (EVs) as a basic model, indicating a typical case in the future. An in-depth analysis of the voltage, current, reactive, and active power is carried out for a microgrid. By coordinating control of diesel generation, renewable energy source (RES) generation, power exchange, and EV generation, the system frequency of a microgrid can be managed by regulating load demand with V2G devices. The proposed microgrid with V2G effectively manages energy and reduces the uncertain and variable nature of RES power generation with enhanced performance. System parameter variations have been investigated for various operating scenarios, and it has been discovered that error is confined to less than 5%.

**Keywords:** electric vehicle (EV); renewable energy sources (RESs); microgrid; vehicle to grid (V2G); photovoltaic (PV); wind

**Citation:** Sarda, J.; Raj, Y.; Patel, A.; Shukla, A.; Kachhatiya, S.; Sain, M. A Vehicle-to-Grid System for Controlling Parameters of Microgrid System. *Sensors* **2023**, *23*, 6852. https://doi.org/10.3390/s23156852

Academic Editors: Behnam Mobaraki and Jose Turmo

Received: 9 July 2023
Revised: 26 July 2023
Accepted: 31 July 2023
Published: 1 August 2023



## 1. Introduction

One of the significant issues in the energy sector in the most developed nations is the decarbonization of transportation fleets [1–4]. There is a strong trend toward integrating RESs with EVs and considering V2G in the smart grid as a solution to reduce dependency on petroleum products and meet energy demand. The NITI Aayog's 2030 electric transportation strategy for India [5] represents a substantial market opportunity. The report analyzes EV sales by market segment, battery needs, required public charging infrastructure, and investments required through 2030 to facilitate India's EV transition. The study simulates three alternative transition scenarios in addition to the 2030 goal. If this ambition is realized, India's EV industry might play a significant role in the country's post-COVID-19 economic recovery. Along the entire value chain, including in the existing and new industries, it can create jobs and economic value. India's 2030 e-mobility vision, which calls for 70% of all commercial vehicles, 30% of private vehicles, 40% of buses, and 80% of two- and three-wheeler sales to be electric, amounts to 26 million EVs. The EVs are anticipated to make up 43% of all new car sales under the high adoption scenario, 10% over the objective up to 112 million units. In a scenario with limited adoption, this may drop to 23% for new car sales, 40% below the desired level, up to 61 million units. Figure 1 shows the growth of EV sales under different scenarios.

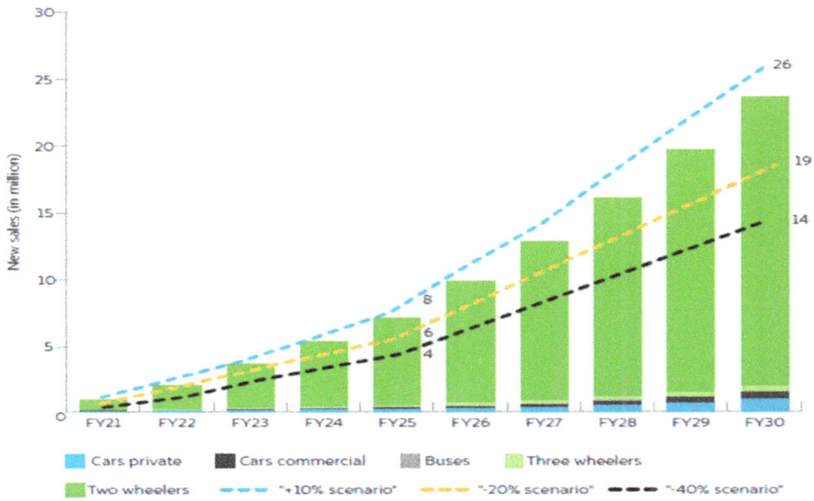


**Figure 1.** EV sales grow under different scenarios.

The adoption of an electric fleet of vehicles offers various benefits for the environment and the economy, but widespread EV adoption could cause significant power system fluctuations, especially during times of high demand [6]. Only 5% of EVs are typically driven daily, while the other 90% are idle and stored in lots [7]. To encourage the active participation of EV owners in revenue generation, different countries have provided different kinds of incentives, shown in Table 1.

**Table 1.** Current zero-emission light-duty vehicle (ZEV) incentives in selected countries [8].

| | | Canada | China | European Union | India | Japan | USA |
|---|---|---|---|---|---|---|---|
| Incentives vehicle | Fiscal incentives | ✓ | ✓ | ✓ | ✓ | ✓ | ✓ |
| Regulations charger | Hardware standards | ✓ | ✓ | ✓ | ✓ | ✓ | ✓ |
| | Building regulations | ✓ | ✓ | ✓ | ✓ | --- | ✓ |
| Incentives charger | Fiscal incentives | ✓ | ✓ | ✓ | ✓ | ✓ | ✓ |

The number of fossil-fuel products used in conventional power plants to produce electricity is reduced in the first place. In the second place, the use of internal combustion (IC) automobiles moves to EVs, reducing the amount of gasoline. Numerous studies have used simulation modules, particularly the Monte-Carlo Simulation [9], while others have created novel optimization models to assess the current power network's suitability for adopting a fleet of electrified vehicles in the future [10,11]. Reducing peak demand and boosting grid reliability are two of the V2G system's most significant goals [12,13]. International Energy Agency has developed a framework for grid integration of electric vehicles to encourage the active participation of EV owners in grid stability, shown in Figure 2.

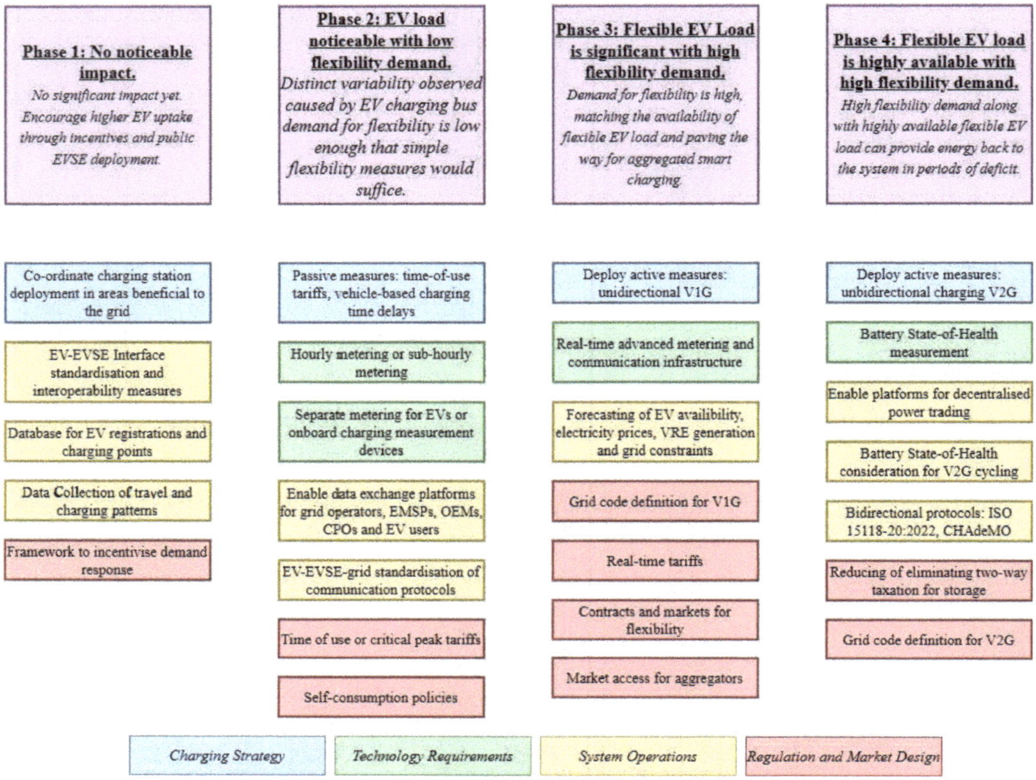


**Figure 2.** Framework for grid integration of EVs [14].

Drude, Pereira Jr., and Ruther [15] used a MATLAB simulation to examine how the V2G in metropolitan locations affected the energy demand profile in various charge and discharge modes. To create a flatter power demand curve, Khemakhem, Rekik, and Krichen [16] thought of using regulated charge for plug-in electric vehicles (PEVs) as a supervisory method to create a flatter power demand curve. By using electric car aggregators with V2G capabilities, Lopez et al. [17] presented an optimization model for power-load shifting in smart grids. To maximize the performance of EVs and smooth out the sporadic nature of RESs and energy cost reduction, Mehrjerdi and Rakhshani [13] introduced nonlinear stochastic programming. Additionally, this approach reduces the number of times the car batteries are charged and discharged, preventing battery deterioration. Trorja et al. [18] examined how RES penetration, emissions, and balancing power plant performance are affected by energy storage systems and V2G. Hoehne and Chester [19] studied the effects of discharging and charging cycles on emissions and found that optimal EV charging lowered the number of emissions in many cases.

Power networks based on conventional power plants have a consistent power supply since they typically do so throughout the day. The strength of this study, compared to previous research, by providing fresh and cutting-edge indices for monitoring these characteristics, is assessing the effects of the V2G system on the emission, cost, and reliability of the grid for different types of energy sources. However, because RESs are intermittent, power networks that rely on them as their main source of produced energy have an erratic supply pattern. Another significant addition to this study is the stochastic power supply, where two innovative indices are introduced for assessing the system's dependability in the context of two different power grid types—the grid with less penetration of RESs and the grid with more penetration of RESs. Reliability and environmental and economic elements of this subject are all considered in this study as a thorough analysis is provided on the effects of EVs, the V2G system and its integration with RESs on the electricity grid, particularly wind and solar generating. Many scenarios are specified, and a Monte-Carlo simulation provides them with the results of the system performance measurements. Based on the power grid's characteristics, the recommended methodology, simulation, and indices may be used in many places and provide excellent tools for policymakers.

There are some potential challenges with the scaling up of EV adoption discussed by [20], which are listed below:

1. Charging Infrastructure: Tie [21] advises constructing a thorough, national charging infrastructure before introducing EVs since it is a critical problem regarding the adoption of EVs.
2. Charging Stations: Several studies have determined that a lack of infrastructure directly impacts customers' intent to buy an EV [22–24]; this affects market sales and ranks at the top of the list of objections against the widespread use of EVs [25].
3. Repair and Maintenance Workshops: Quak, Nesterova and Rooijen [26] claim that existing EV owners are dissatisfied with the lack of EV maintenance facilities and workshops compared to those for ICEVs.
4. Driving Ranges and Charging Times: The restricted capacity and driving range of the batteries as well as their high cost, are two of the critical problems with BEVs. Although some new models have ranges up to 400 km, and subsequent models are projected to have ranges beyond this, the battery size of many existing models restricts their driving range to 250 km [27].

This work is structured as follows: Section 2 discusses the architecture of the V2G interconnected microgrid. The test system is introduced, and results are discussed in Section 3. Finally, Section 4 concludes.

## 2. Architecture of Projected V2G Interconnected Microgrid

A 24-h simulation may be performed quickly using a MATLAB power system network phasor model. A microgrid consists of four sections: the system's primary power source is a DG, a PV coupled with a wind, which provides RES, and a V2G arrangement, which serves as an additional load in the grid. In the spring or autumn, during the days when energy use is low, the size of a microgrid is to serve a community made up of 2000 households. The default model considers 400 EVs, meaning a 1:5 ratios between the EVs and homes, which would be considered a good future configuration.

Figure 3 depicts a schematic representation of the planned microgrid. The electricity provided by RESs, V2G, and consumed power are all balanced out by the diesel generator. Between the grid and the diesel generator, there is a 53 MVA isolation transformer DGT is placed. A 9 MW wind power facility, is linked by a transformer WPPT, which has ratings of 12 MVA and 0.575/25 kV. Between the grid and V2G, there is a 53 MVA transformer is placed. Residential loads with a 20 MW capacity and an Asynchronous Machine (ASM) with a 0.15 MW rating are linked to the distribution side, which runs at 600 V. The battery charge power generating units and grid control are integrated with the V2G, which has a 16 MW rating.

### 2.1. Diesel Generator

The DG helps maintain equilibrium between the amount of energy produced and used. Its synchronous unit's rotor speed could be utilized to determine the grid frequency deviation.

Figures 4 and 5 represent the DG and governor model block diagrams. The governor controls the rotational speed of the diesel engine to maintain a certain electrical frequency under various situations of electrical demand. The governor model is the Woodward diesel governor, which features a 5% frequency droop.

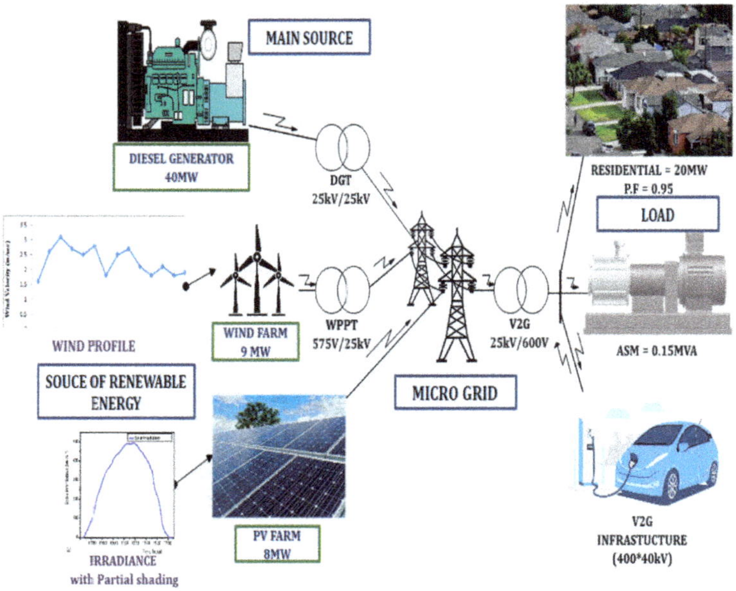


**Figure 3.** A simple structure of the V2G model.

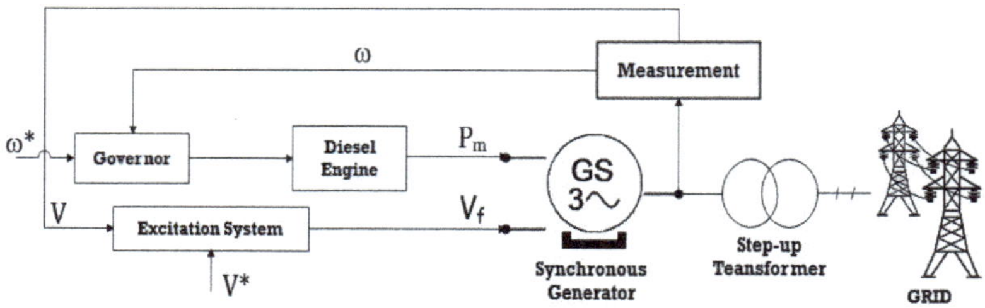


**Figure 4.** DG block diagram.

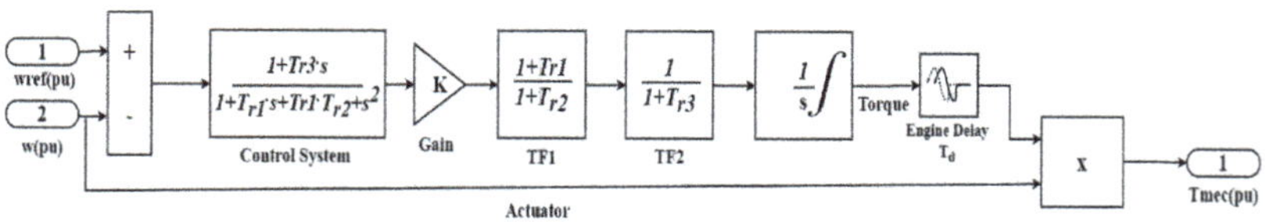


**Figure 5.** Block Diagrams for Governor and Diesel Generator Models.

Based on the d-q rotor reference frame, a 3-phase synchronous machine represents a diesel generator. The following expression is used in the planning of the transfer function ($H_{df}$) of the diesel engine governor system:

$$H_d = \frac{K_g(1 + sT_{r3})}{(1 + sT_{r1} + s^2 T_{r1} T_{r2})} \tag{1}$$

The equation shown below defines the actuator transfer function ($H_a$):

$$H_d = \frac{K_g(1 + sT_{a4})}{s(1 + sT_{a5})(1 + sT_{a6})} \tag{2}$$

where, $T_{r1}$, $T_{r2}$ and $T_{r3}$ are the regulator time constants; $T_{a4}$, $T_{a5}$ and $T_{a6}$ are the actuator time constants; $K_g$ is the regulator gain; and $T_d$ is engine time delay. Here, $T_{r1}$, $T_{r2}$ and $T_{r3}$ are regarded as 0.01, 0.02, and 0.2, respectively; $T_{a4}$, $T_{a5}$ and $T_{a6}$ are assumed to be equal to 0.25, 0.009, and 0.0384, respectively; $T_d$ is set at 0.024; mechanical power is assumed to have a starting value of 0.1361. The DG excitation system is realized by combining an exciter and an IEEE Type-1 voltage regulator of a synchronous machine.

### 2.2. PV Farm

The initial source of RES in this microgrid is a PV farm, which produces energy in accordance with three variables: the area it distributes electricity, the panel efficiency, and the irradiance data. Solar insolation is actuated in real-time using a simulation of irradiance. Additionally, 300 s of partial shadings are replicated for 12 h. Four variables are used in the PV subsystem calculation section to determine the current, which is subsequently inserted into the grid using AC sources, as shown in Figure 6. These are the phase-to-phase AC voltages, the solar radiation's properties, the solar panels' total useable area and their efficiency. The line voltage is converted towards the phase voltage based on the input values:

$$U_A = \frac{1}{3}\left(U_{AB} - a^2 U_{BC}\right) \tag{3}$$

where, a is a complex Fortescue operator, $U_A$ is a phase voltage in volt, and $U_{AB}$ and $U_{BC}$ are line-to-line voltages in volt.

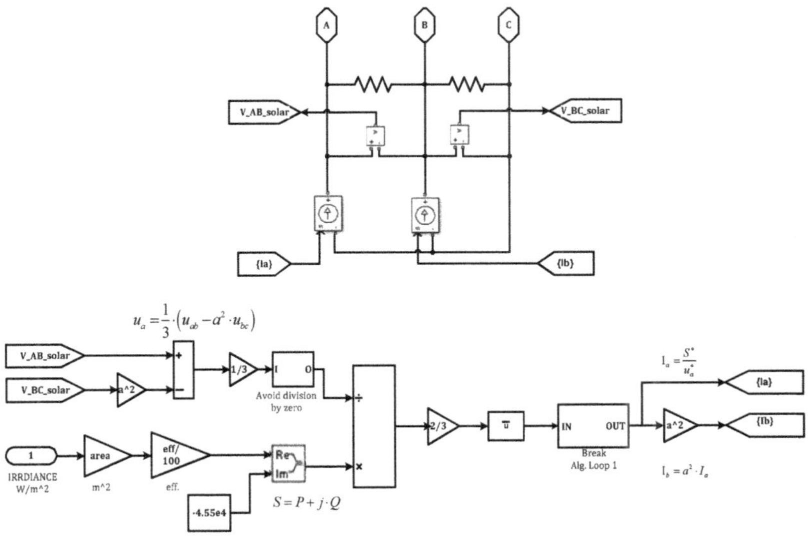


**Figure 6.** Block Diagrams for PV Generator.

By using Equation (4), the active power is determined:

$$P = SR_i * AREA * EFF \tag{4}$$

where *AREA* denotes the entire useable area of solar panels, *SRi* denotes the instantaneous amount of solar radiation, and *EFF* denotes the efficiency of the *PV* generator. Following the expression of the phase current:

$$I_A = \frac{S^*}{U_A^*} \tag{5}$$

where $S$ denotes the apparent power, and $U_A$ is a phase voltage.

A complex vector, also known as the modern complex Fortescue's operator (a), is used to determine the value of the $I_B$ for the examination of the steady-state performance of rotating machines.

$$I_B = a^2 I_A \tag{6}$$

## 2.3. Wind Farm

The next is a streamlined representation of a wind farm that generates wind-generated electricity. The wind farm begins to generate electricity as the wind speed reaches a particular point. Once the wind speed reaches the maximum permitted speed, the wind farm disconnects from the grid and remains disconnected until the wind decreases below the maximum. Table 2 shows the DG, PV, and wind parameters.

**Table 2.** Parameters of DG, PV, and Wind.

| Technical Parameter | Value |
|---|---|
| Diesel Generator | 40 MW |
| Nominal Frequency | 60 Hz |
| PV Farm Power | 8 MW |
| PV Farm Efficiency | 20% |
| PV Farm Area | 8000 $m^2$ |
| Wind Farm Area | 9 MW |
| Nominal Wind Speed | 13.5 $m/S^2$ |
| Maximum Wind Speed | 15 $m/S^2$ |

## 2.4. Vehicle to Grid

The V2G simulates a general collection of EVs. The mask may be adjusted to switch between the model's five distinct profiles by modifying the plug-in and lookup tables of state of charge (SOC). The user may choose how many EVs will follow each sort of available port. The user may also select the power converter efficiency, rated volume, and rated power. The power output of each EV is 40 kW, and the V2G's power rating is 16 MW consequently. To regulate the frequency of plug-in EVs (PHEVs), an aggregator is needed. The V2G aggregator monitors built into the grid's network continually monitor the fleet of EVs. The whole profit depends on how many vehicles are V2G capable. Figure 7 shows the combined flow chart for the regulation and charging modes of V2G.

The energy that can be supplied to the fleet of EVs is included in the power. The operator can send signals or orders to the aggregator, which can then relay them to the fleet of EVs. As a result, it is possible to evaluate the PHEV fleet regulation capacity and choose the best bidding strategy. A specific PHEV's involvement in regulating frequency via the bidirectional energy exchange will be decided and directed by the aggregator. The charge controller has a function when there is excess power in the grid or an over-frequency scenario, as well as when a significant industrial load is rejected or a source is brought back on. Two situations have been researched for charging the EVs while in charging mode. Plug-in first, then SOC second.

There are five groups made up of the whole EV fleet. Every group has a stochastic charging profile based on the amount of time needed to charge and dependent on the availability of charging stations. The SOC and plug-in time were used to profile EVs. The way to identify whether the EVs are in the charging state or the regulation condition is shown in Figure 6. Two limiters have also been implemented within the SOC controller architecture to restrict output variations brought on by plug-in state and SOC initialization. Figure 6 shows the Flow chart for the regulation and charging modes of V2G. The charger controller has the following features as described below:

I. The State Estimation (SE) has been set between 95% and 85%; outside that range, the charging process will stop to guarantee high-quality power output.
II. The EVs are in charging mode when SE is lower than 85% and in regulation mode when it is more than 95%.

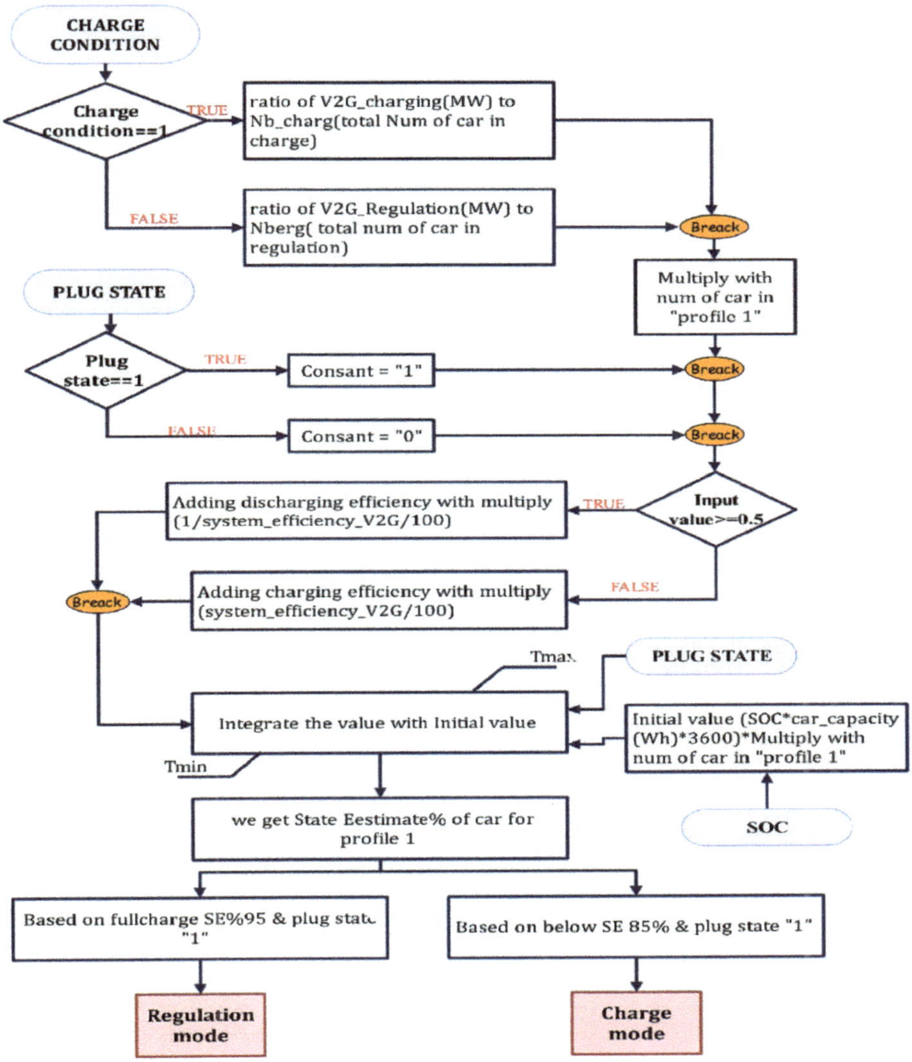


**Figure 7.** Flow Chart for Regulation and Charging Modes of V2G.

Figure 8 summarises the V2G technology utilised for the study and illustrates the operating structure. To regulate the frequency of the PHEVs, an aggregator is needed. The V2G aggregator monitors built into the electricity grid's network continually monitor the fleet of cars. The total number of vehicles with V2G capabilities determines the aggregate profile. The energy that can be supplied to the fleet of cars is included in the power. The operator can send signals or orders to the aggregator, which can then relay them to the fleet of cars. As a result, it is possible to evaluate the PHEV fleet's regulation capacity and choose the best bidding strategy. The frequency regulation employing the bidirectional energy exchange will be decided and directed by the aggregator for a specific PHEV. The V2G controller uses the battery's charge level and real-time frequency to make choices for the BMS. Charger/discharger block/sequence control is also provided by the V2G controller. Additionally, the battery's healthy SOC is tracked by the BMS. Detailed statements of the controls utilised for the investigation are in [14].

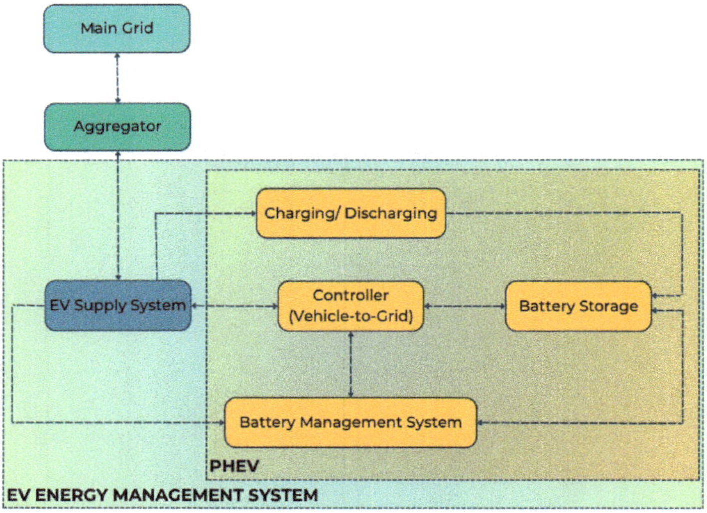


**Figure 8.** Framework of V2G operation.

*2.5. Load*

An asynchronous machine (ASM) and residential loads are utilized to model an industrial load effect on a system. These loads could be unique, like in ventilation systems. The load under consideration has a 20 MW overall capacity. The mechanical torque and rotor speed have a square relationship, which controls the ASM. At 3 h into the simulation, ASM is turned on. The ASM has a rating of 0.15 MW and 60 Hz frequency. Additionally, the networks on diesel generator bus, wind power generation bus, and load bus each have a 1 kW rating. Table 3 shows the Load parameters.

**Table 3.** Load Parameters.

| Technical Parameter | Value |
|---|---|
| V2G Rated Power | 40 MW (per Car) |
| Rated Capacity | 85 kWh (per Car) |
| V2G Efficiency | 90% |
| Total Cars | 400 |
| Domestic Load | 20 MW |
| Power Factor | 0.95 |
| Time-Step | 60 min |
| Asynchronous Machine Load | 0.15 MVA |

## 3. Results Discussion

The simulation of the model is for 24 h. The highest intensity of solar radiation happens in the middle of the day and follows a regular distribution range. Throughout the day, there are several peaks in the wind's strength and brief bursts of minor amplitude. A typical pattern that closely approaches a pattern of residential load consumption is employed to simulate domestic load. The daytime consumption is minimal, and the evening hours are when it peaks, which progressively declines during the nighttime. During the day, the grid parameters would be affected by the following three circumstances:

I. The start of ASM was at the onset of the third hour.
II. At midday, some partial shadowing will be noticed, which impacts how much solar electricity is produced.

III. A wind farm trips every 22 h when the wind speed is higher than the highest allowed wind speed.

The study's primary goal is to balance the system power to reduce changes in system characteristics like voltage and frequency and limit the amount of electricity imported from the utility network. The difference between the amount of energy produced by RES and ASMs and the amount of energy demanded by domestic, commercial, and EV load, as indicated below, determines the need for power balancing.

$$P_{bal} = (P_{gen} + P_w + P_{pv} + P_{V2G}) - (P_{load} + P_{G2V}) \quad (7)$$

$P_{gen}$, $P_w$, $P_{pv}$, and $P_{V2G}$ stand for the power generated by asynchronous machines, wind, solar and EVs discharge. Demand for residential, commercial, and PEV charging is represented by $P_{load}$ and $P_{G2V}$. On the DG bus, whether to take power from or transfer power to the primary grid depends on whether the power balance is negative or positive.

The study focuses on reducing costs while simultaneously supplying consumers with high-quality electricity since the primary goal is to minimize energy conversation with the power grid and changes in system characteristics. This section provides a full analysis of the simulation results for modifications to the system's characteristics, including voltage, current, active power, reactive power, and frequency for all units. This section also discusses the SOC outcomes for each of the five carpools. The results are analyzed to demonstrate the viability of the suggested design for the microgrid interfaced with V2G.

Some assumptions of the methodology are listed below:

- The simulation assumes that the microgrid runs continuously for the entire 24 h.
- The DG may have a linear frequency droop characteristic, according to the model. As a result, the frequency difference from the reference value directly affects how quickly the governor adjusts the speed.
- The model may assume that the sensors that measure the electrical frequency and the actuator that modifies the fuel supply to the engine react instantly and without any mistakes.
- The model assumes perfect transformers, inverters, and voltage regulators, as well as other lossless and ideal components. These components do not consider real-world losses like transformer or converter losses.
- Without considering realistic restrictions like charging/discharging rates and battery degradation, the V2G system is believed to have complete bidirectional power interchange with electric cars.

### 3.1. Diesel Generator Parameters

This section describes the outcomes of several factors for the typical DG. The governor rotor creates power in line with the rotor speed since it is directly connected to the generator shaft. The speed of the rotor is changed between hours 0.02 to 0.1, 2.8 to 2.9, 11.4 to 11.6, 21.9 to 22.05, and 22.2 to 22.4, to manage the power in the grid due to fluctuations in the operational situations caused by various profiles considered in the study. Figure 9 illustrates a DG that uses rotor speed variation to reduce the irregularity of the generation load balance.

Figure 10 displays the current that the DG supplies into the grid. The amount of current varies depending on the amount of electricity generated by RES and the amount of power exchanged with the V2G system. As a result, various spikes and a smaller current magnitude are seen in this figure.

Figure 11 displays the active power and reactive power that the DG provides to the grid. The amount of active power varies depending on the power produced by wind farms, PV farms, and the power exchanged through the V2G system. Reactive power generation from RES and the conversation of reactive power with the V2G affect the amount of power in different ways. As a result, various spikes and a smaller active and reactive power magnitude are seen in Figure 11.

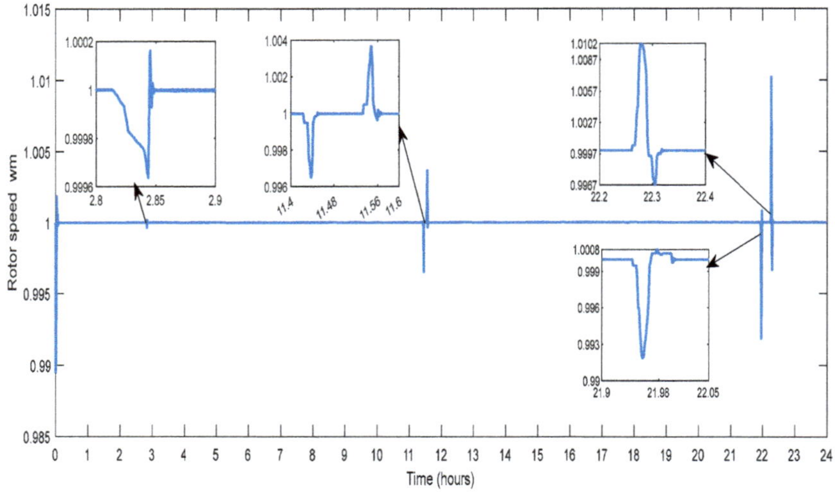


**Figure 9.** Rotor Speed of DG.

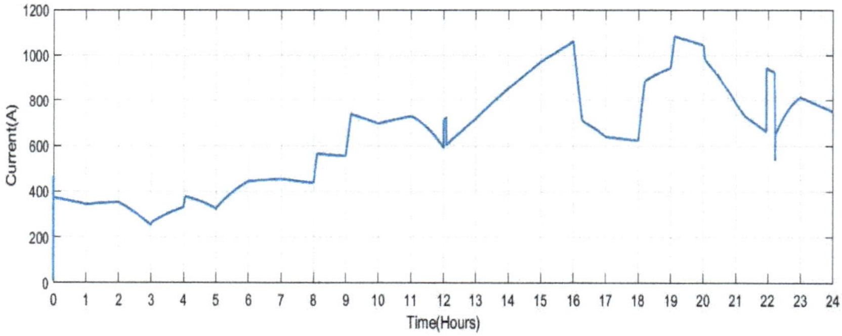


**Figure 10.** Current Measured on the Diesel Generator Bus.

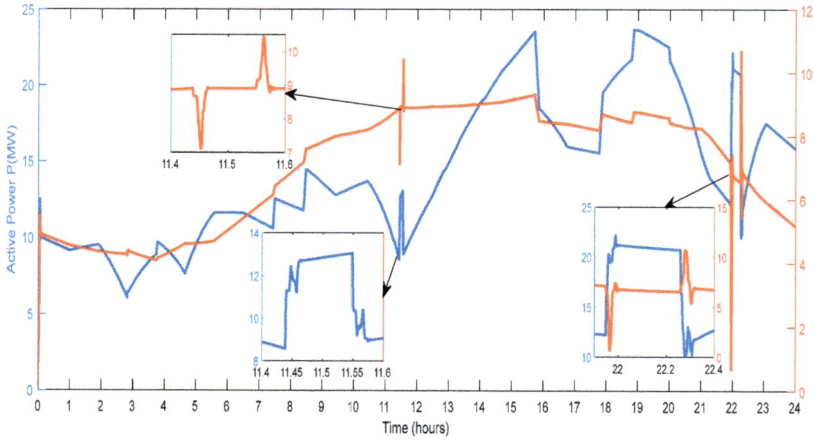


**Figure 11.** Active and Reactive Power of DG.

### 3.2. Wind Generator Parameters

This section describes the parameters recorded on the test system bus and changes in the wind turbine's input wind speed. Deviations in wind speed over a 24-h period are regarded as inputs to the wind turbine to generate electricity, which is shown in Figure 12. From this graph, it can be seen that the minimal speed of wind is 7 m/s$^2$ and that, depending on the time of day, wind speeds can range between 7 and 15 m/s$^2$. This fluctuating wind speed will provide fluctuating power, simulating real-time fluctuations in wind power output.

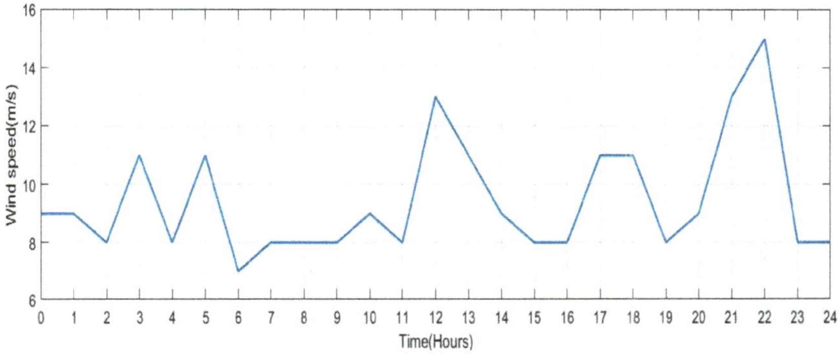


**Figure 12.** Wind Speed for 24 h.

Also, the wind farm phase voltage and current variation are shown in Figure 13. According to changes in wind speed, the current strength increases or decreases. The amount of current the wind generator feeds into the electric grid grows as the wind speed increases. Additionally, as the wind speed diminishes, the utility grid receives less current from the wind generator. As a result, changes in wind speed affect the current that the wind generator produces.

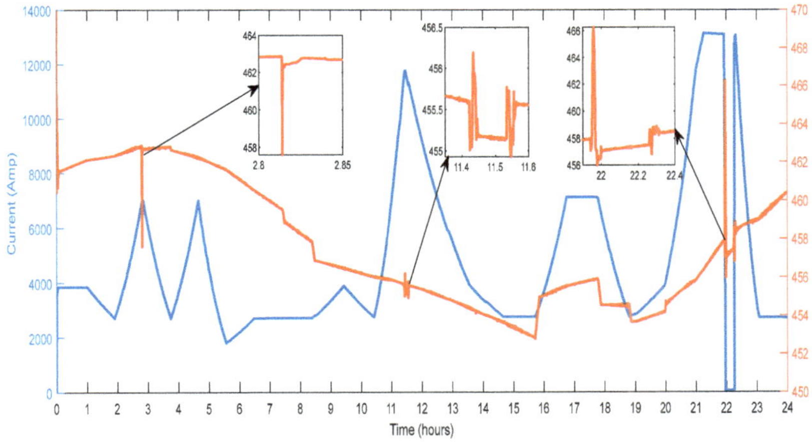


**Figure 13.** Wind Farm Phase Voltage and Current.

Figure 14 displays the active and reactive power of the wind farm provided to the grid. According to changes in wind speed, the power magnitude rises and falls. When the wind blows high, the wind generator will produce more energy to feed into the grid. The quantity of active power the wind generator can send to the utility grid decreases with low wind speed. As a result, changes in wind speed will affect the amount of active power the wind generator produces. Reactive power magnitude varies during three hours: 2.8 to 2.84, 11.4 to 11.6, and 21.9 to 22.4. This aids in controlling changes to the system parameters, particularly the microgrid voltage.

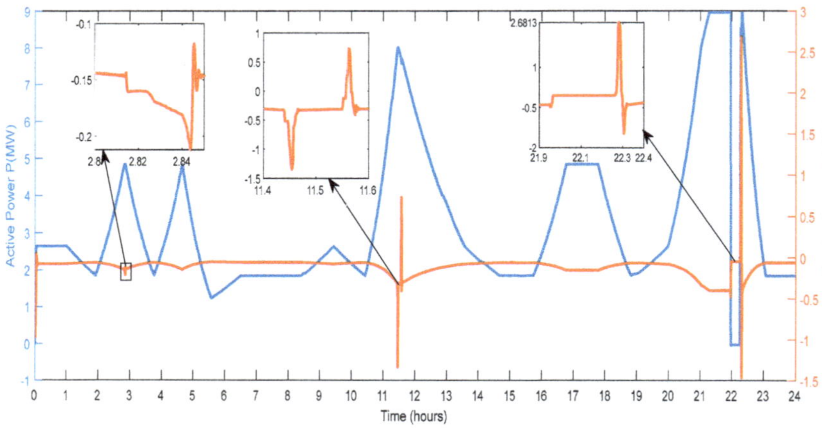


**Figure 14.** Active and Reactive Power of Wind Farm.

### 3.3. *Solar Generator Parameters*

This section describes the parameters recorded on the test system bus and changes in the input solar irradiation. Figure 15 shows changes in solar irradiation for one day that are taken for producing power from PV. The maximum irradiation is noted to be 520 (w/m$^2$). A lowered value of the solar irradiation simulates partial shadowing for a duration of 300 s at the hour 11.5. This fluctuating solar irradiance will provide fluctuating power, simulating real-time fluctuations in solar power generation. At hour 5.7, solar irradiation begins to increase, and at hour 16.8, it reaches zero.

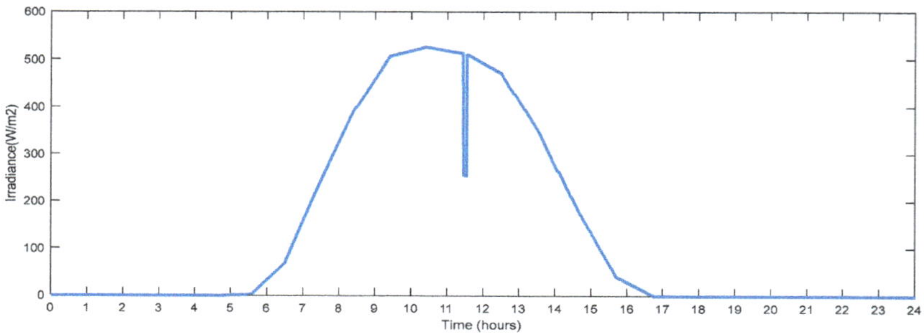


**Figure 15.** Solar Irradiance with Partial Shading.

Figure 16 displays the voltage that was captured on bus B10. The voltage varies over nine time periods.

Figure 17 displays the current supplied to the grid by the PV generator. According to changes in irradiance, the current magnitude increases and decreases. The current provided to the grid by the PV generator will grow as the irradiance level rises. In addition, when the irradiance declines, so does the current so that the PV generator supplies to the grid. Thus, changes in irradiance will affect the current generated by the PV generator—additionally, the current drops after the solar PV panels are partially shaded.

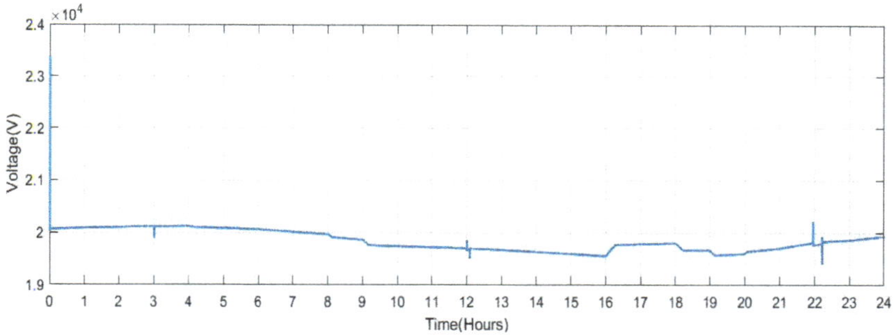


**Figure 16.** Voltage Measured at Solar PV Bus.

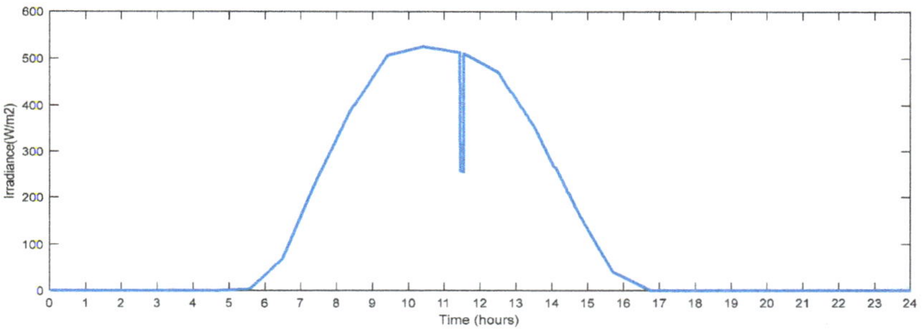


**Figure 17.** Current at PV Bus.

Figure 18 displays the active and reactive power that the PV generator sent to the grid. Depending on the changes in irradiance, the amount of power generated increases and decreases. The amount of active power the PV generator generates and sends to the utility grid rises as the degree of irradiation increases. Additionally, as the irradiance drops, so decreases the power that the PV generator sends to the grid. Therefore, changes in irradiance will impact the amount of active power the solar PV generator can produce. After the solar PV plates were partially shaded, power output also decreased. In general, the power curve characteristics resemble those of irradiance. The microgrid and PV generator exchange reactive power between hours 11.4 to 11.6.

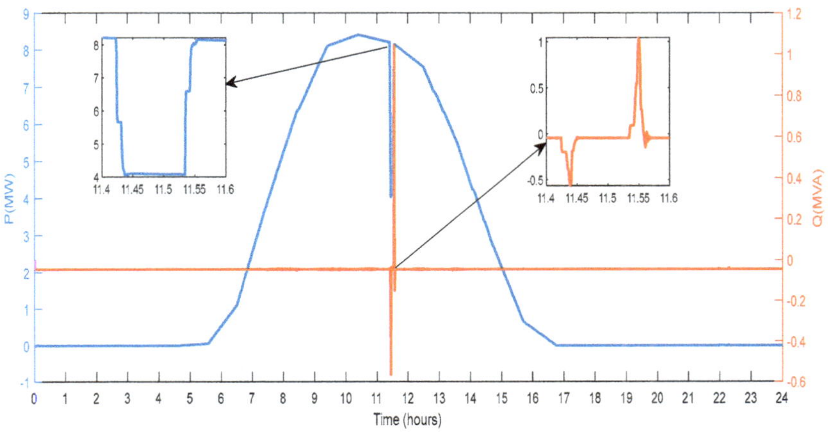


**Figure 18.** Active and Reactive Power at PV Bus.

### 3.4. Consumer Load Parameters

The parameters connected with the consumer loads recorded on the system are described in detail in this section. Figure 19 shows the voltage and current recorded on the system's bus. Between the periods of hours 21.8 and 22.4, a maximum voltage variation of 4.98% is seen in Figure 19. Between the hours of 2.8 and 2.85, there is a voltage sag of magnitude 0.413%. Transient components between hours 11.4 and 11.6 are detected to have magnitude amplitude variations of 0.39%. The transient components have amplitude variations of 3.73% and are seen between hours 22.2 and 22.40. Depending on changes in the load demand, the current's magnitude rises and falls. The grid is used to draw a current in the 15,000 A to 30,000 A range, shown in Figure 19.

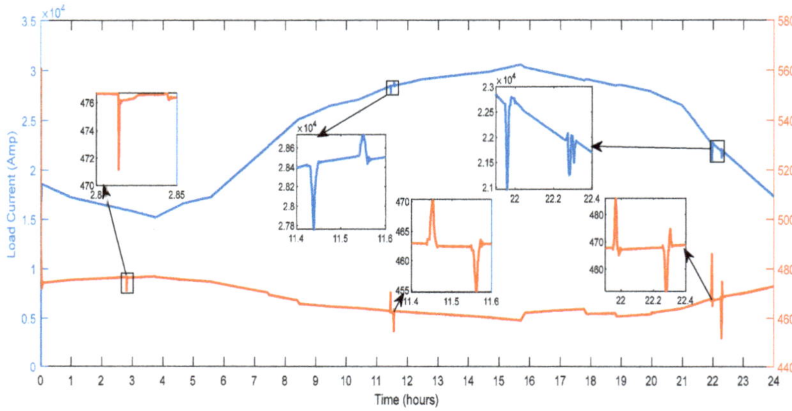


**Figure 19.** Voltage and Current of Residential Load.

Additionally, as load demand fluctuates, local variances are seen. Hours 2.8 to 2.9, 11.4 to 11.6 and 21.8 to 22.4 show abrupt fluctuations in the demand of load. By employing the current balancing with the V2G, variations in load demand are reduced.

Figure 20 shows the active and reactive power that the load draws from the microgrid. The load draws between 10 MW and 20 MW of active power from the microgrid, shown in Figure 20. Hours 2.8 to 2.9, 11.4 to 11.6 and 21.8 to 22.4 show abrupt fluctuations in the demand of load. The current balance employing the V2G has reduced deviations in the load demand. There are abrupt fluctuations in the demand for reactive power load at hours 11.4 to 11.6 and 21.8 to 22.4.

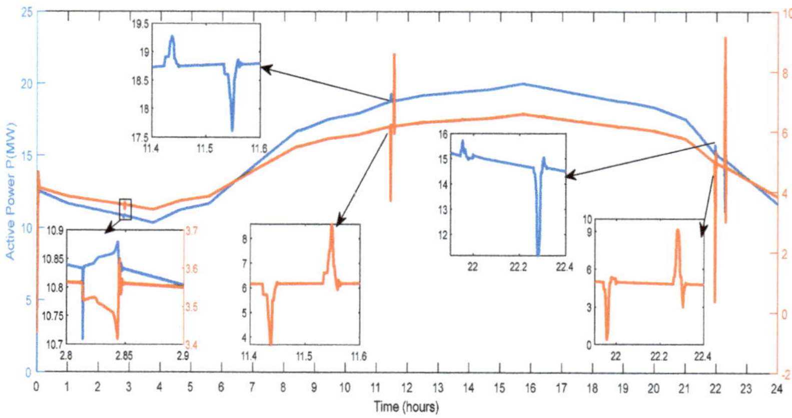


**Figure 20.** Active and Reactive Power of Residential Load.

### 3.5. Industrial Load Parameters

This section describes the outcomes of parameters related to the ASM verified on bus B9 of the microgrid. Figure 21 depicts the voltage and current on the test bus B9. An asynchronous machine load of 0.15 MVA is turned on every three hours. It is perceived that on the ASM, the large value of inrush current, which is equal to 452 A received at the instant of switching. More deviations in the current are also detected at hours 11.4 to 11.6 and 21.8 to 22.4.

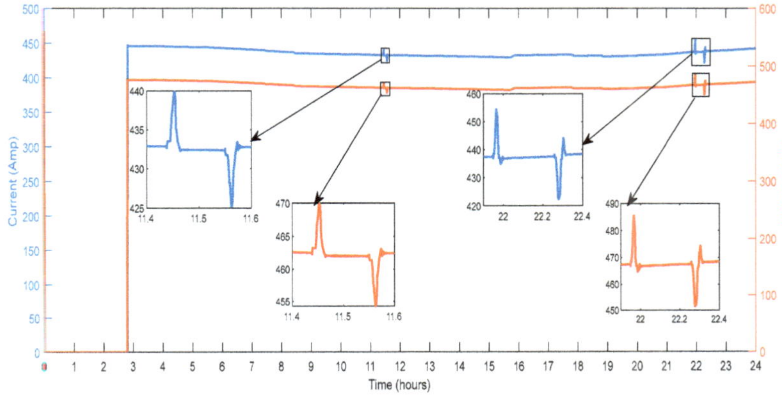


**Figure 21.** Voltage and Current of Asynchronous Machine.

Figure 22 displays the ASM's active and reactive power draw from the grid. It is perceived that a large value of active power is received during switching on the ASM. Consequently, the ASM is consumed continuous active power of 0.32 MW. Additional deviations in the active power are detected at the hour 11.4 to 11.6 and 21.8 to 22.4. Figure 22 displays the ASM's reactive power draw from the grid. It is perceived that a large value of active power is received during switching on the ASM. Consequently, the ASM is consumed continuous reactive power of 0.07 MVA. Additional deviations in the reactive power are detected at 11.4 to 11.6 and 21.8 to 22.4.

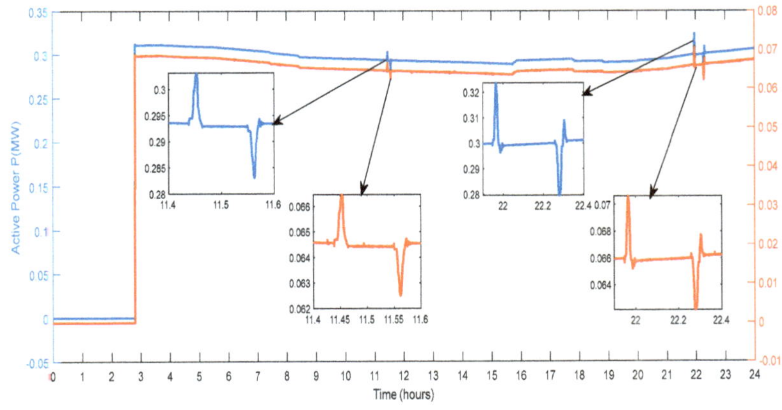


**Figure 22.** Active and Reactive Power of Asynchronous Machine.

### 3.6. V2G Regulation Parameters

This section describes the outcomes of the parameters related to the regulation of V2G on bus B8 as shown in Figure 23.

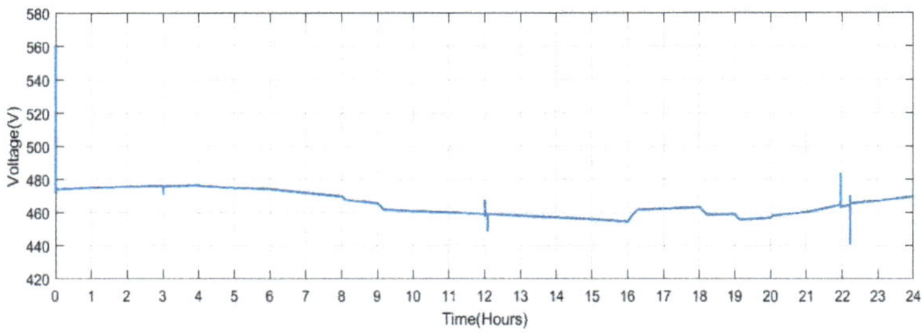


**Figure 23.** V2G Voltage at Bus B8.

Figure 24 shows the current received for regulations of the deviation by the V2G. It is seen that the current is received at the hour 11.4 to 11.6 and 21.8 to 22.4, with high magnitude.

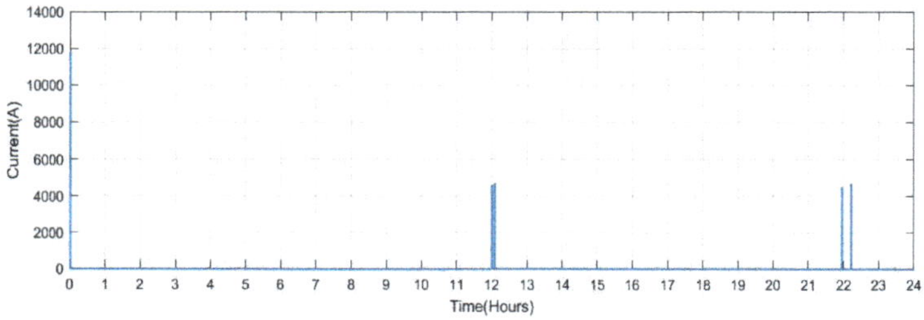


**Figure 24.** V2G Current at Bus B8.

Figure 25 shows the exchange of active and reactive power for the regulation of the deviations by the V2G. The active and reactive power is swapped at the hour 11.4 to 11.6 and 21.8 to 22.4, with high-magnitude peaks, where the first peak shows that the V2G draws the active power and another magnitude shows that the active power is delivered to the V2G to diminish the deviations.

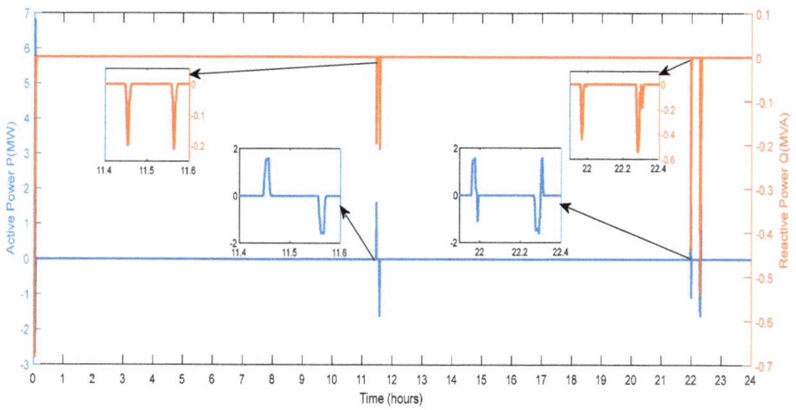


**Figure 25.** Active and Reactive Power of V2G Regulation.

### 3.7. V2G Charging Parameters

The specifications for V2G for the charging of vehicle batteries are described in this section. Seven-time periods of voltage variations are present. The five scenarios are taken for the study, where a probable duration for the car's charging is also stated.

Scenario 1: Persons who commute to work have the option of charging their vehicles there (140 vehicles altogether).

Scenario 2: Persons that commute to work and charge their vehicles there might expect a longer journey time. (100 vehicles altogether).

Scenario 3: Persons who commute to work do not charge their vehicles there (40 vehicles altogether).

Scenario 4: Persons who remain at home (80 vehicles altogether).

Scenario 5: Persons doing the night shift (40 vehicles altogether).

Figure 26 displays the active and reactive power calculated during charging V2G vehicles. The power drained from the grid is marked as "negative". As a result, batteries use the grid for the four time periods that were previously mentioned. For charging batteries, vehicles consumed reactive power from the grid during hours 11.4 to 11.6 and 21.8 to 22.4. When the batteries are connected to charging, the reactive power is given by the batteries, and when the batteries are disconnected, the reactive power is obtained from the grid.

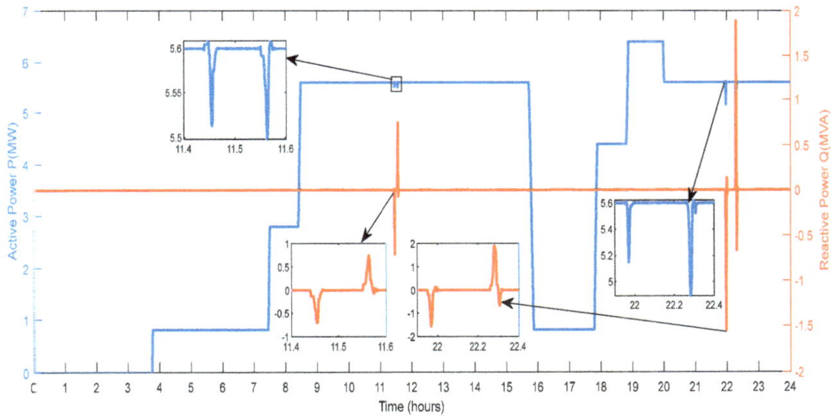


**Figure 26.** Active and Reactive Power of V2G Charging.

In Table 4, the suggested system's sensitivity analysis is displayed. When measuring frequency and voltage under various V2G vehicle scenarios, it has been shown that the inaccuracy is below 5%. Because of this, the suggested model for V2G and microgrid is not overly sensitive to changes in operational situations and successfully runs in various operating situations.

**Table 4.** Sensitivity Analysis under various V2G scenarios.

| | Scenario | Number of EVs | | | | |
|---|---|---|---|---|---|---|
| | | **Profile 1** | **Profile 2** | **Profile 3** | **Profile 4** | **Profile 5** |
| | 1 | 140 | 120 | 100 | 80 | 120 |
| | 2 | 100 | 100 | 80 | 120 | 120 |
| | 3 | 40 | 60 | 40 | 60 | 40 |
| | 4 | 80 | 80 | 100 | 60 | 80 |
| | 5 | 40 | 40 | 80 | 80 | 40 |
| Change in Parameter | Voltage | 3.87% | 3.47% | 3.53% | 3.69% | 3.49% |
| | Frequency | 0252% | 0.232% | 0.247% | 0.260% | 0.234% |
| Error pertaining to Scenario 1 | Voltage | 0.00% | 3.72% | 3.67% | 3.81% | 3.20% |
| | Frequency | 0.00% | 2.90% | 3.15% | 3.37% | 2.35% |

Table 5 shows the SOC value and plug state for each car profile. Figures 27 and 28 depict Scenario 1: this scenario indicates persons commuting to work with a facility charging their vehicles there (no. of vehicles:140). As soon as the car is unplugged, the max SOC set to 0.9 changes to 1 and has a lowered value of 0.817 with respect to time (hour). Its decreasing rate is 8.3%. However, when the car is plugged back in, its increasing rate is 10%.

**Table 5.** SOC value and plug state for each car profile.

| Car Profiles | SOC | Plug State |
|---|---|---|
| 1 | [0.9 0.9 0.9 0.9 0.9 0.9 0.9 1 1 0.834 0.9 0.9 0.9 0.9 0.9 0.9 1 1 0.834 0.9 0.9 0.9 0.9 0.9] | [1 1 1 1 1 1 1 0 0 1 1 1 1 1 1 1 0 0 1 1 1 1 1 1] |
| 2 | [0.9 0.9 0.9 0.9 0.9 0.817 0.734 0.651 0.751 0.851 0.9 0.9 0.9 0.9 0.9 0.9 0.817 0.734 0.651 0.751 0.851 0.9 0.9 0.9] | [1 1 1 1 1 0 0 0 1 1 1 1 1 1 1 1 0 0 0 1 1 1 1 1] |
| 3 | [0.9 0.9 0.9 0.9 0.9 0.9 0.817 0.734 0.651 0.651 0.651 0.651 0.651 0.651 0.651 0.568 0.485 0.402 0.502 0.602 0.702 0.802 0.9 0.9] | [1 1 1 1 1 1 0 0 0 0 0 0 0 0 0 0 0 0 1 1 1 1 1 1] |
| 4 | [0.9 0.9 0.9 0.9 0.9 0.9 0.9 0.9 0.9 0.9 0.9 0.9 0.9 0.9 0.9 0.9 0.9 0.9 0.9 0.9 0.9 0.9 0.9 0.9] | [1 1 1 1 1 1 1 1 1 1 1 1 1 1 1 1 1 1 1 1 1 1 1 1] |
| 5 | [0.9 0.817 0.734 0.651 0.751 0.851 0.9 0.9 0.9 0.9 0.9 0.9 0.9 0.9 0.9 0.9 0.9 0.9 0.9 0.9 0.817 0.734 0.651 0.751] | [0 0 0 0 1 1 1 1 1 1 1 1 1 1 1 1 1 1 1 1 0 0 0 0] |

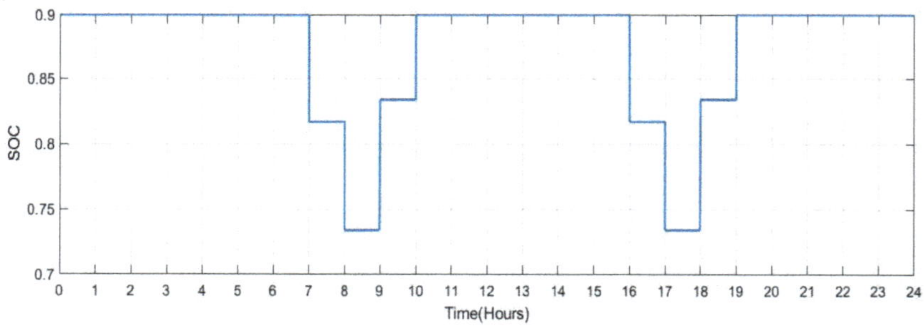


**Figure 27.** SOC Data of Scenario 1: [0.9 0.9 0.9 0.9 0.9 0.9 0.9 1 1 0.834 0.9 0.9 0.9 0.9 0.9 0.9 1 1 0.834 0.9 0.9 0.9 0.9 0.9].

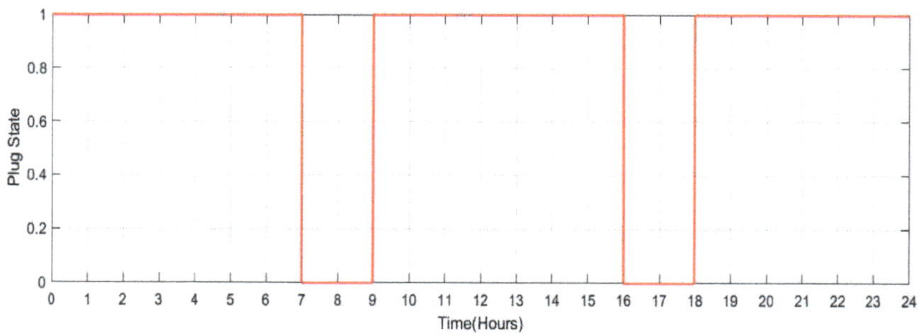


**Figure 28.** Plug State of Car Scenario 1: [1 1 1 1 1 1 1 0 0 1 1 1 1 1 1 1 0 0 1 1 1 1 1 1].

Figure 29 shows the SOC of batteries for Scenario 2: individuals who commute to work with a facility for charging their vehicles but with a lengthier journey (no. of vehicles: 100) during a one-day period. When people make a long trip from their house to their place of work in the morning between the hours of 5:00 to 8:00, it is seen that the SOC of the batteries is lower. Similarly, the SOC also drops when individuals leave work in the late afternoon or evening between 16:00 and 19:00. When batteries are charged at work and reach highs of more than 90. SOC rises during this time. Figure 30 shows the plug state of car Scenario 2.

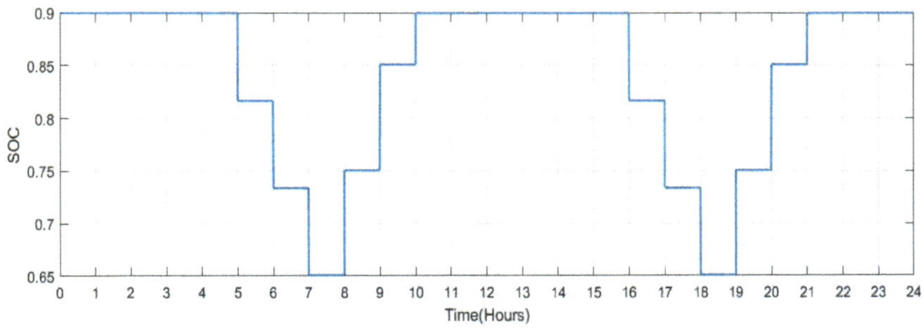


**Figure 29.** SOC Data of Scenario 2: [0.9 0.9 0.9 0.9 0.9 0.817 0.734 0.651 0.751 0.851 0.9 0.9 0.9 0.9 0.9 0.9 0.817 0.734 0.651 0.751 0.851 0.9 0.9 0.9].

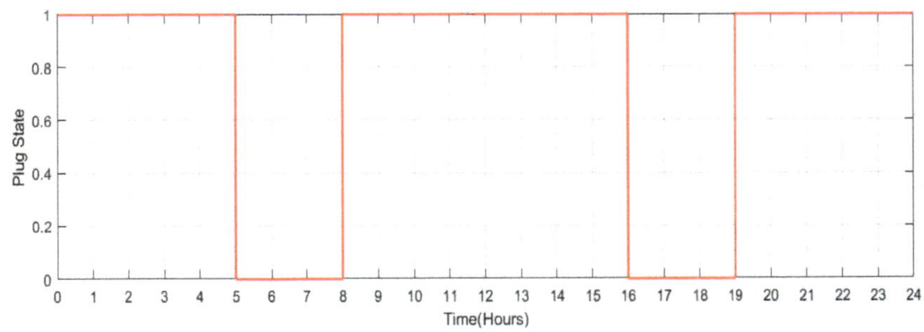


**Figure 30.** Plug State of car Scenario 2: [1 1 1 1 1 0 0 0 1 1 1 1 1 1 1 1 0 0 0 1 1 1 1 1].

Figure 31 shows the SOC of batteries for Scenario 3 over a 24-h period when people arrive at work but are unable to charge their vehicles (number of vehicles: 10). When travelling from home to work in the morning and returning home from hours 15:00 to 18:00 in the evening, the SOC of the batteries drops in this scenario by 8.3%. Figure 32 shows the plug state of car Scenario 3.

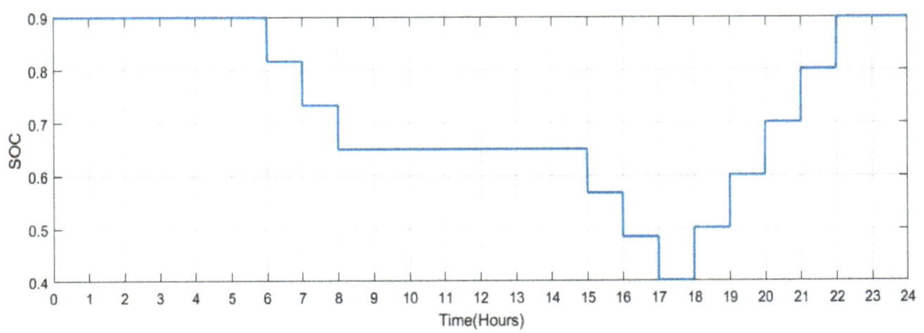


**Figure 31.** SOC Data of Scenario 3: [0.9 0.9 0.9 0.9 0.9 0.9 0.817 0.734 0.651 0.651 0.651 0.651 0.651 0.651 0.651 0.568 0.485 0.402 0.502 0.602 0.702 0.802 0.9 0.9].

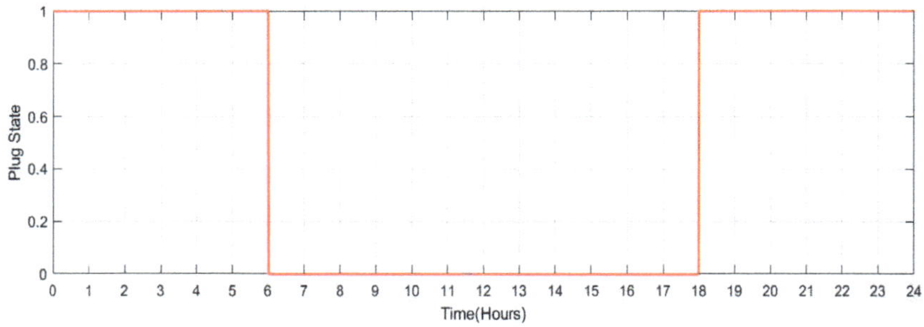


**Figure 32.** Plug State of car Scenario 3: [1 1 1 1 1 1 0 0 0 0 0 0 0 0 0 0 0 0 1 1 1 1 1 1].

Figure 33 shows the SOC of the batteries for Scenario 4 throughout a 24-h period when the individuals remain at home (number of vehicles: 20). It has been noted that the batteries' SOC has been nearly consistent throughout. However, when cars are operated in grid regulation mode, minor variances are seen. Figure 34 shows the plug state of car Scenario 4.

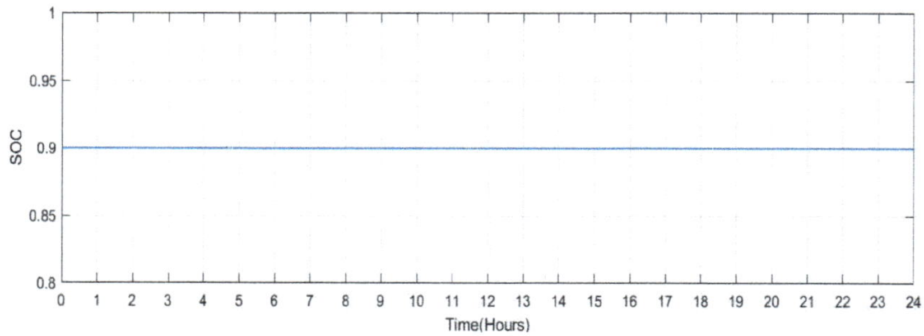


**Figure 33.** SOC Data of Scenario 4: [0.9 0.9 0.9 0.9 0.9 0.9 0.9 0.9 0.9 0.9 0.9 0.9 0.9 0.9 0.9 0.9 0.9 0.9 0.9 0.9 0.9 0.9 0.9 0.9].

Figure 35 shows the SOC of the batteries for Scenario 5 during a 24-h period while individuals are working the night shift (10 total vehicles). When people commute from home to work during the night shift, between the hours of 20 and 4 the following day, the SOC of the batteries decreases in this situation. During daylight hours, when individuals are at home, the likelihood of SOC rises. Because the batteries are not charged at work, the SOC is maintained at low levels throughout the night. Figure 36 shows the plug state of car Scenario 5.

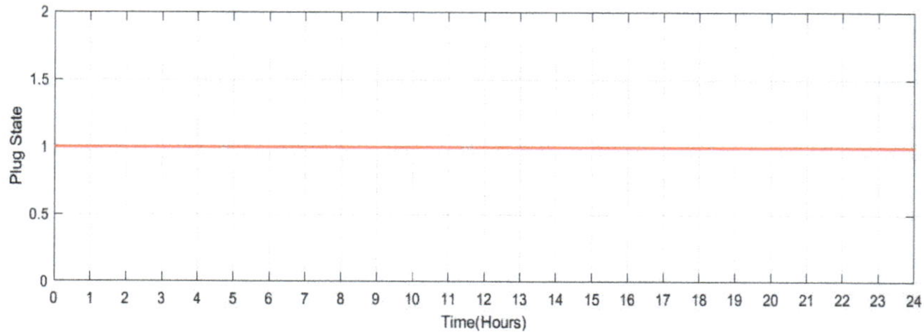


**Figure 34.** Plug State of car Scenario 4: [1 1 1 1 1 1 1 1 1 1 1 1 1 1 1 1 1 1 1 1 1 1 1 1].

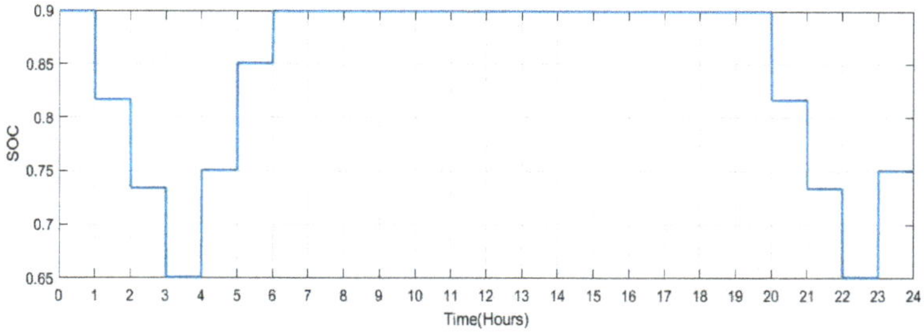


**Figure 35.** SOC Data of car Scenario 5: [0.9 0.817 0.734 0.651 0.751 0.851 0.9 0.9 0.9 0.9 0.9 0.9 0.9 0.9 0.9 0.9 0.9 0.9 0.9 0.9 0.817 0.734 0.651 0.751].

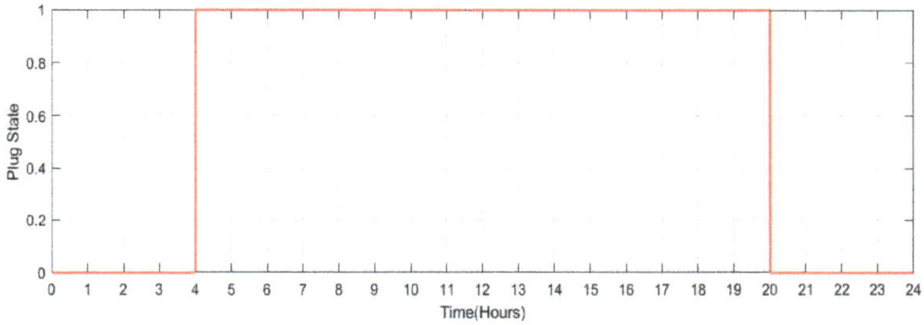


**Figure 36.** Plug State of car Scenario 5: [0 0 0 0 1 1 1 1 1 1 1 1 1 1 1 1 1 1 1 1 0 0 0 0].

Figure 37 depicts the active power of the generating sources and the residential load, industrial load, and V2G load. Throughout the 24 h the power will be balanced, as shown in Figure 37. Here we have the previously discussed scenario of partial shading at noon when generated power is suddenly reduced so that a diesel generator will provide power. When wind speed goes above the speed limit, that wind generator will be cut-off at the hour 22:00, then the diesel generator offers total power to the load.

The limitations of the proposed methodology are listed below:

- The accuracy of the model's representation of real-world behaviour may not have been validated against actual microgrid data or field measurements.
- The V2G model does not provide specific details about different types of EVs (e.g., plug-in hybrid electric vehicles, battery electric vehicles), each of which might have unique charging and discharging characteristics.
- The model does not consider environmental factors, such as temperature and weather, which can influence charging and discharging performance.

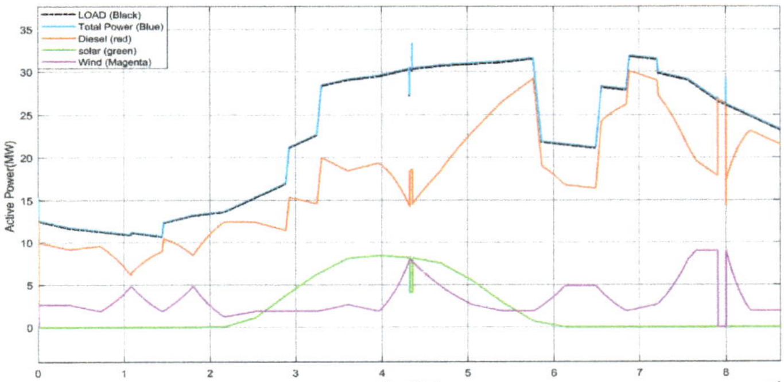


**Figure 37.** Active Power of V2G.

## 4. Conclusions

Although V2G operation may shorten the battery life of vehicles, it is anticipated to be cost-effective for grid operators and car owners. Ancillary services like voltage and frequency control and spinning reserves can be made possible by the V2G concept. To satisfy the demand and control the parameters of the microgrid, a V2G system interfaced with a microgrid has been proposed for a 24-h cycle. A DG, RESs, a V2G system, and the loads are the four parts that make up a microgrid. A microgrid is built to reflect a community of a thousand homes daily with low demand in the spring or fall. To realize 1:5 ratios between EVs and homes, which may be a potential situation in the near future, a total of 400 EVs have been demonstrated as basic models. The suggested microgrid and V2G system successfully ran, satisfying the load demand. By feeding energy from the EV batteries to the grid, V2G has been used to control changes in the microgrid parameters. All the data have been analyzed, including voltage, active and reactive power, current, fluctuations in wind speed and irradiance, and battery SOC, and it has been concluded that the proposed V2G and microgrid architecture successfully controls all of these variables.

The microgrid and V2G characteristics have been controlled through trial and error. It is observed that the V2G operates effectively in various operating scenarios, and the measurement error of frequency and voltage for various V2G vehicle scenarios is below 5%, showing that the suggested V2G model and microgrid are not particularly sensitive to the changes in operating conditions. To efficiently manage the loads linked to the grid, optimization techniques may be utilized in the future to enhance the functionality and efficacy of the energy resources.

Future increases in EV adoption would result in a greater overall strain on the grid. This illustrates a future scenario where the system operator could invest in a well-organized aggregation of EVs rather than new generating units. To encourage owners of EVs to actively contribute to grid stability and provide an additional source of revenue, new incentives would be required.

**Author Contributions:** J.S., Y.R. and S.K.; Methodology, J.S. and A.P.; Writing—original draft preparation, J.S. and Y.R.; Writing—review and editing, J.S., A.P., A.S. and M.S.; Supervision, J.S., A.P., A.S. and M.S. All authors have read and agreed to the published version of the manuscript.

**Funding:** This research received no external funding.

**Institutional Review Board Statement:** Not applicable.

**Informed Consent Statement:** Not applicable.

**Data Availability Statement:** The data presented in this study are available on request from the corresponding author.

**Acknowledgments:** This research was done with the help of Charotar University of Science and Technology, GLA University and Dongseo University.

**Conflicts of Interest:** The authors declare no conflict of interest. The funders had no role in the study's design; in the collection, analyses, or interpretation of data; in the writing of the manuscript, or in the decision to publish the results.

## References

1. *EV Everywhere Grand Challenge: Road to Success*; US Department of Energy: Washington, DC, USA, 2014.
2. Galus, M.D.A.G.; Vayá, M.G.; Krause, T. The role of electric vehicles in smart grids. In *WIREs Energy and Environment*; Wiley Online Library: Hoboken, NJ, USA, 2012; pp. 1–17.
3. *E-Mobility: Closing the Emissions Gap*; World Energy Council: London, UK, 2016.
4. Tarroja, B.; Shaffer, B.; Samuelsen, S. The importance of grid integration for achievable greenhouse gas emissions reductions from alternative vehicle technologies. *Energy* **2015**, *87*, 504–519. [CrossRef]
5. *India's Electric Vehicle Transition*; CEEW The Council: New Delhi, India, 2019.
6. Chen, Z.; Carrel, A.L.; Gore, C.; Shi, W. Environmental and economic impact of electric vehicle adoption. *Environ. Res. Lett.* **2021**, *16*, 045011. [CrossRef]

7. Sovacool, B.K.; Hirsh, R.F. Beyond batteries: An examination of the benefits and barriers to plug-in hybrid electric vehicles (PHEVs) and a vehicle-to-grid (V2G) transition. *Energy Policy* **2009**, *37*, 1095–1103. [CrossRef]
8. IEA. Global EV Outlook 2021, IEA, Paris. 2021. Available online: https://www.iea.org/reports/global-ev-outlook-2021 (accessed on 1 February 2023).
9. Hashemi-Dezaki, H.; Hamzeh, M.; Askarian-Abyaneh, H.; Haeri-Khiavi, H. Risk management of smart grids based on managed charging of PHEVs and vehicle-to-grid strategy using Monte Carlo simulation. *Energy Convers. Manag.* **2015**, *100*, 262–276. [CrossRef]
10. DeForest, N.; MacDonald, J.S.; Black, D.R. Day ahead optimization of an electric vehicle fleet providing ancillary services in the Los Angeles Air Force Base vehicle-to-grid demonstration. *Appl. Energy* **2018**, *210*, 987–1001. [CrossRef]
11. Zheng, Y.; Shang, Y.; Shao, Z.; Jian, L. A novel real-time scheduling strategy with near-linear complexity for integrating large-scale electric vehicles into smart grid. *Appl. Energy* **2018**, *217*, 1–13. [CrossRef]
12. Mehrjerdi, H.; Bornapour, M.; Hemmati, R.; Ghiasi, S.M.S. Unified energy management and load control in building equipped with wind-solar-battery incorporating electric and hydrogen vehicles under both connected to the grid and islanding modes. *Energy* **2018**, *168*, 919–930. [CrossRef]
13. Mehrjerdi, H.; Rakhshani, E. Vehicle-to-grid technology for cost reduction and uncertainty management integrated with solar power. *J. Clean. Prod.* **2019**, *229*, 463–469. [CrossRef]
14. International Energy Agency. Grid Integration of Electric Vehicles A Manual for Policy Makers. March 2022. Available online: https://iea.blob.core.windows.net/assets/21fe1dcb-c7ca-4e32-91d4-928715c9d14b/GridIntegrationofElectricVehicles.pdf (accessed on 1 July 2023).
15. Drude, L.; Junior, L.C.P.; Rüther, R. Photovoltaics (PV) and electric vehicle-to-grid (V2G) strategies for peak demand reduction in urban regions in Brazil in a smart grid environment. *Renew. Energy* **2014**, *68*, 443–451. [CrossRef]
16. Khemakhem, S.; Rekik, M.; Krichen, L. A flexible control strategy of plug-in electric vehicles operating in seven modes for smoothing load power curves in smart grid. *Energy* **2017**, *118*, 197–208. [CrossRef]
17. López, M.A.; De La Torre, S.; Martín, S.; Aguado, J.A. Demand-side management in smart grid operation considering electric vehicles load shifting and vehicle-to-grid support. *Int. J. Electr. Power Energy Syst.* **2015**, *64*, 689–698. [CrossRef]
18. Tarroja, B.; Zhang, L.; Wifvat, V.; Shaffer, B.; Samuelsen, S. Assessing the stationary energy storage equivalency of vehicle-to-grid charging battery electric vehicles. *Energy* **2016**, *106*, 673–690. [CrossRef]
19. Hoehne, C.G.; Chester, M.V. Optimizing plug-in electric vehicle and vehicle-to-grid charge scheduling to minimize carbon emissions. *Energy* **2016**, *115*, 646–657. [CrossRef]
20. Alotaibi, S.; Omer, S.; Su, Y. Identification of Potential Barriers to Electric Vehicle Adoption in Oil-Producing Nations—The Case of Saudi Arabia. *Electricity* **2022**, *3*, 365–395. [CrossRef]
21. Tie, L. Status analysis and development planning for the network of charging stations. *Eur. J. Elec. Eng.* **2018**, *20*, 485.
22. Singh, V.; Singh, V.; Vaibhav, S. A review and simple meta-analysis of factors influencing adoption of electric vehicles. *Transp. Res. Part D Transp. Environ.* **2020**, *86*, 102436. [CrossRef]
23. Egbue, O.; Long, S. Barriers to widespread adoption of electric vehicles: An analysis of consumer attitudes and perceptions. *Energy Policy* **2012**, *48*, 717–729. [CrossRef]
24. Narassimhan, E.; Johnson, C. The role of demand-side incentives and charging infrastructure on plug-in electric vehicle adoption: Analysis of US States. *Environ. Res. Lett.* **2018**, *13*, 74032. [CrossRef]
25. Funke, S.Á.; Sprei, F.; Gnann, T.; Plötz, P. How much charging infrastructure do electric vehicles need? A review of the evidence and international comparison. *Transp. Res. Part D Transp. Environ.* **2019**, *77*, 224–242. [CrossRef]
26. Quak, H.; Nesterova, N.; van Rooijen, T. Possibilities and barriers for using electric-powered vehicles in city logistics practice. *Transp. Res. Procedia* **2016**, *12*, 157–169. [CrossRef]
27. Lewis, M. Electric Vehicles Technology Brief, IRENA. 2017. Available online: https://www.irena.org/-/media/Files/IRENA/Agency/Publication/2017/IRENA_Electric_Vehicles_2017.pdf (accessed on 21 May 2022).

sensors

MDPI

*Review*

# A Survey of Techniques for Discovering, Using, and Paying for Third-Party IoT Sensors

**Anas Dawod [1,*], Dimitrios Georgakopoulos [1], Prem Prakash Jayaraman [1] and Ampalavanapillai Nirmalathas [2]**

[1] Department of Computing Technologies, Swinburne University of Technology, Melbourne 3122, Australia; dgeorgakopoulos@swin.edu.au (D.G.); pjayaraman@swin.edu.au (P.P.J.)

[2] Department of Electrical and Electronic Engineering, The University of Melbourne, Melbourne 3010, Australia; nirmalat@unimelb.edu.au

* Correspondence: adawod@swin.edu.au

**Abstract:** The Internet of Things (IoT) includes billions of sensors and actuators (which we refer to as IoT devices) that harvest data from the physical world and send it via the Internet to IoT applications to provide smart IoT services and products. Deploying, managing, and maintaining IoT devices for the exclusive use of an individual IoT application is inefficient and involves significant costs and effort that often outweigh the benefits. On the other hand, enabling large numbers of IoT applications to share available third-party IoT devices, which are deployed and maintained independently by a variety of IoT device providers, reduces IoT application development costs, time, and effort. To achieve a positive cost/benefit ratio, there is a need to support the sharing of third-party IoT devices globally by providing effective IoT device discovery, use, and pay between IoT applications and third-party IoT devices. A solution for global IoT device sharing must be the following: (1) scalable to support a vast number of third-party IoT devices, (2) interoperable to deal with the heterogeneity of IoT devices and their data, and (3) IoT-owned, i.e., not owned by a specific individual or organization. This paper surveys existing techniques that support discovering, using, and paying for third-party IoT devices. To ensure that this survey is comprehensive, this paper presents our methodology, which is inspired by Systematic Literature Network Analysis (SLNA), combining the Systematic Literature Review (SLR) methodology with Citation Network Analysis (CNA). Finally, this paper outlines the research gaps and directions for novel research to realize global IoT device sharing.

**Keywords:** cost-sharing; description; discovery; Internet of Things (IoT); integration; IoT device; payment; query; sensor; sharing

**Citation:** Dawod, A.; Georgakopoulos, D.; Jayaraman, P.P.; Nirmalathas, A. A Survey of Techniques for Discovering, Using, and Paying for Third-Party IoT Sensors. *Sensors* **2024**, *24*, 2539. https://doi.org/10.3390/s24082539

Academic Editor: Behnam Mobaraki

Received: 22 February 2024
Revised: 7 April 2024
Accepted: 10 April 2024
Published: 15 April 2024



## 1. Introduction

The Internet of Things (IoT) combines billions of Internet-accessible IoT devices (e.g., sensors, RFIDs, wearables, smartphones, smart meters, industrial machines, vehicles, etc.) that are capable of sensing the physical world and sending their data observations (which we refer to as IoT data) to IoT applications. These IoT applications use collected IoT data to make decisions, enabling the development of smart IoT services and products that solve problems that were previously very difficult to solve. Deploying, managing, and maintaining IoT devices for the exclusive use of an IoT application is inefficient and involves significant costs, time, and effort that often outweighs the benefits [1]. Enabling the global sharing of third-party IoT devices that have been deployed, maintained, and owned by different parties (which we refer to as IoT device providers, or simply providers) significantly reduces the cost, timeframe, and effort of realizing IoT applications. Although this development can springboard cheaper and faster developments of novel smart IoT services and products [2], sharing third-party IoT devices (termed simply IoT devices for the remains of the paper) that are deployed and managed by different providers remains unrealized due to a lack of techniques to support global (global refers to the ability to share IoT devices deployed across the globe without the need to own them (similar to the Internet)) sharing (i.e., discovering, using, and paying) for IoT devices.

To illustrate the need for sharing IoT devices globally, we provide the following climate change example. The negative impact of climate change on agriculture can be mitigated by using IoT devices that provide the information needed to determine how various plants perform under changing environmental conditions across the world [3]; however, the procurement, deployment, and maintenance of IoT devices that are needed to monitor micro-climate, soil humidity, solar radiation, and crop performance are difficult to implement at this scale and incredibly expensive. Alternatively, using existing IoT devices, that have been deployed by farmers and agribusinesses for their own purposes to collect the data needed for climate change mitigation, can minimize the effort, cost, and timeframe for responding to the effects of climate change. This can be enabled through global IoT device sharing.

To achieve the global sharing of IoT devices, there is a need to provide the following objectives:

1. Discovering IoT Devices: This objective allows IoT applications to discover the IoT devices that they need from devices that have been made available by various providers. In our earlier example, this would enable an IoT application for climate change mitigation to discover devices already deployed IoT by farmers and agribusinesses, which can be potentially used to take mitigation actions that reduce the impact of climate change in agriculture.
2. Using IoT Devices: This objective allows IoT applications to use the IoT data of already deployed IoT devices instead of procuring, deploying, and maintaining their own devices. In our earlier climate change mitigation example, this would minimize the effort, cost, and timeframe of determining the impact of climate change on crops, and would enable mitigation actions to be taken, e.g., increase irrigation, use crops that require less water or tolerate worsening environmental conditions.
3. Paying IoT Devices: This objective allows IoT applications to share the cost of procuring, deploying, and maintaining IoT devices by paying the providers for their use (i.e., in our climate mitigation example, these would be farmers, agribusinesses, or other third parties). This is similar to the "pay-as-you-go" model seen in cloud computing [4].

While meeting the above objectives, it is equally important to address the following challenges for the global sharing of IoT devices via the Internet.

1. The Internet is an open system (i.e., no specific individuals and/or organizations own it or control it [5]). The Internet provides the infrastructure to allow for interconnection with devices owned by specific entities. However, the infrastructure itself is not owned by a specific entity; instead, thousands of collective people and organizations (entities) across the globe own the Internet [6]. Hence, we must develop a global sharing of IoT devices that are not owned or controlled by specific entities. We use the terms IoT-ownership or IoT-owned. The IoT-owned sharing of devices encourages IoT devices to be owned by different providers, but the infrastructure is not owned by any specific entity;
2. The Internet can be scaled up by supporting an ever-increasing number of users. Similarly, the global sharing of IoT devices requires the ability to scale up for potentially billions of IoT devices and IoT applications, while new ones are continuously added; we term this scalability;
3. The Internet allows computers that are manufactured by different companies and use different operating systems to communicate and share heterogeneous data that makes it interoperable [7]. Therefore, the global sharing of IoT devices requires support for IoT devices that have diverse sensing capabilities, heterogeneous hardware, different protocols, are manufactured by a variety of vendors, and generate heterogeneous data. In our climate change mitigation example, some available IoT devices may provide ground temperature measurements in Celsius, and other IoT devices may

provide only combined ambient temperature and humidity measurements, while an application that uses such IoT data needs to construct a soil temperature map in Fahrenheit for temperatures collected from around the world during the summer months. Basic interoperability is achieved when at a minimum the descriptions of IoT data that are generated by available IoT devices are shared with the IoT applications when they are in the stage of discovering the IoT devices they need;

4. There is a requirement to develop the global sharing of IoT devices that is comprehensive. That means it should support all or most aspects of sharing (more specifically discovering) IoT devices. In our climate change mitigation example, one IoT application may look to discover IoT devices based on their data type, another may look for the geographical area of IoT devices, and another may look for the cost of using IoT devices and payment options, while another one looks for a way to use these IoT devices. Therefore, comprehensive discovery is needed to support the global sharing of IoT devices and to support all or most IoT application needs.

The main contributions of this paper are as follows:

- Providing a review of the most relevant articles and research outcomes supporting the sharing of IoT devices. This is achieved by following a SLNA-based methodology (Systematic Literature Citation Network Analysis);
- Proposing a taxonomy of challenges that helps classify and review related work in the area of sharing IoT devices and outlines related work gaps.
- Reviewing the related work for sharing IoT devices classified by the proposed taxonomy;
- Providing related work gap analysis and directions for novel research that can help other researchers advance the development of global sharing IoT devices.

To the best of our knowledge, this paper presents the first survey of techniques for discovering, using, and paying for third-party IoT devices. The survey area is important due to the benefits of reducing the cost, time, and effort of creating new IoT applications by discovering, using, and paying for third-party IoT sensors instead of procuring, deploying, and maintaining new sensors. The remainder of this paper is organized as follows: In Section 2, we discuss the survey methodology we used to collect and filter papers that are relevant to sharing IoT devices. In Section 3, we present a taxonomy of the challenges in devising the global sharing of IoT devices. Section 4 presents the existing techniques that can contribute to the research and development of the global sharing of IoT devices, while Section 4 provides a related work gap analysis and presents directions for novel research toward the developing the global sharing of IoT devices. Section 5 concludes the paper.

## 2. Survey Methodology

To identify and review the most relevant and cited related work on the sharing of IoT devices, we present our methodology, which is inspired by a Systematic Literature Network Analysis (SLNA). It combines a Systematic Literature Review (SLR) and Citation Network Analyses (CNAs) [8]. The SLR is used to identify ideas and select terms for the initial selection of the most relevant papers in the field. The CNA distinguishes a backbone in a citation network, which assists researchers in recognizing how the frame of knowledge has progressed over time [8]. In Section 2.1, we outline the methodology used for identifying and selecting the most relevant papers. Then, in Section 2.2, we present how we used the CNA [9] to analyze the knowledge progression in the field of study [10]. Figure 1 shows the structure of our survey methodology.

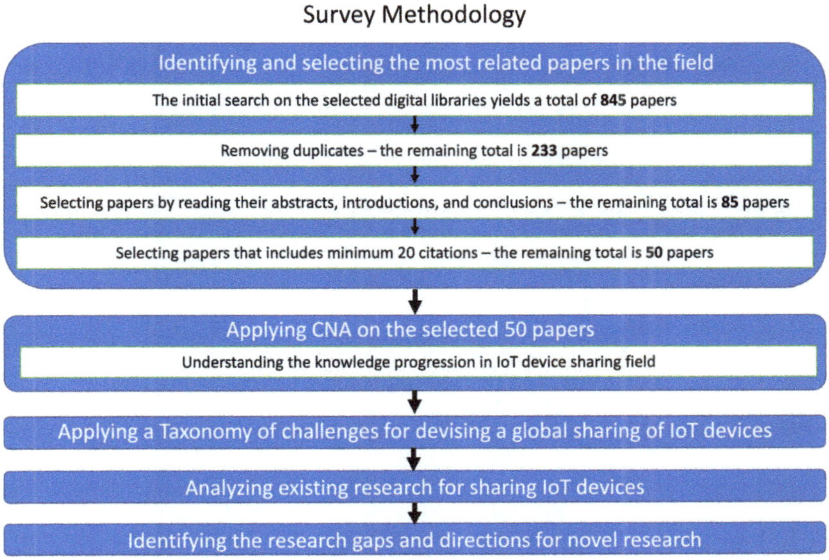


**Figure 1.** Structure of our survey methodology which is inspired by Systematic Literature Network Analysis (SLNA).

### 2.1. *Identifying and Selecting the Most Relevant Papers*

To identify the most relevant papers on sharing IoT devices, we used a combination of the terms listed in Table 1 to search the digital libraries of Google Scholar, Microsoft Academic, IEEE Xplore, Scopus, Emerald, and ScienceDirect. Please note that the search terms and expressions in Table 1 are the global IoT device sharing objectives and challenges we discussed in Section 1. Other aspects (e.g., power consumption) that are important for sharing IoT devices are out of scope of this paper. To search these digital libraries, we used the search strings in Table 1, combining multiple terms, to keep the search focused on sharing IoT devices. The "*" sign in Table 1 refers to a search for various forms of the word. The search on the selected digital libraries of the period between 2010 to 2023 yielded a total of 845 peer-reviewed publications (with duplicates, including journals, conference proceedings, and books). Peer-reviewed publications were used as they provide better quality than non-peer-reviewed ones. We selected the period between 2010 and 2023 as the modern IoT research and applications era started in mid-2010 [11]. After removing duplicates, we reduced the total number of related articles to 233. Next, we filtered this collection by reading the abstract, introduction, and conclusions of each paper. This stage reduced the number of papers to 85. We then filtered the papers based on their number of citations, as this reflects each paper's contribution to and influence on this field of study [8]. To apply a fair citation filtration of recent papers, papers published in the last two years were excluded from the citation filtration, papers published in the last five years had to have at least five citations to be included in the review, and papers published more than five years ago had to have at least 20 citations to be included in the review. This stage removed 35 papers, reducing the total number of papers to 50. We included these 50 papers in the Citation Network Analysis that is discussed next in Section 2.2.

**Table 1.** The keywords, databases, and strings for finding papers.

| Search Terms | Databases | Search Strings |
|---|---|---|
| IoT, Internet of things, Internet of everything, IoT device, Device, Sensor, IoT sensor, IoT data, IoT service, IoT product, IoT resource, IoT source, Discovery, Use, Reuse, Cost-sharing, Pay, Trustworthy, Scalable, Variety, Heterogeneity, Comprehensive | Google scholar, Microsoft academic, IEEE Xplore, Scopus, Emerald, and ScienceDirect | ("sensor*" OR "device*" OR "IoT sensor*" OR "IoT device*" OR "IoT data*" OR "IoT service*" OR "IoT resource*" OR "IoT product*" OR "IoT source*") AND ("discover*" OR "use*" OR "reuse*" " OR "cost-share*" OR "pay*" AND ("IoT" OR "internet of things" OR "internet of everything"), ("scale*" OR "scale-up", OR "trustworthy", OR "trust*", OR "variety*" OR "heterogeneity" OR "comprehensive") AND ("sensor*" OR "device*" OR "IoT sensor*" OR "IoT device*" OR "IoT data*" OR "IoT service*" OR "IoT resource*" OR "IoT product*" OR "IoT source*") AND ("discover*" OR "use*" OR "reuse*" OR "cost-share*" OR "pay*") |

*2.2. Applying CNA to Analyze Knowledge Progression*

In this section, we applied CNA to the 50 selected papers. CNA reveals knowledge progression over time [8]. To identify the citation sources for each paper in the related work collection, we used Scopus to source the citation information (e.g., title, authors, year, and references). Next, we used VOSviewer [12], which is a specialized tool for visualizing a citation network, to perform a citation analysis of the selected papers. The results of the citation analysis of these highly cited papers is shown in Figure 2.

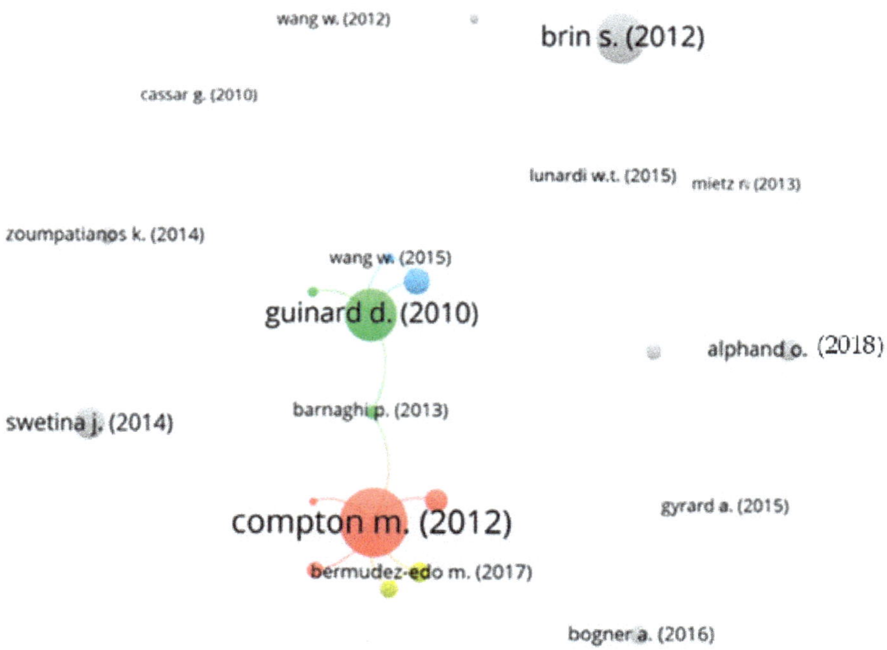


**Figure 2.** Citation analysis of the selected papers.

Note that the VOSviewer software [12] has some limitations, which prevented us from comparing all of the selected papers. These limitations include supporting a limited number of digital libraries (e.g., Google Scholar is not supported) and having limited citation formats. For those reasons, the works of Le-Phuoc et al. [13] and Harris et al. [14]

were not included in the CNA. In Figure 2, papers are represented by circles; the size of these circles represents the number of citations of each paper. The lines between the circles represent the citation links between papers. Two major clusters are connected through Barnaghi et al. [15], as can be seen in Figure 2. These clusters represent the well-established field of IoT device discovery (majority of papers) and IoT devices use (minority of papers). This cluster was started in 2010 through papers on (1) the basics of describing IoT devices to discover them, and (2) techniques to integrate and use IoT devices and their data. We can clearly see that the work produced by Compton et al. [16] is the most cited in the field of sharing IoT devices, as it helped to establish the base of semantic descriptions of IoT devices. Compton et al. [16], along with Guinard et al. [17], represent the backbone of the discovery and use of IoT devices.

Other papers (coloured grey) have no citation link with any other paper in this analysis, as can be seen in Figure 2. These papers mainly provide techniques for querying, indexing, and integrating IoT devices without following the bases established by other researchers.

The CNA analysis helped us understand that the knowledge progression in the field of sharing IoT devices focuses on the (1) semantic discovery of IoT devices, which is enabled by describing IoT devices semantically to support semantic queries, and (2) integration of IoT devices to use their data.

### 2.3. A Taxonomy of Challenges for Devising the Global Sharing of IoT Devices

In this section, we propose a taxonomy of challenges that we use to review and classify work related to sharing IoT devices in order to devise the global sharing of IoT devices. Table 2 presents the proposed taxonomy.

**Table 2.** Taxonomy of challenges of devising the global sharing of Internet of Things devices.

| Taxonomy Categories | | Challenges of Global Sharing of IoT Devices | | | |
|---|---|---|---|---|---|
| | | Scalable | IoT-Owned | Interoperable | Comprehensive |
| Discovering IoT devices | Describing IoT devices and their data | N/A | N/A | Description of heterogenous IoT devices, their data, and supporting heterogenous IoT applications | Description of all or most aspects of IoT devices, including the cost/payment of using IoT devices and integration details |
| | Registering IoT devices | Registration of the large and ever-increasing number of IoT devices | Registration of any IoT device globally without owning the descriptions of the IoT device | Registration of heterogenous IoT devices | N/A |
| | Querying IoT devices | Query of the large and ever-increasing number of IoT devices | Query IoT devices globally without any control from any entity | Query heterogenous IoT devices and their data | Query IoT devices comprehensively including cost/payment and integration capabilities |
| | Indexing IoT devices | Index of the large and ever-increasing number of IoT devices | Index IoT devices without any control from any entity | Index heterogenous IoT devices | N/A |
| Using IoT devices | Integrating IoT devices | Integration of the large and ever-increasing number of IoT devices | Integrate IoT devices globally without any control from any entity | Integrate heterogenous IoT devices and obtain their data | N/A |
| Paying IoT devices | | Payment of the large and ever-increasing number of IoT devices | Payment of IoT devices globally without any control from any entity | Dealing with heterogenous payment options to pay for using IoT devices | N/A |

This section may be divided by subheadings. It should provide a concise and precise description of the experimental results, their interpretation, as well as the experimental conclusions that can be drawn.

1. Describing IoT devices and their data: This category includes existing techniques (e.g., ontologies, data models, and schemas) that have been proposed to describe IoT devices and their data. IoT device and data descriptions are standards set of information about the software, hardware, system, and measurement capabilities, which can be used to describe sensors on IoT devices and their data (e.g., type, name, data unit, and location). Describing IoT devices is the key to discovering them and sharing them because it provides the necessary information that allows IoT applications to find the required IoT devices and IoT data. Simple descriptions of IoT devices prevent IoT applications from finding the required IoT devices and IoT data. For example, assume that a basic value name pairs technique describes IoT devices like "Type": "temperature", "location": "Australia", "name": "lab temperature IoT device". This basic description does not provide enough information to differentiate a temperature IoT device from an "ambient air temperature IoT device" or a "system/machine temperature IoT device". This prevents IoT applications from finding the required IoT devices. To help devise a global, IoT-owned, scalable, and interoperable sharing of IoT devices, a technique for describing IoT devices and their data should address the following challenges: (1) describe heterogeneous IoT devices and their data and support interoperability, and (2) provide comprehensive information to describe all possible aspects, including ownership of IoT devices to pay owners (providers) for using their IoT devices, the cost of using IoT devices to support paying for IoT devices, and the integration IoT devices detail to support IoT device use.
2. Registration of IoT Devices: This category includes existing techniques that have been proposed to manage and store the description of IoT devices to make them discoverable by IoT applications. To help devise a global, IoT-owned, scalable, and interoperable sharing of IoT devices, a technique for IoT device registration should address the following challenges: (1) scale up to the large and ever-increasing number of IoT devices, and (2) allow any device across the globe to be registered (i.e., IoT-owned) without owning the description of the IoT device.
3. Querying of IoT Devices: To discover IoT devices, IoT applications are required to query IoT devices and retrieve their descriptions based on IoT application needs. To help devise a global, IoT-owned, scalable, and interoperable sharing of IoT devices, a technique for IoT device query should address the following challenges: (1) scalability to query the large and ever-increasing number of IoT devices, (2) ability to query heterogeneous IoT devices and their data to support interoperability, (3) IoT-ownership to query IoT devices globally without control from any entity, and (4) comprehensiveness to query all or most possible aspects for finding IoT devices that match all or most IoT application needs. However, querying a large number of IoT devices and data descriptions (potentially billions) will increase the time required to find IoT devices and may impact the scalability of the global sharing of IoT devices. Therefore, there is a need for IoT device indexing that indexes IoT devices and data descriptions to support fast querying and, therefore, support the scalability of the global sharing of IoT devices. To help devise a global, IoT-owned, scalable, and interoperable sharing of IoT devices, a technique for IoT device indexing should address the following challenges: (1) be scaled to index the large and ever-increasing number of IoT devices, (2) be able to index heterogeneous IoT devices and their data to support interoperability, and (3) be IoT-owned to index IoT devices globally without control from any entity.
4. Integrating IoT Devices: To use IoT devices, there is a need for a technique (e.g., protocol and middleware) that can integrate IoT devices with IoT applications by establishing a communication channel between them, allowing IoT applications to use IoT data generated by IoT devices. To help devise a global, IoT-owned, scalable, and interoperable sharing of IoT devices, a technique for IoT device integration needs

to address the following challenges: (1) support the integration of heterogeneous IoT devices (i.e., use different software, vendors, and integration protocols) and their data to support interoperability, (2) scale up to integrate a large number of heterogeneous IoT devices and their data, and (3) be IoT-owned to integrate IoT devices and their data without control from any entity.

5. Paying IoT Devices: This category includes existing techniques that have been proposed to enable a pay-as-you-go model to pay providers for sharing their IoT devices and data. To enable paying IoT devices, there is a need for a pay-as-you-go model that allows IoT applications to (1) pay for IoT device providers as long as they are using their IoT devices and data, and (2) be compensated in case the IoT data are not fully received due to any issue with the IoT device, such as low battery. To help devise a global, IoT-owned, scalable, and interoperable sharing of IoT devices, a technique for pay-as-you-go IoT device payment needs to (1) be scalable to manage payment and compensation transactions for a large and ever-increasing number of IoT devices, (2) support heterogeneous payment options such as cryptocurrencies, PayPal, and bank cards, and (3) support IoT-owned payment, i.e., allowing any IoT application to pay any IoT device globally with any payment option without any control from any entity.

## 3. Related Work for Sharing IoT Devices

This section provides a review of the existing techniques that contribute to the development of the global sharing of IoT devices. The existing techniques are classified based on our taxonomy provided in Section 3 and are reviewed based on their ability to address the challenges that are represented in the taxonomy.

### 3.1. Related Research for Describing IoT Devices and Their Data

Describing IoT devices and their data can be achieved via using techniques that use an ontology, a data model, or a schema. Most relevant papers use ontologies to describe IoT devices and their data. The most commonly used ontologies in the related work are Semantic Sensor Network (SSN) [16] and Sensor Observation Sampling Actuator (SOSA) [18], which are World Wide Web Consortium (W3C) recommendations that were developed by the W3C semantic sensor network incubator group (SSN-XG) [16,19,20]. SSN/SOSA are semantic-based ontologies that present a standard machine-readable semi-autonomous or even autonomous system to find, assemble, process, analyze, and proceed on IoT devices. The SSN ontology can be seen from four main perspectives, which are (1) the sensor perspective, which focuses on what senses, how it senses, and what is sensed; (2) the observation perspective, which focuses on observation data and related IoT device descriptions; (3) the system perspective, which focuses on deployments and systems that IoT devices belong to; and (4) the feature and property perspective, which focuses on what senses a particular property or what observations have been made about a property. SSN/SOSA can describe heterogeneous IoT devices and their data. Many researchers have used SSN/SOSA ontologies (modified or as is) in their work, such as Le-Phuoc et al. [13], Perera et al. [21], Kamilaris et al. [22], and Wang et al. [23].

The Machine-to-Machine Measurements (M3) ontology [24] has been proposed to solve some of the limitations of the existing standards (e.g., SSN [16], oneM2M [25], and ETSI M2M [26]). M3 is a semantic-based ontology, and it can describe heterogeneous IoT devices and their data. Also, it provides a unified description for sensors, data, and measurements, but it is limited to measurement name, device name, and unit of measurement. The authors have provided a link for the complete unified description, but it is not reachable. Other researchers who have used the M3 ontology for their work, such as the authors of [27,28], have modified the M3 ontology to fit their objectives for providing a unified semantic search engine for smart cities and the IoT. MyOntoSens [29] provides a semantic open data model for IoT devices and their data. The data model is designed to deal with the heterogeneity of IoT devices and their data. Although this data model was proposed mainly for the Wireless

Sensor Networks (WSNs) domain, the model was designed based on other ontologies such as SSN [16], which makes it suitable for the IoT as well.

IoT-Lite [20] provides an instantiation of SSN [16] that allows for the interoperability and discovery of IoT devices by using lightweight semantics. IoT-Lite focuses on three key concepts, which are objects, devices, and services. It provides fast processing times for semantic queries by reducing the complexity of concepts and relationships. Barnaghi et al. [15] have proposed data modelling and annotation to describe the observation and measurement of streaming data. The data stream schema includes value, time, location, type, and unit of IoT data. Although this semantic model describes data observation and measurements, it does not cover describing IoT devices. Bharti et al. [30] have proposed a semantic resource (IoT device) ontology that describes a physical device and its characteristics. This ontology has four main concepts, which are (1) physical characteristics, including IDs, energy, and processing time; (2) working features, including accuracy, latency, and measurement; (3) deployment, including geographical location; and (4) interests, including device property. The Open Geospatial Consortium's (OGCs) Sensor Web Enablement (SWE) group have proposed the Sensor Model Language (SensorML) for describing and discovering IoT devices [31]. The ontologies/models proposed in [15,20,30,31] cannot deal with heterogeneous devices and their data.

Bogner et al. [32] use their own schema to describe internet-connected objects (i.e., cars, bikes, and scooters) via IoT devices. This schema describes the cost of sharing these objects via IoT devices. However, the description focuses on the object, not on IoT devices.

Based on this review of the related work describing IoT devices and their data, only the ontologies/models proposed in [16,18,24,29] have the ability to describe heterogeneous IoT devices and their data. However, no ontologies/models in the related work provide the ability to support interoperability, i.e., provide a description of IoT devices and their data that allows IoT applications to understand heterogeneous IoT devices and their data automatically. To illustrate, each IoT device sends data in different sequences and formats. For example, consider two weather station IoT devices sending data to an IoT application: the first IoT device sends data in the following sequence and format (sensor 1 sends a value and its unit; sensor 2 sends a value only; and sensor 3 sends a value, its unit, and a time stamp of captured value); the second IoT device sends data in the following sequence and format (sensor 2 sends a value only; sensor 3 sends a value and the duty-cycle of capturing values; and sensor 1 sends a value and a unit). In this case, the IoT application does not understand the received data from these weather station IoT devices. These techniques need a standard or unified measurement that describes the syntax and format of the IoT data they send. While Barnaghi et al. [15] and Gyrard et al. [24] have proposed a schema towards unified sensor measurements, their schema does not consider the structure of an IoT device that includes several sensors, which requires other attributes to differentiate sensors in one IoT device. Also, none of the related work provides a comprehensive description of IoT devices and their data as can be seen in Table 3. For example, none support describing (1) the cost of using IoT devices, (2) the details of integrating IoT devices, and (3) the ownership of IoT devices.

**Table 3.** The related research that addresses describing Internet of Things devices and data challenges.

| Interoperable | Comprehensive | Fully Support Descriptions of IoT Devices and Their Data |
|---|---|---|
| Refs. [16,18,24,29] support heterogenous IoT devices, but show limited support for interoperability | None | None of the related work fully supports describing IoT devices and their data |

### 3.2. *Related Research for IoT Devices Registration*

The registration of IoT devices involves managing and storing the descriptions of IoT devices.

Perera et al. [21] have proposed registering IoT devices in a MySQL database [33]. The authors have claimed that this database is limited to joining 61 tables only, which later on impacts querying IoT device descriptions as full IoT device description properties cannot be retrieved. To solve this issue, the authors have suggested using different storage methods or applying several queries at the same time. The authors have focused on evaluating the scalability of querying IoT devices registered in the MySQL database without consideration of evaluating the scalability of storing IoT device descriptions in the database. Lunardi et al. [34] have managed and stored IoT device descriptions using a PostgreSQL database [35]. The authors have evaluated their work by registering up to 1 million devices (i.e., IoT devices) without any significant issues with the database.

Barnaghi et al. [15] have proposed using distributed repositories that support managing and storing semantic IoT device descriptions. The authors have used a clustering algorithm to distribute the IoT device descriptions among diverse clusters in order to enable faster queries for retrieving the IoT device descriptions. However, their evaluation does not cover the scalability of managing and storing data streams. Wang et al. [36] have proposed managing and storing IoT device descriptions using semantic-based distributed gateways in order to provide scalable queries and support a range of different queries. The authors have evaluated managing and storing up to 10,000 IoT device descriptions. Paganelli and Parlanti [37] have proposed distributed repositories to manage and store IoT device descriptions. However, their evaluation does not cover the scalability of registering IoT devices.

GeoCENS [38] have used a hybrid peer-to-peer network for registering IoT devices and providing IoT device discovery. Their network consists of two major types of sensor web servers: (1) powerful servers that are maintained by large organizations, such as NASA, which are less likely to disappear or provide degraded services (i.e., very low volatility), and (2) less powerful servers that are managed by people or small organizations that are more likely to join and leave the network (i.e., high volatility). Their work has been used by different projects, which means that the GeoCENS network has been successfully deployed, but their paper has no evaluation of the scalability of managing and storing large numbers of IoT device descriptions. Although the storage of IoT device descriptions is considered an IoT-owned storage as no one specifically owns or controls it, the registration technique is generally controlled by the GeoCENS solution, which is not IoT-owned.

The authors of IOT Chain [39] proposed using a peer-to-peer blockchain to register IoT devices. It has been proposed as a decentralized consensus mechanism to manage IoT device descriptions, and it uses a private Ethereum blockchain to store IoT device descriptions. From an evaluation of the solution, the authors have estimated that the blockchain-based IoT could be influential and could produce a significant revolution. Wen et al. [40] have proposed a Blockchain-based Supply Chain System (BSCS) to provide a traceable and trustworthy mechanism to register IoT devices and store their data. Rahman et al. [41] have proposed using a permissioned Ethereum blockchain to provide a secure and privacy-assured data storage method for medical data. This system aims to (1) store the patients' information along with data provided by related medical IoT devices, such as human body temperature; and (2) share the data with related stakeholders via a permissioned smart contract for access control. IOT Chain, BSCS, and Rahman et al. [41] use IoT-owned registration techniques for managing and storing IoT device descriptions and data. However, IOT Chain and BSCS have no evaluation of the scalability of managing and storing a large number of IoT devices' metadata. Rahman et al. [41] have evaluated a blockchain with up to 5000 transactions and two mining nodes. However, this evaluation does not show the scalability of the system when dealing with a large number of IoT devices and transactions. Also, data storage is expected to grow exponentially as it stores the data of medical IoT devices and not just their descriptions and the patients' information. Bogner et al. [32] have proposed using an Ethereum-based decentralized application (DAPP) for managing and storing descriptions of the objects (e.g., cars, bikes, and apartments, which are like IoT devices in our perspective) in an IoT-owned decentralized ledger. The storage of IoT device descriptions is considered to be IoT-owned, but managing IoT device descriptions is not. Also, the Ethereum blockchain is not scalable enough to register a large and ever-increasing number of IoT devices.

HCL-BaFoG [42] is a fog computing architecture that collects IoT devices' data and stores them. HCL-BaFoG utilizes permissioned blockchain functionalities to provide secure IoT data sharing for clients (IoT applications). Although HCL-BaFoG stores IoT data and shares it, it does not have IoT device registration functionality. Given the permissioned architecture of HCL-BaFoG, it cannot be scaled up for IoT.

Based on this review of related work, most do not support a scalable and heterogenous registration of IoT devices, while Wang et al. [36] and Lunardi et al. [34] provide evaluations that show the registering of up to 1,000,000 IoT devices as can be seen in Table 4. With the increasing number of IoT deployments, there is a significant scope to expand these evaluations to billions of IoT devices. Also, the related work provides very limited support for IoT-owned registration. IOT Chain [39] and Wen et al. [40] have proposed a decentralized consensus mechanism that manages IoT device descriptions without control from any specific entity. However, it does not support storing the descriptions of heterogeneous IoT devices and it is not scalable for registering a large number of IoT devices.

**Table 4.** The related research that addresses Internet of Things device registration challenges.

| Scalable | IoT-Owned | Interoperable | Full Support of Registering IoT Devices |
|---|---|---|---|
| Refs. [34,36] evaluate registration of up to 1,000,000 IoT devices | [39,40] | None | None of the related work fully supports registering IoT devices |

### 3.3. Related Research for Querying IoT Devices

Querying IoT devices involves using attributes such as Data Type: Temperature and Location: Melbourne City to find IoT devices (more specifically, finding IoT devices and data descriptions) that match an IoT application's needs. An example of such a query can be detecting the battery level of electric cars located less than 5 km from the Melbourne city center; the IoT device query technique should find the IoT device description that has a type: battery level, system: electric car, and geospatial: circular area with a radius of 5 km from Melbourne city center.

Google search [43] uses a scalable hypertextual search technique that can provide high-precision results. However, this technique is not designed for querying IoT device descriptions as they are much more complicated than websites.

Simurgh [44] have used a two-phase query technique based on the syntax of IoT device descriptions. The first phase is responsible for finding the matching IoT devices via a syntax-based query. The second phase is responsible for finding the required Application Program Interface (API) for the selected IoT devices in order to integrate them. Lunardi et al. [34] have proposed an IoT device query technique that can check synonyms and find other words with similar meanings and stop words that are not relevant to the query. Also, this technique provides a geographical interface for IoT applications to help them select the best match for IoT devices. GeoCENS search [38] have proposed using a hybrid peer-to-peer network (See Section 3.2 to allow for the distributed query of IoT devices. This technique allows finding even non-GeoCENS OGC web services (OWSs) (i.e., IoT devices) that are not registered in the hybrid peer-to-peer network. GeoCENS utilizes SOS [45] and SensorML [45] to enable the querying of IoT devices. This technique is scalable for a large number of stored IoT devices. The IoT device querying proposed by [34,38,44] uses matching keywords techniques, which lack flexibility for answering IoT device queries [17].

Simple Protocol and RDF Query Language (SPARQL) [14] is an RDF query language that has the ability to manipulate and retrieve RDF data. SPARQL supports a variety of queries and can return accurate results that match the IoT application's needs. Also, SPARQL is scalable for querying a large number of registered IoT devices. There are several papers that have proposed using SPARQL as a query technique, such as Perera et al. [21], Wang et al. [34], Gyrard and Serrano [27], Barnaghi et al. [15], Le-Phuoc et al. [13], Mietz et al. [41], and Kamilaris et al. [22].

Based on the above review, most techniques in the related work have the ability to query heterogeneous IoT devices and their data. However, none of the related work supports comprehensive queries, mainly due to the lack of ontology describing IoT devices as IoT device query techniques cannot search for information that has not been described. The IoT device query techniques proposed in [34,38,43,44] are not IoT-owned as they are centralized and/or controlled by specific individuals or organizations. The current implementations of SPARQL [14] are on centrally owned infrastructure. Also, according to the review of related work, the IoT device query techniques proposed in [34,38,44] did not provide evidence for their scalability. Lunardi et al. [34] provide an evaluation for querying up to 1,000,000 IoT devices. Wang et al. [36] and Perera et al. [21] have experimented for up to 500,000 IoT devices to evaluate the querying of IoT devices using SPARQL as can be seen in Table 5. These evaluations were based on centralized databases and not on an IoT-owned infrastructure.

**Table 5.** A summary of the review of related work for querying Internet of Things devices.

| Scalable | IoT-Owned | Interoperable | Comprehensive | Full Support of Querying IoT Devices |
|---|---|---|---|---|
| Ref. [34] evaluates querying up to 1,000,000 IoT devices<br>Refs. [21,36] evaluate querying up to 500,000 IoT devices | None | None | None | None of the related work fully supports querying IoT devices |

As we explained earlier in Section 3, there is a need for indexing IoT devices and data descriptions to support fast queries and, therefore, support the scalability of IoT device query techniques and the global sharing of IoT devices.

Wang et al. [36] have proposed using a geospatial indexing technique to reduce the search space for a faster IoT device query. They aimed to divide IoT devices based on geographic area, where the IoT device discovery was divided into distributed gateways and each gateway stored the description of IoT devices that are connected to it. Linked Sensor Middleware (LSM) [13] uses a spatial indexer for indexing IoT devices. LMS provides a basic indexing technique based on the type and location of IoT devices [21]. These techniques are limited to spatial indexing and cannot index IoT devices based on other properties (e.g., IoT device type or data units). The experiment conducted by Wang et al. [36] was limited to one small geographical area and a maximum of 1000 sensor services.

Perera et al. [21] have proposed using a Comparative-Priority-based Weighted Index (CPWI) technique for indexing IoT devices based on an IoT application's needs. This technique indexes IoT devices during query time by calculating the similarity between the IoT application's needs and the stored IoT device and data descriptions via a priority-based, weighted Euclidean distance in a multi-dimensional space mechanism. They have also applied heuristic filtering and a filtering-based relational expression to minimize the total amount of IoT devices and data descriptions that are required to be processed. This technique is able to index heterogeneous IoT devices. Perera et al. [21] have conducted an experiment to index up to 1,000,000 IoT devices with an indexing time that did not exceed 1000 ms.

Cassar et al. [46] have proposed using Probabilistic Latent Semantic Analysis (PLSA) and Latent Dirichlet Allocation (LDA) for indexing IoT devices based on a comparison between the IoT application's query and the IoT device and data descriptions. There was no experiment to show the scalability of this technique.

Liang and Huang [38] have used a space-filling technique to index the sensor data (i.e., IoT data). Also, Tree-based (e.g., LOST-Tree) techniques were used to index IoT data by using spatial information. Barnaghi et al. [15] have used geospatial and temporal information to index IoT data. They have used Geohash tagging to provide geospatial information. Singular Value Decomposition (SVD) has also been used to reduce the dimensionality of Geohash vector data before applying a k-means clustering algorithm to allocate IoT data among repositories. Zoumpatianos et al. [47] have claimed that the main bottleneck for IoT device discovery is the indexing technique. Therefore, they have proposed shifting part of the time that is required for indexing IoT data from the IoT device registration to the IoT device query to provide faster IoT device discovery. The authors did this by indexing the parts of IoT data that are related to the IoT application's query during query time. The authors introduced a mechanism that produces a tree of iSAX [48], which represents each of the data series (i.e., IoT data); the real data remains in the raw files and is loaded once a matched query appears [47]. The techniques proposed by Zoumpatianos et al. [47], Barnaghi et al. [15], and Liang and Huang [38] are designed to index data streams, which are not fully applicable to indexing IoT devices. Further, their evaluation does not cover large amounts of data (i.e., a large number of IoT devices and data descriptions) to show the scalability of these techniques.

Based on the review of the related work, most do not support the scalable and heterogenous indexing of IoT devices, while Perera et al. [21] provide an evaluation of indexing up to 1,000,000 IoT devices. With the increasing number of IoT deployments, there is a significant scope to expand these evaluations into billions of IoT devices. Also, none of the related work provides IoT-owned indexing techniques for IoT devices, as can be seen in Table 6.

**Table 6.** The related research that addresses indexing Internet of Things devices challenges.

| Scalable | IoT-Owned | Interoperable | Full Support of Indexing IoT Devices |
|---|---|---|---|
| Ref. [21] evaluates indexing up to 1,000,000 IoT devices | None | None | None of the related work fully supports indexing IoT devices |

### 3.4. Related Research for Integrating IoT Devices

A common technique for integrating IoT devices is using a wrapper that can virtually integrate any IoT device [19]. A wrapper is a piece of code developed individually to integrate each IoT device. These wrappers are usually developed by the developers of the system/solution that offers IoT device integration, or by the provider of the IoT device. Several solutions/systems have used wrappers to integrate IoT devices, including LSM [13], OpenIoT [49], and (GSN) middleware [50].

Application Programming Interfaces (APIs), which are software that allow for IoT applications to interact with IoT devices [51], can provide an interface for deployed IoT devices to be integrated with IoT applications. One API can provide an interface for integrating several IoT devices. Several solutions/systems have used APIs to integrate IoT devices, including Simurgh [44] and Agri-IoT [22]. Although using a wrapper or API technique can virtually support the integration of any IoT device, it is not scalable due to the need to develop a wrapper or API for each IoT device or group of IoT devices. Also, IoT-ownership depends on the system/solution that uses these wrappers or APIs.

Other IoT device integration techniques are providing information like Uniform Resource Locators (URLs) for IoT applications and are expecting IoT applications to use their own IoT platform or middleware to integrate IoT devices. Paganelli and Parlanti [37] have proposed providing a Universal Resource Identifier (URI) for the repository that has the required information to integrate IoT devices. Wang et al. [36] have provided unique identifications for IoT devices, such as endpoints that allow for IoT applications to integrate with IoT devices. The scalability, IoT-ownership, and ability to support heterogeneous IoT devices depend on the IoT platform and/or the middleware used by IoT applications.

Simurgh [44] uses APIs and low-level programming libraries to integrate IoT devices and obtain their data. These APIs are developed by Simurgh developers, volunteer developers, or IoT device providers. Simurgh contains an API access management layer that provides the necessary information to manage the integration of IoT devices.

Perera et al. [52] have proposed a Context-Aware Sensor Configuration Model (CASCoM) to provide automation in the integration of IoT devices for IoT middleware, such as GSN [50]. CASCoM has the ability to integrate IoT devices without the need for a wrapper or API, and it is easy to use for non-IT experts. However, CASCoM needs all the information related to IoT devices and data processing components to be stored in a repository. Perera et al. [52] have conducted an experiment to integrate up to 10,000 IoT devices. The experiment shows that CASCoM can be up to 250 times faster (respectively) in comparison to some existing techniques, such as GSN.

Madureira et al. [53] have proposed the Internet of Things Protocol (IoTP), which supports IoT data aggregation from the network layer among heterogeneous IoT devices to improve interoperability. IoTP does not rely on routing protocols; thus, it is a more generic protocol and supports interoperability. IoTP requires another layer to manage the aggregation of IoT data. The authors have ran an experiment with a total of 50 IoT devices to evaluate IoTP by measuring several metrics including transmitted data, network efficiency, and average delay.

Based on the review of related work, all of the techniques except for [53] do not provide support for interoperable integration due to either the need for an API or wrapper for almost each different IoT device. The technique proposed in [53] partially supports interoperable integration by allowing data aggregation for heterogeneous IoT devices. However, it requires an extra layer to manage the integrated data. The related work lacks an application layer protocol that fully supports the integration of any IoT device with any IoT application. Also, the related work provides very limited support for the scalable integration of IoT devices. Perera et al. [52] have used up to 10,000 IoT devices, and Madureira et al. [53] have used up to 50 IoT devices to evaluate their integration techniques as can be seen in Table 7. Further, none of the related work provides any support for the IoT-owned integration of IoT devices.

**Table 7.** The related research that addresses integrating Internet of Things devices challenges.

| Scalable | IoT-Owned | Interoperable | Fully Support Integrating of IoT Devices |
|---|---|---|---|
| Refs. [36,52] uses up to 10,000 IoT devices to evaluate their integration techniques Ref. [53] uses up to 50 IoT devices to evaluate the integration technique | None | None | None of the related works fully support integrating IoT devices |

### 3.5. Related Research for Paying IoT Devices

IoT device payment techniques involve supporting a pay-as-you-go model for sharing IoT devices in a method similar to cloud computing [4]. Such a model allows IoT applications to pay IoT device providers as long as they use their IoT devices.

Ether cryptocurrency [54] is proposed by several solutions, such as [32] as a payment technique to pay for using IoT devices. However, based on the performance of the Ethereum blockchain [54], it is not scalable for paying for a large number of IoT devices. Also, it does not support a pay-as-you-go-model.

There are other techniques that can be used to provide payment for IoT devices, such as PayPal [55] and cryptocurrencies like Bitcoin [56] and MIOTA [57]. However, there is no paper that proposes using PayPal [55] or Bitcoin [56] to support paying IoT devices. Also, PayPal [55], Bitcoin [56], and MIOTA [57] require development to support a pay-as-you-go model for using IoT devices.

Existing IoT device sharing solutions/marketplaces do not support a pay-as-you-go model for using IoT devices.

Tzianos et al. [58] provide preliminary work on supporting a payment mechanism for IoT device sharing by using the IOTA wallet [57] to enable IoT clients to purchase data packets. However, Tzianos et al. [58] did not provide any model or mechanisms to process payment, calculate required payment, nor support pay-as-you-go payment for IoT devices.

Özyilmaz et al. [59] support IoT device sharing via a blockchain-based marketplace. Özyilmaz et al. [59] have proposed using the Raiden network [60] off-chain payment channel to support micro-payments for IoT device providers. Rahman et al. [61] have proposed a blockchain-based marketplace for sharing IoT devices. Rahman et al. [61] use EOS permissioned blockchain [62] to support paying IoT devices for their data. However, Özyilmaz et al. [59] and Rahman et al. [61] did not provide information for a pay-as-you-go model. Also, they did not provide a mechanism to calculate the required payment.

All of the discussed solutions, including [54,56–59,61], do not fully support paying IoT devices through a pay-as-you-go model as can be seen in Table 8.

**Table 8.** A summary of the review of related work for paying Internet of Things devices.

| Scalable | IoT-Owned | Interoperable | Fully Support Paying for IoT Devices |
|---|---|---|---|
| [57,58] | [54,56–59,61] | None | None of the related work fully supports paying for IoT devices nor supports pay-as-you-go |

## 4. Gap Analysis and Directions for Novel Research

In the previous section, we reviewed the related work that can contribute to the development of IoT device sharing and enable its objectives, which are IoT device discovery, use, and pay. This review has helped us identify existing gaps in the related work IoT device sharing. In this section, we analyze the related work gaps that have hindered the development of global IoT device sharing. Also, we provide directions for novel research that help researchers advance the development of global IoT device sharing. From reviewing the related work in Section 4 and Tables 3–8, we can see that most of the existing techniques do not address the challenges of achieving global IoT device sharing. From the reviewed techniques that are based on using, or on part of blockchain technology such as [30,37–40,55–57,59], we can say that blockchain technology is a possible approach to establishing global and IoT-owned IoT device sharing. Blockchains are a decentralized ledger that can allow for IoT applications to query IoT devices and data descriptions along with other needed information (e.g., endpoint) stored in them. Also, it can run layers or modules on top of it in a decentralized structure to support IoT device discovery, using and paying functionalities for IoT applications. Blockchain is not owned or controlled by individuals; it runs on globally distributed computers (i.e., nodes) and is controlled by a consensus. It can thus support the IoT-owned sharing of IoT devices. However, in our survey, there is no blockchain-based technique for indexing IoT devices. Also, the analyzed blockchain techniques did not provide proof of scalability for sharing a large number of IoT devices. Based on that, in this section, we describe the identified gaps that hindered the development of the global sharing of IoT devices and our vision to address these gaps. Figure 3 shows our envisioned architecture for IoT device sharing.

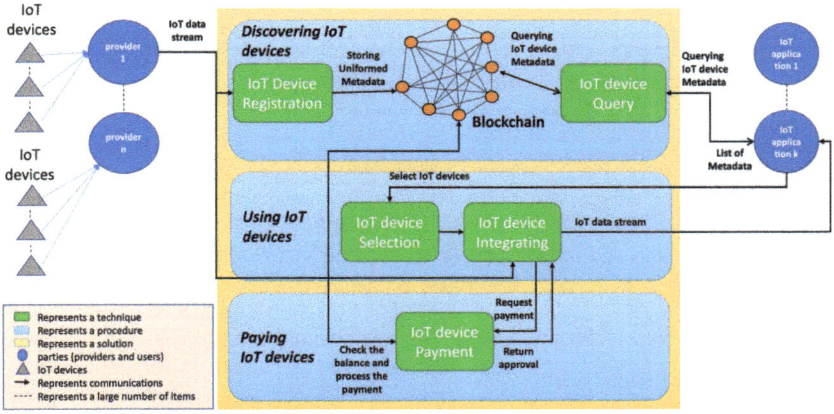


**Figure 3.** Our envisioned architecture for sharing IoT devices.

### 4.1. Describing of IoT Devices and Their Data

Existing techniques for describing IoT devices and data require further development to support the global sharing of IoT devices.

1. Semantic technology is highly relevant to describing IoT devices and their data as it provides the foundations for the development of sharing IoT device solutions. Semantic technology is proposed by the IoT community for providing IoT devices and data descriptions to enable interoperability [63]. The Semantic Sensor Network (SSN) ontology is the leading semantic description of IoT devices and their data and provides the basis of most of the well-known ontologies in this field. However, SSN is not comprehensive as it does not cover the ownership of IoT devices, location, and the measurement format of IoT data. Also, there is other information that is not covered by SSN, and this is required to support the integration and payment of IoT devices (e.g., payment ID for IoT devices, cost of using IoT device, Uniform Resource Identifier (URI), integration protocol, and special token for fetching IoT data from IoT devices). Thus, there is a need for a technique which describes IoT devices and their data and that (1) supports heterogeneous IoT devices and their data as well as heterogeneous IoT applications, and (2) provides a comprehensive description that facilitates all or most of the possible aspects, such as the description of payment and the integration of IoT devices. Early work on this area is published in [64]. The authors have proposed an SSN-based ontology that supports semantic queries, integration, and the payment of IoT devices. However, further development is required to fully support a comprehensive description and heterogeneous IoT application.
2. The problem with semantic description is that it requires both significant effort and considerable expertise in RDF-based ontology definitions and use [65]. We envision a machine learning-based description technique that is able to analyze the data stream of IoT devices to generate automatic semantic descriptions for IoT devices and their data. Early work in this area has been published in [66]. The author has proposed using k-fold cross-validation (CV) to train a novel metadata-assisted cascading ensemble classification framework (MACE) for annotating heterogeneous IoT data autonomously. However, this model does not generate the metadata needed for IoT devices.
3. Storing semantic information in a Blockchain requires a specialized blockchain that has an RDF triple store for the efficient query of IoT devices and data descriptions. Early work in this area is published in [1,2]. The authors have implemented an RDF triple store inside each of the distributed nodes of the blockchain in order to store semantic descriptions of IoT devices and data as triples. They have called this a

semantic-based blockchain. However, further development is required to achieve an efficient triple store for a scalable semantic-based blockchain.

The techniques for describing IoT devices, automatically generating semantic metadata, and storing metadata in a blockchain can be called IoT device registration, as can be seen in Figure 1.

### 4.2. Querying IoT Devices

Existing techniques for querying and indexing IoT devices require further development to support the global sharing of IoT devices.

1. Querying IoT device techniques lacks scalability due to the complexity of semantic data and the large number of IoT devices and data descriptions. Therefore, we envision a distributed IoT device query technique that is deployed on each blockchain node. Each query can be processed by multiple nodes that each query part of the RDF-store concurrently to achieve a fast and scalable query technique. Further, deploying IoT device query techniques on the blockchain can make it IoT-owned as no entity owns or controls the query technique.
2. Semantic query languages, which are required to support semantic description such as SPARQL, are complex and require experts to write queries. Also, they are not comprehensive enough to query all or most of the possible aspects of IoT devices, mainly for as they lack a comprehensive ontology. Therefore, we envision the development of an easy-to-use and comprehensive IoT device query language that is able to provide the constructs needed to query IoT devices and support all or most IoT application needs.
3. Existing techniques for indexing IoT devices are not IoT-owned and lack scalability. We envision developing a novel blockchain-based indexing technique that provides fast access to semantically clustered descriptions of IoT devices and their data stored in a blockchain. Early work on this area is published in [67]. The authors have proposed providing (1) a lookup table index for semantic information among blocks and (2) a block-level recursive model index for blocks to improve the query efficiency. However, this work requires development to address the semantic description of IoT devices.

### 4.3. Integrating IoT Devices

Existing techniques for integrating IoT devices require further development to support the global sharing of IoT devices.

1. The integration of IoT device techniques lacks scalability because (1) they either rely on developing APIs or wrappers for almost each IoT device manually or require a set of information for each IoT device to create the wrapper automatically; and (2) there are no standards for developing these wrappers and APIs, as every provider has its own developed one. Therefore, we envision a machine learning-based integration technique that cooperates with the machine learning-based description technique discussed earlier in Section 4.1 to provide an automatic integration of IoT devices.
2. The integration of IoT device techniques lacks interoperability, which describes the ability for heterogeneous IoT applications to integrate the same IoT device and for one IoT application to integrate heterogeneous IoT devices. Therefore, we envision an application layer protocol that can cooperate with (1) existing Internet and IoT protocols, such as MQTT [68], CoAP [69], and HTTP [70]; (2) the machine learning-based description technique discussed earlier in Section 4.1; and (3) the machine learning-based integration technique discussed earlier in this section in order to (a) integrate heterogeneous IoT devices and collect their data, and (b) interpret collected data to formats and structures understandable by heterogeneous IoT applications.

### 4.4. Paying IoT Devices

Very few techniques in the related work support paying for shared IoT devices, and none of them support heterogeneous payments or a pay-as-you-go model. To support paying for globally shared IoT devices, we need to achieve the following: (1) identify the cost of reusing IoT devices per unit of time and/or per unit of IoT data (this part can be performed in IoT device and data descriptions as explained above in the first point); (2) provide a unique payment ID for IoT devices; (3) allow for IoT applications to query IoT devices based on their cost; (4) allow for IoT applications to pay for using IoT devices; (5) compensate IoT applications in case IoT data are not fully collected; and (6) support IoT-owned and heterogenous payments, such as cryptocurrencies, bankcards, and PayPal. Early work in this area is published in [64]. The authors have proposed an initial pay-as-you-go model that achieves some of the above requirements. However, the authors have not evaluated the model. Also, further development is required to (a) support the compensation of IoT applications in case IoT data are not fully shared, and (b) support heterogeneous payment.

### 4.5. Scalability

Using a blockchain-based structure for the global sharing of IoT devices supports IoT-ownership, but it may not be scalable to managing the data of a large number of IoT devices and IoT applications because it requires some time and effort to generate blocks (i.e., contain data). Therefore, there is a need for the following:

1. Limiting the size of information stored in the blockchain to ensure scalability for dealing with a large number of IoT devices. The authors in [71] have proposed a self-managed marketplace for sharing IoT devices using a blockchain. The authors limited the information stored in the blockchain by preventing IoT devices from storing their data in the blockchain and by only storing the IoT devices and data descriptions that are required for discovery, use, and pay. However, there is a need to reduce the size of this information by compressing it or storing part of it in off-chain storage.
2. Developing a new consensus mechanism for blockchain that ensures (1) scalability, (2) suitability to be run by IoT devices, and (3) the majority of nodes that generate blocks belong to the Internet of Things (e.g., IoT devices and gateways).

## 5. Conclusions

Global sharing of IoT devices is emerging as an important area, with a plethora of IoT application developments and deployments. The true value of IoT will be realized when it can be operated similarly to the internet, i.e., IoT device providers can share their IoT devices with consumers developing IoT applications and be rewarded for doing the same. In this paper, we have conducted an SLNA-based review of the related work regarding advances in techniques that support the development of the global sharing of IoT devices. We have proposed a taxonomy of challenges that helps to review and classify techniques for IoT device discovery, use, and pay. By reviewing the related work in IoT device sharing, we have found that existing techniques that enable discovering, using, and paying for IoT devices have come a long way but are still in the infancy of being able to support the global sharing of IoT devices. Existing techniques do not sufficiently cater to the challenges imposed by the IoT, such as IoT-ownership, scalability, and the heterogeneity of IoT devices and the data they generate. Our analysis leads to the identification of directions for novel research, including the development of a scalable and semantic blockchain-based marketplace for enabling automated data sharing from third-party IoT devices to IoT applications.

**Author Contributions:** In this paper, authors are contributed as follows. Conceptualization, A.D., D.G., P.P.J. and A.N.; methodology, A.D., D.G. and P.P.J.; writing—original draft preparation, A.D.; writing—review and editing, A.D., D.G., P.P.J. and A.N.; visualization, A.D. All authors have read and agreed to the published version of the manuscript.

**Funding:** This research was supported by the Australian Research Council (ARC) No. DP220101420.

**Conflicts of Interest:** The authors declare no conflict of interest.

## References

1. Dawod, A.; Georgakopoulos, D.; Jayaraman, P.P.; Nirmalathas, A. Advancements towards Global IoT Device Discovery and Integration. In Proceedings of the 2019 IEEE International Congress on Internet of Things (ICIOT), Milan, Italy, 8–13 July 2019; pp. 147–155.
2. Dawod, A.; Georgakopoulos, D.; Jayaraman, P.P.; Nirmalathas, A. An IoT-owned service for global IoT device discovery, integration and (Re) use. In Proceedings of the 2020 IEEE International Conference on Services Computing (SCC), Beijing, China, 18–24 October 2020; pp. 312–320.
3. Chandler, S. How the Internet of Things Will Help Fight Climate Change. AI. Available online: https://www.forbes.com/ai/?sh=5ff582a07052 (accessed on 1 December 2023).
4. Armbrust, M.; Fox, A.; Griffith, R.; Joseph, A.D.; Katz, R.; Konwinski, A.; Lee, G.; Patterson, D.; Rabkin, A.; Stoica, I.; et al. A view of cloud computing. *Commun. ACM* **2010**, *53*, 50–58. [CrossRef]
5. Open Systems. Gale Encyclopedia of E-Commerce. Available online: https://www.encyclopedia.com/economics/encyclopedias-almanacs-transcripts-and-maps/open-systems (accessed on 1 December 2023).
6. Strickland, J. Who Owns the Internet? Available online: https://computer.howstuffworks.com/internet/basics/who-owns-internet1.htm (accessed on 15 November 2023).
7. Cyphers, B.; Doctorow, C. A Legislative Path to an Interoperable Internet. Available online: https://www.eff.org/deeplinks/2020/07/legislative-path-interoperable-internet#:~:text=The%20Internet%20is%20interoperability.,to%20talk%20to%20one%20another (accessed on 15 November 2023).
8. Colicchia, C.; Strozzi, F. Supply chain risk management: A new methodology for a systematic literature review. *Supply Chain Manag. Int. J.* **2012**, *17*, 403–418. [CrossRef]
9. Hummon, N.P.; Dereian, P. Connectivity in a citation network: The development of DNA theory. *Soc. Netw.* **1989**, *11*, 39–63. [CrossRef]
10. Zhao, G.; Liu, S.; Lopez, C.; Lu, H.; Elgueta, S.; Chen, H.; Boshkoska, B.M. Blockchain technology in agri-food value chain management: A synthesis of applications, challenges and future research directions. *Comput. Ind.* **2019**, *109*, 83–99. [CrossRef]
11. Oppitz, M.; Tomsu, P. Internet of Things. In *Inventing the Cloud Century*; Springer: Berlin/Heidelberg, Germany, 2018; pp. 435–469.
12. VOSviewer. Leiden University. 2019. Available online: http://www.vosviewer.com/ (accessed on 1 November 2023).
13. Le-Phuoc, D.; Quoc, H.N.M.; Parreira, J.X.; Hauswirth, M. The linked sensor middleware–connecting the real world and the semantic web. *Proc. Semant. Web Chall.* **2011**, *152*, 22–23.
14. Harris, S.; Seaborne, A.; Prud'hommeaux, E. SPARQL 1.1 query language. W3C Recommendation. 2013. Available online: https://www.w3.org/TR/sparql11-query/ (accessed on 15 July 2023).
15. Barnaghi, P.; Wang, W.; Dong, L.; Wang, C. A linked-data model for semantic sensor streams. In Proceedings of the 2013 IEEE International Conference on Green Computing and Communications and IEEE Internet of Things and IEEE Cyber, Physical and Social Computing, Beijing, China, 20–23 August 2013; pp. 468–475.
16. Compton, M.; Barnaghi, P.; Bermudez, L.; García-Castro, R.; Corcho, O.; Cox, S.; Graybeal, J.; Hauswirth, M.; Henson, C.; Herzog, A.; et al. The SSN ontology of the W3C semantic sensor network incubator group. *Web Semant. Sci. Serv. Agents World Wide Web* **2012**, *17*, 25–32. [CrossRef]
17. Guinard, D.; Trifa, V.; Karnouskos, S.; Spiess, P.; Savio, D. Interacting with the soa-based internet of things: Discovery, query, selection, and on-demand provisioning of web services. *IEEE Trans. Serv. Comput.* **2010**, *3*, 223–235. [CrossRef]
18. Haller, A.; Janowicz, K.; Cox, S.J.; Lefrançois, M.; Taylor, K.; Le Phuoc, D.; Lieberman, J.; García-Castro, R.; Atkinson, R.; Stadler, C. The modular SSN ontology: A joint W3C and OGC standard specifying the semantics of sensors, observations, sampling, and actuation. *Semant. Web* **2019**, *10*, 9–32. [CrossRef]
19. Georgakopoulos, D.; Jayaraman, P.P. Internet of things: From internet scale sensing to smart services. *Computing* **2016**, *98*, 1041–1058. [CrossRef]
20. Bermudez-Edo, M.; Elsaleh, T.; Barnaghi, P.; Taylor, K. IoT-Lite: A lightweight semantic model for the Internet of Things. In Proceedings of the 2016 Intl IEEE Conferences on Ubiquitous Intelligence & Computing, Advanced and Trusted Computing, Scalable Computing and Communications, Cloud and Big Data Computing, Internet of People, and Smart World Congress (UIC/ATC/ScalCom/CBDCom/IoP/SmartWorld), Toulouse, France, 18–21 July 2016; pp. 90–97.
21. Perera, C.; Zaslavsky, A.; Liu, C.H.; Compton, M.; Christen, P.; Georgakopoulos, D. Sensor search techniques for sensing as a service architecture for the internet of things. *IEEE Sens. J.* **2013**, *14*, 406–420. [CrossRef]
22. Kamilaris, A.; Gao, F.; Prenafeta-Boldú, F.X.; Ali, M.I. Agri-IoT: A semantic framework for Internet of Things-enabled smart farming applications. In Proceedings of the 2016 IEEE 3rd World Forum on Internet of Things (WF-IoT), Reston, VA, USA, 12–14 December 2016; pp. 442–447.
23. Wang, W.; Barnaghi, P.; Cassar, G.; Ganz, F.; Navaratnam, P. Semantic sensor service networks. In Proceedings of the SENSORS, 2012 IEEE, Taipei, Taiwan, 28–31 October 2012; pp. 1–4.

24. Gyrard, A.; Datta, S.K.; Bonnet, C.; Boudaoud, K. Standardizing generic cross-domain applications in Internet of Things. In Proceedings of the 2014 IEEE Globecom Workshops (GC Wkshps), Austin, TX, USA, 8–12 December 2014; pp. 589–594.
25. Swetina, J.; Lu, G.; Jacobs, P.; Ennesser, F.; Song, J. Toward a standardized common M2M service layer platform: Introduction to oneM2M. *IEEE Wirel. Commun.* **2014**, *21*, 20–26. [CrossRef]
26. Alaya, M.B.; Banouar, Y.; Monteil, T.; Chassot, C.; Drira, K. OM2M: Extensible ETSI-compliant M2M service platform with self-configuration capability. *Procedia Comput. Sci.* **2014**, *32*, 1079–1086. [CrossRef]
27. Gyrard, A.; Serrano, M.; Jares, J.B.; Datta, S.K.; Ali, M.I. Sensor-based linked open rules (s-lor) an automated rule discovery approach for IoT applications and its use in smart cities. In Proceedings of the 26th International Conference on World Wide Web Companion, Perth, Australia, 3–7 April 2017; pp. 1153–1159.
28. Gyrard, A.; Serrano, M. A unified semantic engine for internet of things and smart cities: From sensor data to end-users applications. In Proceedings of the 2015 IEEE International Conference on Data Science and Data Intensive Systems, Sydney, Australia, 11–13 December 2015; pp. 718–725.
29. Nachabe, L.; Girod-Genet, M.; El Hassan, B. Unified data model for wireless sensor network. *IEEE Sens. J.* **2015**, *15*, 3657–3667. [CrossRef]
30. Bharti, M.; Kumar, R.; Saxena, S.; Jindal, H. Optimal resource selection framework for Internet-of-Things. *Comput. Electr. Eng.* **2020**, *86*, 106693. [CrossRef]
31. Botts, M.; Percivall, G.; Reed, C.; Davidson, J. OGC® sensor web enablement: Overview and high level architecture. In *International Conference on GeoSensor Networks*; Springer: Berlin/Heidelberg, Germany, 2006; pp. 175–190.
32. Bogner, A.; Chanson, M.; Meeuw, A. A decentralised sharing app running a smart contract on the ethereum blockchain. In Proceedings of the 6th International Conference on the Internet of Things, Stuttgart Germany, 7–9 November 2016; pp. 177–178.
33. Oracle Corporation. MySQL. Available online: https://www.mysql.com/ (accessed on 20 November 2023).
34. Lunardi, W.T.; de Matos, E.; Tiburski, R.; Amaral, L.A.; Marczak, S.; Hessel, F. Context-based search engine for industrial IoT: Discovery, search, selection, and usage of devices. In Proceedings of the 2015 IEEE 20th Conference on Emerging Technologies & Factory Automation (ETFA), Luxembourg, 8–11 September 2015; pp. 1–8.
35. The PostgreSQL Global Development Group. PostgreSQL: The World's Most Advanced Open Source Relational Database. Available online: https://www.postgresql.org/ (accessed on 20 November 2023).
36. Wang, W.; De, S.; Cassar, G.; Moessner, K. An experimental study on geospatial indexing for sensor service discovery. *Expert Syst. Appl.* **2015**, *42*, 3528–3538. [CrossRef]
37. Paganelli, F.; Parlanti, D. A DHT-based discovery service for the Internet of Things. *J. Comput. Netw. Commun.* **2012**, *2012*, 107041. [CrossRef]
38. Liang, S.; Huang, C.-Y. Geocens: A geospatial cyberinfrastructure for the world-wide sensor web. *Sensors* **2013**, *13*, 13402–13424. [CrossRef]
39. Alphand, O.; Amoretti, M.; Claeys, T.; Dall, S.; Duda, A.; Ferrari, G.; Rousseau, F.; Tourancheau, B.; Veltri, L.; Zanichelli, F. IoTChain: A blockchain security architecture for the Internet of Things. In Proceedings of the 2018 IEEE Wireless Communications and Networking Conference (WCNC), Barcelona, Spain, 15–18 April 2018; pp. 1–6. [CrossRef]
40. Wen, Q.; Gao, Y.; Chen, Z.; Wu, D. A blockchain-based data sharing scheme in the supply chain by IIoT. In Proceedings of the 2019 IEEE International Conference on Industrial Cyber Physical Systems (ICPS), Taipei, Taiwan, 6–9 May 2019; pp. 695–700.
41. Rahman, M.Z.U.; Surekha, S.; Satamraju, K.P.; Mirza, S.S.; Lay-Ekuakille, A. A collateral sensor data sharing framework for decentralized healthcare systems. *IEEE Sens. J.* **2021**, *21*, 27848–27857. [CrossRef]
42. Cech, H.L.; Großmann, M.; Krieger, U.R. A fog computing architecture to share sensor data by means of blockchain functionality. In Proceedings of the 2019 IEEE International Conference on Fog Computing (ICFC), Prague, Czech Republic, 24–26 June 2019; pp. 31–40.
43. Brin, S.; Page, L. Reprint of: The anatomy of a large-scale hypertextual web search engine. *Comput. Netw.* **2012**, *56*, 3825–3833. [CrossRef]
44. Khodadadi, F.; Dastjerdi, A.V.; Buyya, R. Simurgh: A framework for effective discovery, programming, and integration of services exposed in IoT. In Proceedings of the 2015 International Conference on Recent Advances in Internet of Things (RIoT), Singapore, 7–9 April 2015; pp. 1–6.
45. Jirka, S.; Bröring, A.; Stasch, C. Discovery mechanisms for the sensor web. *Sensors* **2009**, *9*, 2661–2681. [CrossRef]
46. Cassar, G.; Barnaghi, P.M.; Moessner, K. Probabilistic Methods for Service Clustering. *SMRR@ ISWC*. 2010. Available online: http://people.csail.mit.edu/pcm/tempISWC/workshops/SMR22010/SMR2Proceedings.pdf#page=4 (accessed on 20 July 2023).
47. Zoumpatianos, K.; Idreos, S.; Palpanas, T. Indexing for interactive exploration of big data series. In Proceedings of the 2014 ACM SIGMOD International Conference on Management of Data, Snowbird, UT, USA, 22–27 June 2014; pp. 1555–1566.
48. Shieh, J.; Keogh, E. i SAX: Indexing and mining terabyte sized time series. In Proceedings of the 14th ACM SIGKDD International Conference on Knowledge Discovery and Data Mining, Las Vegas, NV, USA, 24–27 August 2008; pp. 623–631.
49. Soldatos, J.; Kefalakis, N.; Hauswirth, M.; Serrano, M.; Calbimonte, J.-P.; Riahi, M.; Aberer, K.; Jayaraman, P.P.; Zaslavsky, A.; Žarko, I.P.; et al. Openiot: Open source internet-of-things in the cloud. In *Interoperability and Open-Source Solutions for the Internet of Things*; Springer: Berlin/Heidelberg, Germany, 2015; pp. 13–25.
50. Global Sensor Networks Team. Global Sensor Networks Project. Available online: https://www.epfl.ch/labs/lsir/global-sensor-networks/ (accessed on 24 January 2022).

51. Pazos, N.; Müller, M.; Aeberli, M.; Ouerhani, N. ConnectOpen-automatic integration of IoT devices. In Proceedings of the 2015 IEEE 2nd World Forum on Internet of Things (WF-IoT), Milan, Italy, 14–16 December 2015; pp. 640–644.
52. Perera, C.; Zaslavsky, A.; Compton, M.; Christen, P.; Georgakopoulos, D. Semantic-driven configuration of internet of things middleware. In Proceedings of the 2013 Ninth International Conference on Semantics, Knowledge and Grids, Beijing, China, 3–4 October 2013; pp. 66–73.
53. Madureira, A.L.R.; Araújo, F.R.C.; Sampaio, L.N. On supporting IoT data aggregation through programmable data planes. *Comput. Netw.* **2020**, *177*, 107330. [CrossRef]
54. Buterin, V. A next-generation smart contract and decentralized application platform. *White Pap.* **2014**, *3*, 2.
55. González, A.G. PayPal: The legal status of C2C payment systems. *Comput. Law Secur. Rev.* **2004**, *20*, 293–299. [CrossRef]
56. Nakamoto, S. Bitcoin: A peer-to-peer electronic cash system. 2008. Available online: https://assets.pubpub.org/d8wct41f/31611263538139.pdf (accessed on 10 June 2023).
57. Popov, S. The Tangle. 2016, p. 131. Available online: https://assets.ctfassets.net/r1dr6vzfxhev/2t4uxvsIqk0EUau6g2sw0g/45eae33637ca92f85dd9f4a3a218e1ec/iota1_4_3.pdf (accessed on 12 June 2023).
58. Tzianos, P.; Pipelidis, G.; Tsiamitros, N. Hermes: An open and transparent marketplace for IoT sensor data over distributed ledgers. In Proceedings of the 2019 IEEE International Conference on Blockchain and Cryptocurrency (ICBC), Seoul, Republic of Korea, 14–17 May 2019; pp. 167–170.
59. Özyilmaz, K.R.; Doğan, M.; Yurdakul, A. IDMoB: IoT data marketplace on blockchain. In Proceedings of the 2018 Crypto Valley Conference on Blockchain Technology (CVCBT), Zug, Switzerland, 20–22 June 2018; pp. 11–19.
60. The Raiden Network. Available online: https://raiden.network/ (accessed on 17 January 2024).
61. Rahman, M.U.; Baiardi, F.; Ricci, L. Blockchain smart contract for scalable data sharing in IoT: A case study of smart agriculture. In Proceedings of the 2020 IEEE Global Conference on Artificial Intelligence and Internet of Things (GCAIoT), Virtual, 12–16 December 2020; pp. 1–7.
62. EOS. An Exceptionally Fast and Infinitely Scalable Smart Contract Platform. Available online: https://eosnetwork.com/introducing-eos/ (accessed on 17 January 2024).
63. Ngu, A.H.; Gutierrez, M.; Metsis, V.; Nepal, S.; Sheng, Q.Z. IoT middleware: A survey on issues and enabling technologies. *IEEE Internet Things J.* **2016**, *4*, 1–20. [CrossRef]
64. Dawod, A.; Georgakopoulos, D.; Jayaraman, P.P.; Nirmalathas, A.; Parampalli, U. IoT device integration and payment via an autonomic blockchain-based service for IoT device sharing. *Sensors* **2022**, *22*, 1344. [CrossRef] [PubMed]
65. Georgakopoulos, D.; Jayaraman, P.P.; Dawod, A. SenShaMart: A Trusted IoT Marketplace for Sensor Sharing. In Proceedings of the 2020 IEEE 6th International Conference on Collaboration and Internet Computing (CIC), Atlanta, GA, USA, 1–3 December 2020; pp. 8–17.
66. Montori, F.; Liao, K.; De Giosa, M.; Jayaraman, P.P.; Bononi, L.; Sellis, T.; Georgakopoulos, D. A Metadata-Assisted Cascading Ensemble Classification Framework for Automatic Annotation of Open IoT Data. *IEEE Internet Things J.* **2023**, *10*, 13401–13413. [CrossRef]
67. Yao, Z.; Xin, J.; Hao, K.; Wang, Z.; Zhu, W. Learned-Index-Based Semantic Keyword Query on Blockchain. *Mathematics* **2023**, *11*, 2055. [CrossRef]
68. Hunkeler, U.; Truong, H.L.; Stanford-Clark, A. MQTT-S—A publish/subscribe protocol for Wireless Sensor Networks. In Proceedings of the 2008 3rd International Conference on Communication Systems Software and Middleware and Workshops (COMSWARE'08), Bangalore, India, 5–10 January 2008; pp. 791–798.
69. Shelby, Z.; Hartke, K.; Bormann, C. The constrained application protocol (CoAP). 2014. Available online: https://www.rfc-editor.org/rfc/rfc7252 (accessed on 21 July 2023).
70. Fielding, R.; Gettys, J.; Mogul, J.; Frystyk, H.; Masinter, L.; Leach, P.; Berners-Lee, T. Hypertext transfer protocol--HTTP/1.1. 2070-1721. 1999. Available online: https://www.rfc-editor.org/rfc/rfc2616?data1=dwnsb4B&data2=abmurltv2b(accessed on 21 July 2023).
71. Dawod, A.; Georgakopoulos, D.; Jayaraman, P.P.; Chrysanthis, P.K. A Self-managed Marketplace for Sharing IoT Sensors. In Proceedings of the 2023 IEEE 39th International Conference on Data Engineering Workshops (ICDEW), Anaheim, CA, USA, 3–7 April 2023; pp. 111–117.

sensors

Review

# Non-Terrestrial Networks for Energy-Efficient Connectivity of Remote IoT Devices in the 6G Era: A Survey

**Stefanos Plastras [1,*,†], Dimitrios Tsoumatidis [1], Dimitrios N. Skoutas [1], Angelos Rouskas [2], Georgios Kormentzas [1] and Charalabos Skianis [1]**

[1] Department of Information and Communication Systems Engineering, University of the Aegean, 83200 Samos, Greece; d.tsoumatidis@aegean.gr (D.T.); d.skoutas@aegean.gr (D.N.S.); gkorm@aegean.gr (G.K.); cskianis@aegean.gr (C.S.)

[2] Department of Digital Systems, University of Piraeus, 18532 Piraeus, Greece; arouskas@unipi.gr

* Correspondence: s.plastras@aegean.gr

† Current address: Lymperis Building, 2 Palama & Gorgyras St., 83200 Samos, Greece.

**Abstract:** The Internet of Things (IoT) is gaining popularity and market share, driven by its ability to connect devices and systems that were previously siloed, enabling new applications and services in a cost-efficient manner. Thus, the IoT fuels societal transformation and enables groundbreaking innovations like autonomous transport, robotic assistance, and remote healthcare solutions. However, when considering the Internet of Remote Things (IoRT), which refers to the expansion of IoT in remote and geographically isolated areas where neither terrestrial nor cellular networks are available, internet connectivity becomes a challenging issue. Non-Terrestrial Networks (NTNs) are increasingly gaining popularity as a solution to provide connectivity in remote areas due to the growing integration of satellites and Unmanned Aerial Vehicles (UAVs) with cellular networks. In this survey, we provide the technological framework for NTNs and Remote IoT, followed by a classification of the most recent scientific research on NTN-based IoRT systems. Therefore, we provide a comprehensive overview of the current state of research in IoRT and identify emerging research areas with high potential. In conclusion, we present and discuss 3GPP's roadmap for NTN standardization, which aims to establish an energy-efficient IoRT environment in the 6G era.

**Keywords:** internet of things; IoT; 5G; 6G; internet of remote things; IoRT; unmanned aerial vehicles; UAVs; satellites; non terrestrial networks; NTNs

**Citation:** Plastras, S.; Tsoumatidis, D.; Skoutas, D.N.; Rouskas, A.; Kormentzas, G.; Skianis, C. Non-Terrestrial Networks for Energy-Efficient Connectivity of Remote IoT Devices in the 6G Era: A Survey. *Sensors* **2024**, *24*, 1227. https://doi.org/10.3390/s24041227

Academic Editors: Behnam Mobaraki and Jose Turmo

Received: 17 January 2024
Revised: 6 February 2024
Accepted: 12 February 2024
Published: 15 February 2024



## 1. Introduction

The primary goal of the Internet of Things (IoT) is to create a comprehensive computing environment where everyday objects can sense and interact with their surroundings, bridging the physical and digital realms [1]. The IoT is expected to continue growing further, with forecasts of over 5.5 billion cellular IoT connections by 2027, reflecting its fast expansion [2]. The rapid development of IoT technology opens up numerous applications that have the potential to transform our daily lives [3,4]. These IoT technologies are being leveraged to enhance various aspects of our environment, giving rise to concepts like smart buildings, smart cities, flexible energy infrastructure, and smart agriculture, effectively embedding intelligence within these domains [5,6]. In the realm of smart agriculture, for example, the activation or deactivation of an irrigation system can be precisely controlled based on real-time data collected from humidity sensors deployed in remote fields. This automated approach not only optimizes irrigation efficiency but also enables remote monitoring and management of agricultural operations, while simultaneously providing valuable insights for comprehensive evaluation and analysis [7].

Efficient data transmission is paramount in the expansive realm of the IoT, as it underpins the seamless operation of IoT systems across various domains [8]. The continuous exchange of data and information between devices, data storage systems, and end-users is crucial for the effective functioning of IoT applications. Current IoT architectures employ a diverse range of communication technologies, encompassing short-range solutions tailored for localized environments like homes and industrial settings as well as long-range solutions capable of covering vast geographic areas [9,10]. The introduction of 5G technology marks a pivotal moment in the IoT landscape, addressing the evolving connectivity needs of IoT applications. The standardization of 4G-IoT systems, such as Long-Term Evolution Machine type communication (LTE-M) and NarrowBand Internet of Things (NB-IoT), continued to evolve in 5G standards [11,12].

These advancements collectively contribute to a transformative 5G-IoT ecosystem, unlocking a vast spectrum of possibilities for the future of connected devices and services [13–16]. The IoT is rapidly evolving, paving the way for innovative scenarios and expanding the range of existing IoT applications, laying the foundation for future 6G-IoT systems [17,18]. However, despite the progress made, there are still unique challenges that need to be addressed, particularly in the research area of the Internet of Remote Things (IoRT).

*Internet of Remote Things*

The Internet of Remote Things (IoRT) extends the reach of the Internet of Things to remote and physically isolated regions where terrestrial and cellular networks are absent. This innovative paradigm broadens the scope of the IoT by connecting the farthest and the most remote corners of the world. However, IoRT deployments often encounter significant challenges, such as limited connectivity and harsh environmental conditions [18,19].

Despite these difficulties, there is a growing demand for IoRT applications, as they hold immense potential in various domains, including environmental monitoring, wildlife tracking, disaster relief management, and the delivery of critical healthcare and education services to under-served rural communities [20]. Thus, a number of specialized IoRT environments are formed, as illustrated in Table 1. The Internet of Agricultural Things (IoAT), for instance, has the potential to enhance agricultural productivity and income generation in remote places with arable land [7,21,22]. In oceans and seas, the Internet of Maritime Things (IoMT) is also an excellent example of an IoRT environment [23,24] where, for example, sensors on ships can transmit vital data, facilitating weather forecasts and enhancing maritime safety [25]. Specialized IoRT approaches are also required for the Internet of Underground Things (IoUT), which is applicable to subterranean operations such as mining [26–28], and the Internet of Underwater Things (IoUwT), which is applicable to marine exploration or aquaculture [29–32].

In addition, the Internet of Space, which extends IoT capabilities to satellites and extraterrestrial expeditions, is another prominent example of IoRT deployment [33,34]. In the Internet of Battle Things (IoBT), IoT-enabled devices on military assets, equipment, and personnel provide vital information, optimize deployment strategies, and improve situational awareness, thereby enhancing the efficiency and safety of military operations [35–37] even in the most remote areas.

Similarly, IoRT technologies are essential for gathering data on wildlife behavior, climate, and glacier melt in the far-flung and harsh Arctic, forming the Internet of the Arctic. These IoT systems enable scientists and researchers to study and comprehend the swiftly transforming Arctic conditions, which is essential for environmental monitoring and the development of strategies to mitigate the effects of climate change [38]. Hence, IoRT systems are essential for bridging isolated ecosystems in remote environments, enabling communication, data sharing, and intelligent operations.

This survey investigates the application of non-terrestrial networks to connect remote IoT systems within the framework of the Internet of Remote Things. The primary objectives are to identify the scope of Non-Terrestrial Networks (NTNs), classify technological and energy efficiency challenges, and discuss open issues towards 6G-IoT. Hence, the main contributions of the present study are:

- A thorough analysis of various NTN architectures, including satellite networks operating in different orbits, UAV-based networks, and hybrid NTNs, that can address the limitations of current terrestrial networks. Each NTN type is discussed in detail, outlining its strengths, limitations, and potential applications.

- Identifies and discusses key challenges in integrating NTNs into the IoRT ecosystem. It further outlines seven important research objectives for NTN integration in IoRT, ranging from energy efficiency and cost-effective deployment to enhanced coverage, throughput, data transmission reliability, and timely data acquisition.
- Delving into the research field of NTN-based IoRT, this study dissects three key approaches: satellite, hybrid satellite-UAV, and UAV-based systems. By further analyzing each category through the lens of individual research objectives, it unveils a comprehensive overview of this dynamic landscape.
- Exploring the future of IoRT, this study examines research trends and challenges while also spotlighting promising technologies. Focusing on the 3GPP (3rd Generation Partnership Project) standardization roadmap, it highlights the crucial role of upcoming research and standardization efforts in bridging remote connectivity gaps and paving the way for energy-efficient IoRT environments in the 6G era.

**Table 1.** Internet of Remote Things (IoRT).

| References | IoRT Environments | Description |
|---|---|---|
| [7,21,22] | Internet of Agricultural Things (IoAT) | Remotely monitor and manage agricultural operations, optimizing crop yields, reducing water usage, and enhancing pest control, thereby improving agricultural sustainability and profitability. |
| [23,24] | Internet of Maritime Things (IoMT) | Enhance marine security, optimize shipping routes, and reduce environmental impact by monitoring and tracking vessels, detecting illegal fishing activities, and optimizing fuel consumption. |
| [26–28] | Internet of Underground Things (IoUT) | Monitor and maintain underground infrastructure, including pipelines, tunnels, and power lines, to prevent leaks, collapses, and outages, ensuring public safety and infrastructure reliability. |
| [29–32] | Internet of Underwater Things (IoUwT) | Monitor and protect underwater ecosystems, detect pollution, and perform research on marine life by tracking fish migrations, assessing the health of coral reefs, and monitoring water quality. |
| [33,34] | Internet of Satellite/Space Things (IoST) | Enhance navigation and communication in space, improve network reliability, and enable remote monitoring of satellites by providing real-time data and alerts. |
| [35–37] | Internet of Battle Things (IoBT) | Enhance situational awareness and decision-making in combat conditions by providing real-time data on enemy movements, troop locations, and battlefield conditions. |
| [38] | Internet of Arctic Things (IoAT) | Support search and rescue operations, conduct scientific research on Arctic ecology, and monitor the impact of climate change in the Arctic by tracking and analyzing environmental data. |

## 2. Related Work and Survey Scope

In order to establish clear research objectives for the present study, we conducted a comprehensive review of previous studies in the relevant research literature. Below, we present our key findings that emerged from this review, which are also summarized in Table 2.

The study in [39] provides a comprehensive overview of satellite communication systems for the IoT. However, it overlooks two crucial aspects: energy efficiency considerations in NTNs and the integration of Unmanned Aerial Vehicles (UAVs) in the IoRT framework, which are essential for the successful implementation of IoRT solutions. Similarly, while [40] while provides a comprehensive overview of the Low Earth Orbit (LEO) satellite-based IoRT architecture, the authors do not examine energy efficiency issues or the employment of UAVs in the IoRT environment. The study in [41] proposes integrating IoT, 5G, UAVs, and satellites to address IoT deployment challenges and overcome terrestrial infrastructure limitations, such as limited coverage and capacity. Nevertheless, while it provides a general architectural framework, it does not adequately address the challenges of integrating UAVs, satellites, and terrestrial 5G infrastructure, nor does it address the energy management complexities of NTNs.

The authors of the article [42] explore the potential of Artificial Intelligence (AI) techniques, including Machine Learning (ML) and Deep Learning (DL), for enabling ultra-reliable and low-latency communications (URLLC) and ubiquitous interconnectivity in the NTN-based Industrial Internet of Things (IIoT). However, the study primarily focuses on AI-based solutions and does not fully address the diverse range of challenges presented by IoRT use cases. In [43], the authors propose a simplified approach to analyzing Hybrid Satellite–Terrestrial Networks (HSTNs) by introducing three fundamental cooperative models, providing a survey of the state-of-the-art technologies for each model, and outlining prospective research directions. Nevertheless, their research specifically targets a subset of possible IoRT architectures.

The merits and drawbacks of current satellite-based IoT solutions for remote regions are highlighted in [44], which examines a variety of architectures and technical approaches. This study does not adequately address the diverse energy efficiency challenges associated with NTNs, and while it focuses on satellite-based IoT architectures, it does not consider the potential use of UAVs in the context of IoRT. Kua et al. in [45] provide a comprehensive overview of the potential benefits and challenges of using IoT and space-based technologies for future space exploration missions. While detailed, this study is focused on IoT for space and, therefore, does not consider all the range of IoRT scenarios and the respective challenges they pose. The study in [24] explores the need for hybrid satellite–terrestrial maritime networks, highlighting technologies that enhance efficiency, expand coverage, and support specialized maritime services. This study primarily focuses on the IoT implementations for maritime environments and thus does not comprehensively address the diverse landscape of IoRT scenarios.

The study [46] explores the integration of UAVs into Wireless Sensor Networks (WSNs). The authors discuss the performance and capabilities of UAVs as communication nodes, examine architectural aspects and emerging technologies within UAV-enabled WSNs (U-WSNs), and shed light on crucial factors that influence the design of U-WSNs. The study's scope is limited to UAVs, and thus, it does not investigate satellite systems within the framework of IoRT. In [47], the authors examine the growing potential of NTNs in 5G and beyond networks, particularly when combined with terrestrial networks. The survey covers various aspects, including services, architectures, technological enablers, and challenges associated with NTN integration. Although NTNs are explored in this study, they are not evaluated in the IoRT context; hence, the associated challenges are not investigated.

Furthermore, while the study in [48] provides a comprehensive overview of the rapidly evolving landscape of wireless technologies, its focus on the broader IoT domain limits its ability to delve into the specific details and challenges of IoRT connectivity. The paper's coverage of key performance indicators such as scalability, energy efficiency, reliability, and low latency is valuable, but its broad scope hinders a thorough examination of these factors within the unique context of IoRT applications. Additionally, the paper's discussion of IoRT connectivity scenarios, particularly those involving satellite communication, has limitations due to its overarching focus on IoT as a whole.

Moreover, the study in [49] explores resource management for integrated space–air–ground–sea 6G networks incorporating UAVs and satellites, providing valuable insights. However, the study takes a general approach to these networks and does not specifically address the intricacies of IoRT environments. Moreover, it lacks adequate emphasis on energy-efficient strategies, particularly in utilizing Non-Orthogonal Multiple Access (NOMA) schemes and applying machine learning techniques such as Deep Reinforcement Learning (DRL).

**Table 2.** Related surveys and reviews.

| | Reference | Year | Main Focus | NTNs | | Coverage of IoRT Connectivity Solutions | Coverage of Energy Efficiency Solutions |
|---|---|---|---|---|---|---|---|
| | | | | Satellites | UAVs | | |
| 1 | [39] | 2015 | Overview of satellite technologies for IoT Applications | ✓ | × | Partially | × |
| 2 | [40] | 2017 | Utilization of LEO satellites in IoT | Partially | × | Partially | × |
| 3 | [41] | 2019 | Utilization of satellite and UAV technologies in 5G-IoT networks | Partially | ✓ | Partially | × |
| 4 | [42] | 2020 | AI-based techniques to improve the performance of NTN-based IIoT services | Partially | ✓ | Partially | Partially |
| 5 | [43] | 2021 | Hybrid Satellite–Terrestrial Networks (HSTNs) | ✓ | ✓ | Partially | Partially |
| 6 | [44] | 2021 | Satellite-based IoT solutions for remote/rural areas | ✓ | × | Partially | × |
| 7 | [45] | 2021 | Integration status of IoT and space communication technologies | ✓ | ✓ | Partially | Partially |
| 8 | [24] | 2021 | IoT Maritime Communication Networks (MCNs) | ✓ | ✓ | Partially | Partially |
| 9 | [46] | 2022 | UAV-enabled Wireless Sensor Networks (U-WSNs) | × | ✓ | Partially | Partially |
| 10 | [47] | 2022 | Evolution of NTNs in 5G context | ✓ | ✓ | Partially | Partially |
| 11 | [48] | 2022 | Communication solutions for IoT applications in cellular, wide-area, and non-terrestrial networks | Partially | ✓ | Partially | Partially |
| 12 | [49] | 2023 | Space–air–ground–sea-integrated (SAGSI) networks deployment resource optimization analysis | ✓ | ✓ | Partially | Partially |
| 13 | [50] | 2023 | Satellite communications overview for multi-orbit IoT | ✓ | × | Partially | × |
| 14 | This Survey | 2024 | Non-terrestrial networks, including UAV and satellite technologies, for energy-efficient connectivity of remote IoT Devices | ✓ | ✓ | ✓ | ✓ |

Finally, the study in the work [50] offers a thorough analysis of challenges in satellite IoT networks, focusing on broad connectivity, extensive geographical coverage, mobility, and power consumption limitations. However, its focus remains exclusively on satellite NTNs, omitting any discussion of hybrid satellite-UAV and pure UAV connectivity approaches in IoRT environments. Furthermore, the study lacks a comprehensive analysis of energy-efficient approaches, especially through the utilization of machine learning techniques.

*Survey Scope*

Therefore, as evident from the preceding discussion, a significant research void exists in the relevant literature, defining the domain that this current work aims to address: the realm of connectivity within the IoRT context, with a particular focus on the crucial aspects of energy efficiency and sustainability. Regarding the paper structure, Section 3 provides a detailed description of the characteristics of non-terrestrial networks, laying the foundation for understanding their significance in the context of IoRT. Moving forward, Section 4 identifies and clarifies the key challenges involved in integrating NTNs within IoRT systems. To address these challenges, Sections 5–7 present comprehensive surveys of IoRT systems based on Space-borne NTN Networks, Aerial NTN Networks, and Hybrid NTN Networks, respectively. Each survey investigates the unique objectives, characteristics, and challenges associated with these NTN implementations in IoRT. Section 8 examines future research trends and challenges in IoRT, while Section 9 focuses on the 3GPP standardization roadmap, emphasizing the importance of research and standardization efforts in both bridging remote connectivity gaps and fostering energy-efficient IoRT environments in the 6G landscape. Finally, Section 10 concludes the paper by summarizing our work and identifying future research directions.

## 3. Non-Terrestrial Networks

Current terrestrial networks face limitations in providing extensive wireless coverage to remote regions, adequate availability and reliability, and resilience to natural and man-made disasters. Non-terrestrial base stations, on the other hand, offer several advantages over their terrestrial counterparts. Their ability to be quickly deployed in emergency situations is crucial for disaster response, environmental monitoring, and search-and-rescue operations [51]. Additionally, their elevated positioning above the ground enhances radio link quality, reducing signal attenuation and interference from terrestrial obstacles. In essence as shown in Figure 1, the integration of terrestrial and non-terrestrial networks provides a 3D strategy for network coverage, paving the way for seamless and reliable services in remote areas, high-altitude locations, and deep-sea environments [52–54].

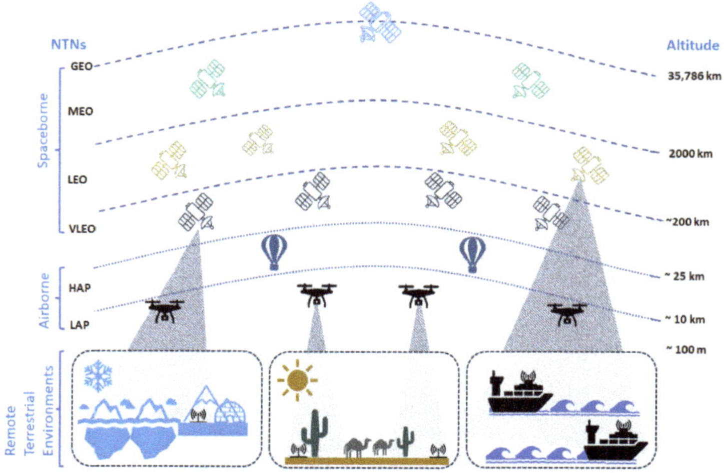


**Figure 1.** Non-terrestrial networks.

- *Space-Borne Networks Based on Satellites*

  Satellites operate in different orbits, each offering distinct advantages and limitations. The choice of orbit depends on the specific requirements of the IoT application. Geostationary Earth Orbit (GEO) satellites maintain a fixed position over the Earth, providing broad coverage but suffering from high latency and signal attenuation. LEO satellites, on the other hand, operate at lower altitudes (500–2000 km), minimizing latency and attenuation but requiring a larger constellation for global coverage [55,56]. Medium Earth Orbit (MEO) satellites offer a balance between coverage and latency, with a smaller constellation than LEO but a higher latency. Very Low Earth Orbit (VLEO) satellites, operating at altitudes below 200 km, promise ultra-low latency and attenuation but require advanced propulsion systems due to increased atmospheric drag and suborbital launch capabilities [57]. LEO and VLEO orbits are commonly utilized for deploying Cubesats, which are small, standardized satellites with a cube-shaped form factor [58,59]. CubeSats are often employed for educational and scientific purposes due to their compact size and lower launch costs. Nanosats, a broader category of small satellites that includes CubeSats but extends to slightly larger satellites, are suitable for a wider range of applications [60].

- *Airborne Networks Based on Unmanned Aerial Vehicles*

  Unmanned Aerial Vehicles (UAVs) offer a promising solution for expanding wireless coverage in challenging environments and beyond terrestrial infrastructure. They can serve as mobile base stations or relay nodes, bridging connectivity gaps and extending coverage to remote or isolated areas. This flexibility addresses the limitations of traditional IoT communication technologies, which often struggle in harsh or geographically constrained conditions [61,62]. UAVs can seamlessly integrate into existing terrestrial networks, effectively utilizing existing base stations without the need for additional infrastructure deployment. The altitude at which UAVs operate determines their role in the network: low-altitude platforms (LAPs) operate at altitudes of up to approximately 17 km, providing localized coverage and relaying data between terrestrial networks and remote areas. High-altitude platforms (HAPs), operating at altitudes of up to 25 km, offer broader coverage and can serve as central nodes for large-scale wireless systems [46].

- *Hybrid Non-Terrestrial Networks*

  Hybrid Non-Terrestrial Networks (Hybrid NTNs) refer to advanced communication infrastructures that seamlessly integrate multiple non-terrestrial technologies to create a unified and resilient network. This integration typically involves combining satellite communication systems with UAVs, either LAPs or HAPs [63]. The goal is to leverage the unique strengths of each component to enhance network performance, coverage, and flexibility. Satellites form the backbone of the network, providing global coverage and enabling long-range data transmission. UAVs and HAPs supplement this backbone infrastructure with dynamic and mobile capabilities, allowing for targeted coverage, rapid deployment, and persistent communication in specific areas. This synergistic combination of global reach and agility ensures the network's ability to adapt to changing communication requirements and effectively serve remote and dynamic environments [64,65].

In conclusion, the utilization of non-terrestrial networks presents significant potential for expanding communication capabilities, enhancing coverage, facilitating the distribution of computing resources, reducing data processing delays, and establishing local IoT networks that can adapt to diverse Quality of Service (QoS) demands [66]. These platforms are characterized by their rapid and flexible development, which further enhances their ability to adapt to remote environments. In this context, the implementation of NTNs can facilitate the realization of different scenarios and services related to the IoRT ecosystem [47]. This involves providing support for demanding applications, including remote infrastructure control, monitoring, data collection, and connectivity, specifically in environments characterized by their remoteness, such as the Arctic, maritime regions, and rural areas.

## 4. Key Challenges in the Integration of NTNs in IoRT

The Internet of Remote Things represents a groundbreaking advancement in IoT technology, extending its reach to connect remote and resource-constrained devices beyond the limitations of traditional terrestrial networks. Leveraging NTNs such as satellites, UAVs, and hybrid combinations of these platforms enables the collection, transmission, and processing of data from remote locations, unlocking a vast array of applications in diverse fields. Figure 2 illustrates a generic IoT system model for interfacing remote IoT systems to backbone communication infrastructure.

Despite significant advancements in Non-Terrestrial Networks, integrating them into the Internet of Remote Things presents notable challenges, as shown in Figure 3. The utilization of space and aerial channels in NTNs, poses unique challenges in implementing underlying communication across diverse settings such as oceans, maritime zones, the Arctic, and rural areas. While enhancing the quality of space-aerial channels holds promise for extensive coverage and increased throughput, it confronts challenges related to signal propagation, atmospheric conditions, and latency. Hence, developing innovative approaches is essential to ensuring uninterrupted and reliable communication [67].

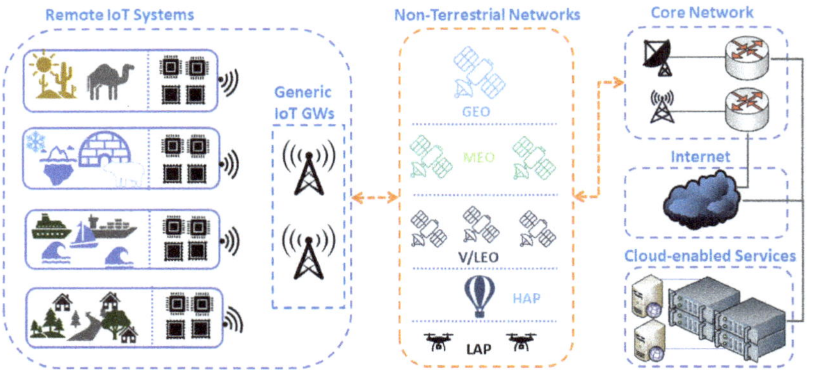


**Figure 2.** Generic IoRT System Model.

Furthermore, data collection in geographically isolated areas presents a distinctive set of challenges in ensuring timely and reliable data transmission and analysis. The interplay between spatial and temporal dynamics in these remote ecosystems can be intricate [68]. For instance, in a remote forest, the distribution of plant species may be influenced by geographic factors such as elevation and soil type as well as temporal factors like seasonal changes in temperature and precipitation. Due to the complexity of remote ecosystems, traditional methods of data collection, transmission, and analysis are often inadequate. Novel approaches are required, and new technologies and protocols are being developed to address these challenges [69].

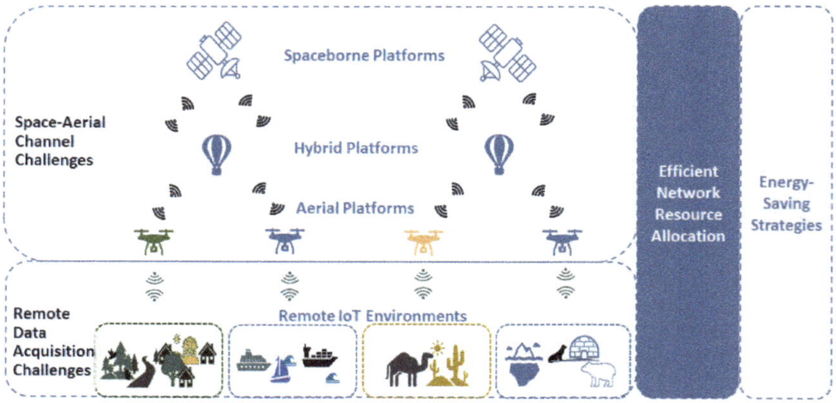


**Figure 3.** Key challenges in NTN-based IoRT systems.

Moreover, efficient resource allocation is always a paramount challenge, driving the development of advanced techniques to optimize resource distribution on satellite, UAV, or hybrid platforms [70]. This necessitates addressing obstacles related to scalability, prioritization and adaptability in resource allocation. Energy efficiency especially poses a critical challenge demanding sustainable and efficient energy management strategies for NTNs operating in remote and challenging environments [71]. Tackling these issues is crucial for the seamless integration of NTNs into the IoRT ecosystem, enabling efficient communication across a wide range of applications in remote and challenging conditions [72]. In our classification of papers on non-terrestrial networks for the IoRT, we identified seven important objectives. While specific objectives might overlap and impact one another, we centered on each work's main research focus.

1. *Energy Efficiency*: NTNs play a crucial role in extending the lifespan and reducing the operational costs of IoT devices, especially in remote or resource-constrained environments. Research efforts aim to optimize NTNs to minimize energy consumption while maintaining network performance.
2. *Cost-Efficient Deployment/Operation*: The deployment and operation of NTNs, particularly in large-scale IoT deployments, must be cost-effective. Researchers explore strategies to reduce infrastructure costs, optimize network utilization, and minimize maintenance expenses.
3. *Enhanced Coverage and Availability*: NTNs provide an opportunity to extend network coverage beyond the limitations of terrestrial networks, ensuring reliable connectivity in challenging terrains and remote areas. Research focuses on improving NTN coverage, overcoming obstacles, and maintaining connectivity in harsh conditions.
4. *Enhanced Throughput/Transmission Rate*: As IoT data volumes continue to grow, NTNs need to handle high-speed data transmission to meet the demands of applications. Researchers investigate techniques to increase NTN throughput and support efficient data transfer.
5. *Enhanced Data Transmission Reliability*: Reliable data transmission is paramount, especially for mission-critical IoT applications. NTN research addresses factors such as interference, fading, and signal attenuation to ensure data integrity and minimize packet loss.
6. *Enhanced and Timely Data Acquisition*: IoT devices generate vast amounts of data at frequent intervals. Timely data acquisition is crucial for proper processing and analysis. NTN research explores methods to optimize data collection, reduce delays, and ensure timely data delivery, often leveraging advanced Medium Access Control (MAC) protocols to optimize channel access.
7. *Enhancement of Edge Computing Capabilities*: Edge computing brings processing power closer to IoT devices, reducing latency and enabling real-time decision-making.

By addressing these objectives, NTNs for IoRT are poised to revolutionize the way data are collected, processed, and utilized, enabling a wide range of applications in remote and isolated environments.

## 5. IoRT Systems Based on Space-Borne NTN Networks

This category of satellite-based IoRT systems is the most prevalent, as it represents the most fundamental approach to establishing remote connectivity. This can be achieved through either the direct connection of remote devices to the satellite or through the establishment of a local network that utilizes a central communication gateway to interface with the satellite. Therefore, in this category, we identified 54 related research works that are classified according to their main research objective into the following subsections.

### *5.1. Enhancement of Energy Efficiency*

In [73], the contribution of satellites to cost-effective solutions for a large number of users and devices in the emerging 5G cellular system is highlighted. Given the broadcast nature of satellite communications, access to remote areas, and support for multiple devices, satellites are positioned to play a significant role in the development of the Internet of Things industry. The purpose of this paper is to investigate Network Coding (NC) techniques within a hybrid satellite/terrestrial network, specifically to improve the reliability of multicast services. The authors propose an energy-optimization method involving adjustments in data symbol repetition and coded packet numbers based on the group's needs and the prevailing satellite link conditions, with a focus on software updates, control messages, and the efficient delivery of content to multiple devices.

The authors of [74] propose a novel random access preamble design and detection method to address the energy-intensive nature of random access for IoT devices in LEO satellite-based NTNs. Their approach eliminates the need for additional signaling and simplifies the detection process. They also present and employ a metric that effectively mitigates the impact of carrier frequency offset and reduces noise-induced timing estimation errors. The proposed metric's statistical analysis reveals that increasing correlation length improves the output Signal-to-Noise Ratio (SNR) power ratio, and the first-path detection threshold is independent of noise statistics. Simulation results in various LEO scenarios demonstrate the proposed method's robustness in achieving considerable improvements, particularly in terms of energy, when compared to existing random-access methods.

Power-efficient scheduling for NTNs to support IoT applications is the focus of the study in [75]. In order to overcome the challenges posed by the typical duty-cycled nature of IoT applications and the limited satellite access time, this article proposes a method for synchronizing sleep and wake-up periods with satellite availability. By reducing the frequency of satellite orbital information updates mandated by Simplified General Perturbation 4 (SGP4), this approach has the potential to substantially enhance the battery life of IoT-NTN devices. Data from a real communication link between a terrestrial device and a LEO CubeSat is utilized to validate the proposed method.

The authors in [76] propose utilizing collaborative LEO satellites to enhance IoRT systems. This approach involves combining uplink signals from multiple LEO satellites, substantially improving SNR, and reducing power requirements for energy-constrained terminals. The paper outlines a comprehensive system design, encompassing frequency planning, waveform selection, collaboration strategies, and terminal design. A coherent combining scheme is introduced to address signal reception impairments, and a modified SUMPLE algorithm is presented for phase difference compensation and estimation. Simulation results validate the effectiveness of the proposed algorithms, particularly in low SNR conditions.

A novel MEC framework for terrestrial-satellite IoT utilizing LEO satellites and terrestrial-satellite terminals is proposed to enable computation outsourcing from IoT Mobile Devices (IMDs) [77]. This framework employs a two-tiered subproblem approach to minimize the weighted-sum energy consumption of IMDs. The lower layer utilizes sequential fractional programming to minimize space segment latency, while the upper layer employs a convex structure and Lagrangian dual decomposition. Based on solutions to these subproblems, an Energy-efficient Computation Outsourcing and Resource Allocation algorithm (E-CORA) is introduced. Simulation results demonstrate that E-CORA effectively reduces the energy consumption of IMDs by optimizing bit offloading, outperforming both full bit offloading and local computing.

Resource allocation remains a critical challenge in satellite IoT deployments, particularly in remote areas. However, existing strategies often fail to adequately address the unique characteristics of low-earth-orbit satellite systems, such as mobility and energy constraints. In [78], the authors propose DeepCA, a Deep Reinforcement Learning-based (DRL) method for energy-efficient channel allocation. DeepCA effectively tackles the challenge of limited channel resources in ground-based nodes, leading to a significant reduction in energy consumption compared to conventional approaches.

The growing demand for IoT connectivity necessitates enhanced spectrum and energy efficiency in satellite-based IoT systems. The work in [79] proposes a novel power control algorithm for IoT terminals in satellite-based networks, leveraging terrestrial base stations for data acquisition and resource management. By employing the Poisson Point Process (PPP) theory, the optimization problem is formulated considering the distribution of IoT devices within the network. The optimal power control scheme is determined by optimizing user distribution and Signal-to-Interference-plus-Noise Ratio (SINR) demand. To address the complexity of the PPP-based objective function, the pattern search procedure is employed. Numerical evaluations of user rates and energy efficiency demonstrate the effectiveness of the proposed power control algorithm.

Furthermore, the study in [80] introduces multi-connectivity to multi-orbit NTNs, allowing user terminals to connect to numerous satellites in order to increase peak throughput. The researchers propose a terminal-aware multi-connectivity scheduling algorithm that optimizes uplink data rates and minimizes energy consumption at the user terminal based on radio availability and propagation data. Operating within a differentiated multi-layer NTN resource scheduling architecture, the efficacy of the algorithm is compared to that of other scheduling methods, evaluating the uplink data rate and energy efficiency. In addition, they present an architectural design for implementable schedulers in multi-orbital satellite networks that can accommodate a variety of terminal classes.

In [81], the authors explore the challenges of enabling massive access in a Beyond-Fifth-Generation (B5G) satellite IoT network using a multibeam architecture. They focus on the issue of channel phase uncertainty caused by the transmission of channel-state information from devices to satellites through gateways. To overcome this challenge, they propose a novel NOMA scheme that effectively supports the wide-scale deployment of IoT devices. Considering the energy limitations of LEO satellites, they introduce two robust beamforming algorithms to minimize total power consumption for non-critical and critical IoT applications while taking channel phase uncertainty into account. Theoretical analysis and simulations demonstrate the effectiveness and robustness of these algorithms in enabling massive access in the satellite IoT.

The authors of [82] propose a DRL algorithm to manage channel allocation and power control in the satellite IoT uplink scenario. The algorithm aims to optimize resource utilization and service quality. Simulations demonstrate that the approach successfully achieves an efficient balance between power efficiency and service quality. Under typical scenarios for satellite IoT, the proposed scheme greatly improves power efficiency in comparison to current methods, with minimal training overhead.

In conclusion, to enhance the energy efficiency of IoRT systems based on space-borne NTN networks, various techniques have been proposed, ranging from utilizing the repetition factor of data symbols to DRL-based methods for dynamic and energy-efficient channel allocation. Table 3 presents a concise summary of the information presented in this section.

**Table 3.** IoRT systems based on space-borne NTN networks: **enhancement of energy efficiency**.

| Ref. | NTN Architecture | Description |
|---|---|---|
| [73] | LEO | Energy optimization utilizing the repetition factor of data symbols on multiple subcarriers of the transmitted OFDM signal and the mean number of coded packets required to satisfy dynamic requirements. |
| [74] | LEO | A novel random access preamble design and detection mechanism for IoT devices in LEO satellite-based NTNs that addresses the energy-intensive nature of random access. |
| [75] | LEO | A method for synchronizing sleep and wake-up periods with satellite availability to substantially enhance the battery life of IoT-NTN devices. |
| [76] | LEO | Combination of uplink signals from multiple LEO satellites to improve SNR and reduce power consumption for energy-constrained terminals. |
| [77] | LEO | Novel MEC framework for terrestrial-satellite IoT accompanied by an energy-efficient computation offloading and resource allocation algorithm named E-CORA. |
| [78] | LEO | Development of DeepCA, a deep reinforcement learning-based method for dynamic and energy-efficient channel allocation. |
| [79] | Generic | Development of a novel power control algorithm for IoT terminals in satellite-based networks that optimizes user distribution and SINR demand. |
| [80] | Generic | Development of a scheduling algorithm for multi-connectivity in multi-orbit NTNs to maximize uplink data rates while minimizing user terminal energy consumption. |

**Table 3.** *Cont.*

| Ref. | NTN Architecture | Description |
|---|---|---|
| [81] | LEO | Development of NOMA-based beamforming algorithms for tackling massive access in B5G IoT networks and minimizing total power consumption for non-critical and critical IoT applications. |
| [82] | LEO | Development of the Deep Reinforcement Learning method (DRL-QoS-RA) for online joint resource allocation to solve the uplink channel allocation and power control problems for satellite IoT systems. |

### *5.2. Cost-Efficient Deployment/Operation*

The study in [83] tackles the challenge of residual Doppler shifts in LEO satellite communications for NB-IoT over LEO NTN systems. The proposed technique combines demodulation reference signal symbols and reduced satellite beam coverage to compensate for residual Doppler shifts. Link-level simulations and link budget analysis for actual three-dimensional orbits demonstrate that the proposed method significantly reduces the number of required satellites while maintaining continuous communication, positively impacting both the total service time and the most persistent service time.

Furthermore, as described in [84], the Direct-to-Satellite Internet of Things (DtS-IoT) concept employs satellite constellations as gateways in orbit for a global IoT network. The research emphasizes the suitability of particular Long-Range (LoRa) network protocol configurations for achieving this objective. As a consequence, the idea of sparse satellite constellations emerges, significantly reducing in-orbit infrastructure while only marginally increasing latency. An algorithmic approach incorporating specific heuristics is proposed to generate quasi-optimal topologies for sparse IoT constellations. The results indicate that LoRa-enabled DtS-IoT services can be enabled on a global scale with substantially fewer satellites than traditional constellations require.

CubeSats offer a novel method for achieving global connectivity by providing cost-efficient and wide geographic coverage. In order to overcome existing constraints and optimize network management, the authors of [85] propose the adoption of Software-Defined Networking (SDN) and Network Function Virtualization (NFV), which would allow for precise control over hardware and network resources. The study provides a thorough and detailed description of the design and components of the system, addressing the distinct challenges presented by the space environment. An initial performance test, which focuses on important measurements such as latency and throughput, highlights the significant potential of this technology in the IoRT domain.

The implementation of a Software-Defined Radio (SDR)-based satellite gateway is proposed in [86] as a solution to the challenges that arise from the diverse communication standards associated with IoRT applications. Utilizing SDR's capacity to decrease hardware expenses and improve adaptability, this approach maximizes efficiency. The proposed architecture was developed utilizing a standalone SDR platform and Commercial Off-The-Shelf (COTS) modules and is able to support the main IoT telecommunication standards.

Satellite communication systems, particularly those utilizing LEO satellite constellations, offer a cost-effective and scalable solution for global IoT applications, enabling ubiquitous connectivity across vast geographical regions with minimal infrastructure deployment. The study in [87] delves into the optimization of uplink transmission scheduling within LEO satellite networks to address the challenge of limited bandwidth resources The proposed algorithm, a novel hybrid of Simulated Annealing and Monte Carlo (SA-MC) techniques, dynamically adapts to varying data traffic demands and network conditions, achieving significant cost reductions and rapid convergence.

CubeSat platform satellites are introduced in [88] as an attempt to provide low-cost, extensive coverage for IoT/Machine-to-Machine (M2M) services in remote areas. In particular, they investigate the challenges of the earth–space communication link in the context of IoT/M2M. Their system model consists of Machine-Type Devices (MTDs), LEO satellites and ground stations (i.e., gateways). The MTDs serve as data sources since they provide uplink signals to satellites, which then transmit the data to the first gateway. However, due to the highly variable uplink channel load and the limited communication cycle of MTD, an ineffective channel reserve system results. Therefore, they propose an energy-efficient multi-access technique for M2M communication in IoRT systems and develop an energy-efficient random access protocol for delay-tolerant applications.

Table 4 summarizes recent research on cost-effective deployment and operation of IoRT systems based on space-borne NTN networks. The studies proposed various solutions, including compensating for residual Doppler shifts, investigating LoRa protocol variations, using CubeSat SDN NFV-based architectures, employing SDR, and developing efficient uplink scheduling algorithms. The proposed solutions aim to minimize satellite hardware requirements, reduce gateway costs, and optimize the utilization of transmission resources, leading to more cost-efficient IoRT systems.

**Table 4.** IoRT systems based on space-borne NTN networks: **cost-efficient deployment/operation**.

| Ref. | NTN Architecture | Description |
|---|---|---|
| [83] | LEO NB-IoT | By compensating for residual Doppler shifts, it minimizes the number of required satellites while maintaining continuous communication. |
| [84] | LEO LoRa | Investigates the potential of LoRa protocol variations for a reduction in the number of required satellites. |
| [85] | Generic | A CubeSat SDN NFV-based architecture is proposed to achieve cost-efficient deployment and operation, enhanced geographic coverage, and improved throughput. |
| [86] | Generic | Employing Software-Defined Radio (SDR) for cost-effective gateway implementation and resolving interoperability issues. |
| [87] | LEO-GEO | Development of an uplink scheduling algorithm that optimizes the utilization of uplink transmission resources in order to achieve a fair distribution of resources and significant cost reductions. |
| [88] | LEO | Application of CubeSat nanosatellites as APs to establish a cost and energy-efficient IoT/M2M network using erasure coding. Using massive numbers of nanosatellites, area coverage is extended. |

### 5.3. Enhanced Coverage and Availability

To address the growing demand for global connectivity in the IoT realm, ref. [89] explores the potential of beam-hopping-based satellite systems. Tailored for IoT applications, these systems offer efficient coverage expansion and adaptability to dynamic traffic patterns, making them well-suited for massive machine-type communication scenarios. By incorporating adaptations for beam scheduling, initial access, and data transmission, the proposed scheme effectively addresses the unique requirements of IoT communication. Numerical analyses reveal the system's superior performance in terms of battery life, link budget, and system capacity, paving the way for optimized resource allocation across diverse IoT use cases.

In an effort to address the issue of unreliable internet connectivity in remote or underserved regions, a solar-powered satellite internet access point is proposed in [90]. By utilizing solar energy, this innovative Access Point (AP) ensures a reliable internet connection, especially in areas where the electricity supply is irregular. It ensures that sensors, IoT devices, and other technological equipment have continuous access to the internet. Moreover, it prioritizes efficiency in terms of costs and ease of maintenance through a meticulous selection of components. Its ability to maintain internet connectivity for a duration of four days without solar power is particularly notable.

In [91], the researchers propose a novel approach for evaluating poor connectivity due to latitude shifts while maximizing the number of IoT devices that can be synchronously accessible despite coverage variations. To achieve this, they developed a novel coverage level measurement method that is used to evaluate the average coverage level within a return circle of an LEO satellite constellation at varying distances. Furthermore, they used a spot beam communication technique with Time Division Multiple Access (TDMA) as a discrete-time mechanism to determine the maximum number of devices within a unit area that can be accessed during a time frame. Finally, they studied the relationship between device density, maximum delay tolerance, and satellite constellation coverage degree.

Integrating LEO satellite constellations with terrestrial networks to support IoRT services requires careful placement of gateways to maximize network coverage, minimize access distance, and maximize revenue. The authors in [92] propose a novel gateway placement algorithm that considers both communication and IoRT service data demands, as well as channel conditions and service demand distribution, and ensures efficient network resource utilization while safeguarding local information confidentiality. Simulation results demonstrate the algorithm's improvement in resource utilization and coverage performance. Additionally, the findings highlight the significance of service demand distribution in gateway placement decisions.

A comprehensive modeling framework for IoT-over-satellite access systems is introduced in [93]. This framework considers a multitude of factors, including satellite orbit, uplink interference from terrestrial IoT devices, atmospheric conditions, gas absorption, and the probability of line-of-sight. The efficacy of the system is evaluated based on the uplink SINR and satellite link time availability during a pass. Focusing on satellites in low-earth orbit, the study employs actual orbital parameters ranging from 300 to 800 km. Additionally, a numerical model for optimizing antenna beamwidth is proposed to maximize link availability while minimizing terrestrial interference and enhancing the IoT signal.

The authors of [94] introduce a Ka-band multibeam satellite industrial IoT framework to overcome the limitations of conventional ground-based IIoT, particularly its limited coverage range. By employing NOMA within each beam, they aim to optimize transmission rates for enhanced QoS. Resource allocation is maximized by optimizing beam power to align theoretical transmission rates with service demands. Moreover, the study proposes a satellite-ground integrated IIoT solution, utilizing terrestrial cellular networks to extend satellite coverage in obstructed areas while incorporating transmission cost reduction strategies. Simulation results validate the benefits of NOMA in satellite IIoT.

Table 5 summarizes the research on enhancing the coverage and availability of IoRT systems based on space-based NTN networks. These advancements include the development of beam-hopping-based satellite systems for IoT, solar-powered satellite access points, algorithms for optimal gateway placement, modeling frameworks for maximizing link availability, and novel multibeam satellite architectures. These innovations have the potential to significantly expand the reach of IoRT applications to remote and underserved areas, paving the way for a more interconnected and globally accessible IoT ecosystem.

**Table 5.** IoRT systems based on space-borne NTN networks: **enhanced coverage and availability**.

| Ref. | NTN Architecture | Description |
|---|---|---|
| [89] | LEO | Development of a beam-hopping-based satellite system for IoT that can support extensive coverage with fewer transmitters and offers flexibility to cope with time-variant traffic distributions. |
| [90] | Generic | Development of solar-powered satellite access points to provide energy autonomy and reliable internet connectivity in areas with irregular electricity supply. |
| [91] | LEO | Study of satellite coverage variations due to latitude shift. |
| [92] | LEO-Cellular | Development of a novel gateway placement algorithm that considers channel conditions and service demand distribution to optimize network resource utilization and coverage. |
| [93] | LEO | Development of a modeling framework for IoT-over-satellite access systems that aims to maximize link availability while minimizing interference. |
| [94] | GEO | Development of a NOMA multibeam satellite IIoT architecture in Ka-band to extend coverage to devices located in remote areas. |

### 5.4. Enhanced Throughput/Transmission Rate

Although the implementation of satellite connectivity for IoRT devices shows potential, it also poses difficulties in terms of spectrum management for 5G satellite networks. In response to these challenges, ref. [95] presents an innovative, dynamic spectrum management scheme that is tailored to this particular context. The effectiveness of the proposed scheme in spectrum management for machine-to-machine and human-to-human communications in 5G satellite networks is demonstrated by simulation results on throughput, delay, spectral efficiency, and fairness index.

NB-IoT, a promising technology for Low-Power Wide-Area Networks (LPWANs), enables widespread deployment of IoT services. However, in remote areas, terrestrial coverage gaps necessitate the integration of LEO satellites as a supplementary network layer. However, the inherent differential Doppler shift associated with satellite communication can degrade system performance. Authors [96] delve into the NB-IoT over LEO satellite architecture, where the satellite provides NB-IoT connectivity to fixed NB-IoT terminal equipment located beyond the reach of terrestrial infrastructure. To address the significant differential Doppler among terrestrial user channels, the authors propose a resource allocation scheme that effectively mitigates excessive Doppler values. Specifically, they propose segmenting the coverage area into smaller sub-regions, ensuring that the differential Doppler within each region remains within permissible limits.

A phased-array antenna framework for IoT-over-satellite applications leveraging nano-satellites is presented in [97] to facilitate global IoT connectivity in remote areas. The proposed approach enables precise beamforming toward targeted IoT devices, boosting received signals while minimizing interference from other terrestrial devices. By efficiently sharing radio spectrum resources, this approach enhances capacity. A practical implementation of this framework involves fabricating an X-band phased array antenna, which has been validated through gain pattern and return loss testing, confirming its feasibility.

In the realm of the satellite-based Internet of Things, efficient access for numerous sensors and short-burst transmissions remains a challenge. To address this issue, ref. [98] proposes a time-slot random access protocol employing Walsh codes specifically tailored for sink node scenarios. A load estimation-based, dynamic Walsh code selection mechanism is introduced to optimize system throughput. Simulation results demonstrate the protocol's effectiveness in enhancing system throughput under high load conditions while effectively managing resource utilization during periods of medium and low demand.

The research in [99] explores a Non-Orthogonal Multiple Access-based multi-user Beamforming (NOMA-BF) scheme for improving spectral efficiency in satellite-based IoT. The proposed scheme has been developed for scenarios in which primary users are prioritized with timely data transmissions and guaranteed data rates, while secondary users are opportunistically served. To reduce inter-beam interference, improve effective channel gains, and optimize inter-beam power allocation, the scheme employs a multi-layered beamforming strategy. Numerical results demonstrate that the proposed scheme enhances spectral efficiency while meeting user data rate requirements.

The research work in [100] emphasizes the potential of the upcoming 6G standard to provide seamless IoT network coverage by 2030. Underscoring the importance of satellite-based communications in meeting IoT service requirements in the 6G era, this study analyzes in depth the multiple access technologies required for effective satellite-based IoT deployment. It includes information-theoretic considerations and key technologies such as NOMA and Random Access (RA). This analysis evaluates how the satellite transmission environment impacts these access technologies. Furthermore, it aims to improve system throughput and robustness under varying traffic conditions by proposing a novel Non-Orthogonal Massive Grant-Free Access (NoMaGFA) scheme that combines the benefits of RA and NOMA for asynchronous transmissions in satellite-based IoT while maintaining low signaling overhead.

In response to the growing demand for spectrum-efficient satellite-based IoT systems, ref. [101] introduces a cognitive satellite-terrestrial framework that leverages multibeam satellite capabilities with full frequency reuse. The proposed scheme maximizes the achievable rate while ensuring outage probability and power consumption constraints. To address the complex optimization problem, a two-level iterative algorithm is developed that incorporates Bernstein-type and large deviation inequality-aided methods for probabilistic constraint handling. Numerical simulations illustrate that the proposed scheme is efficient as well as sensitive to key parameters.

The study in [102] identified NB-IoT as a key technology for supporting massive machine-type communication (mMTC) scenarios within 5G networks. LEO satellites offer the potential to enhance the link budget by reducing propagation signal loss, which is crucial for low-complexity, low-power, and cost-effective IoT devices. However, LEO orbits introduce higher Doppler effects, which can negatively impact the performance of IoT communication. To mitigate these effects, the study proposes an uplink scheduling technique for a mMTC-NB-IoT system based on LEO satellites. Numerical simulations are employed to evaluate the performance of the proposed technique and highlight the limitations imposed by the satellite channel.

Table 6 presents a concise summary of the information presented in this section. Various techniques, such as efficient spectrum management, beamforming, and NOMA, are employed to achieve significant improvements in data transmission rates. These techniques address the challenges posed by large-scale IoT deployments, paving the way for future IoRT systems that can support massive connectivity.

**Table 6.** IoRT systems based on space-borne NTN networks: **enhanced throughput/transmission rate.**

| Ref. | NTN Architecture | Description |
|---|---|---|
| [95] | GEO | Development of a spectrum management scheme for human-to-human and machine-to-machine communications to improve spectral efficiency, throughput and fairness index. |
| [96] | LEO NB-IoT | Development of a resource allocation scheme to mitigate the issue of high differential Doppler among terrestrial user channels. |
| [97] | Nano-satellites | Precise beamforming directed towards specific IoT devices enhances received signals while minimizing interference. |
| [98] | LEO | Development of a slotted random access protocol based on Walsh codes for a satellite IoT scenario with sink nodes that aims to improve the throughput of the system, especially under high load. |
| [99] | GEO | Development of a geographical non-orthogonal multiple access-based multiuser beamforming (NOMA-BF) scheme to improve spectral efficiency in multi-beam satellite-based IoT systems. |
| [100] | LEO | Scheme that combines NOMA and RA technologies for asynchronous transmissions to achieve better throughput performance and an enhanced data transmission rate. |
| [101] | LEO | Development of a robust multigroup multicast beamforming scheme to improve spectrum efficiency and adequately serve a large number of IoT devices. |
| [102] | LEO NB-IoT | Development of an uplink scheduling technique for an LEO satellite-based mMTC-NB-IoT system that mitigates the differential Doppler shift, thereby improving the system's performance in terms of throughput and block error rate. |

### 5.5. Enhanced Data Transmission Reliability

Researchers in [103] examine the importance of satellite Machine Type Communication (MTC) in the maritime IoT and highlight the challenges hindering its implementation. Their focus is on investigating potential interference issues between a satellite MTC system for maritime IoT and co-frequency terrestrial communication systems already in operation. The main obstacle stems from the lack of established regulatory criteria and methods to safeguard terrestrial systems against potential interference from LEO satellite systems or similar technologies. This absence makes it challenging to assess the impact on existing land systems. The researchers contribute significantly by mathematically deriving an electromagnetic Power Flux Density (PFD) mask. This mask helps evaluate and limit satellite space station emissions within the existing regulatory boundaries designed for mitigating interference among terrestrial communication systems in the same frequency band. The principle guiding this approach is ensuring that the interference experienced by a land system due to a space station is no more severe than that from a co-frequency land system permitted by regulations.

Addressing the challenging link between IoT devices and GEO satellites, a novel modulation and signaling scheme based on Chirp-Spread Spectrum (CSS) is proposed in [104]. This scheme, termed Unipolar-coded CSS (UCSS), enables reliable transmission at ultra-low bit rates and supports true random multiple access for a large number of devices, even at high carrier frequencies such as C-band to Ka-band. This breakthrough paves the way for ubiquitous connectivity and mMTC via satellite, even in remote and underserved areas. Furthermore, in [105], the same authors further delve into the challenges and opportunities associated with direct access of IoT devices to GEO satellites. They highlight the potential of UCSS to revolutionize satellite-based IoT communication, enabling reliable and efficient connectivity for a wide range of IoT applications. The successful demonstration of UCSS with a Ku-band testbed validates its feasibility and opens up exciting possibilities for ultra-narrow band mMTC via satellite.

The study in [106] highlights the transformative potential of satellites to significantly enhance the performance of IoRT networks, enabling seamless connectivity and robust data transmission across remote areas. By leveraging a Reinforcement Learning (RL) framework, the authors effectively address the challenge of optimizing resource allocation and IoRT data scheduling in dynamic and unpredictable environments. To overcome the limitations of traditional RL algorithms, they propose customized feature functions tailored to the unique characteristics of satellite-based IoRT networks and employ function approximation techniques to enhance the accuracy of resource allocation decisions. Their proposed state-action-reward-state-action-based RL strategy proves highly effective in simulations. The utilization of a linear combination of proposed features specifically designed for decision-making and the prevention of battery overflow results in effective energy management and reduced energy consumption.

Excessive propagation delay is also a significant challenge in satellite-based IoT (S-IoT). In this particular context, conventional Hybrid Automatic Repeat request (HARQ) strategies are frequently unsuccessful as a result of limited feedback efficiency. To tackle this challenge, a novel Network Code HARQ (NC-HARQ) transmission protocol is introduced in [107]. Enabling efficient multi-hop communication is its primary objective, particularly in S-IoT environments where feedback mechanisms are inadequate or absent. In order to evaluate the performance of the NC-HARQ protocol in terms of Age of Information (AoI), a four-state Markov model is utilized. This model facilitates the derivation of closed-form expressions for the mean AoI in the context of end-to-end two-hop transmission. The results of the simulation illustrate that the NC-HARQ protocol is more effective than several modern HARQ schemes, as indicated by the significantly decreased average AoI levels.

In [108], the authors highlight the very limited choice of available technologies and the challenges of developing data collection and control systems in remote areas. They conduct a critical review and analysis of various protocols and technologies used for IoT data transmission, and they propose a hybrid IoT-satellite network to address these limitations. This network intends to gather data employing a terrestrial LoRa LPWAN, with backhaul connectivity provided via the Iridium satellite system. Different data presentation formats for low-speed satellite channels were evaluated through simulations. Furthermore, the authors introduce GDEP (Gateway Data Encoding and Packaging), a data encoding and packaging technique that is able to reduce the number of Short Burst Data (SBD) containers, improving network efficiency.

In order to improve the efficiency of accessing and utilizing resources in a massive machine-type communication scenario, the authors of [109] suggest a collision avoidance technique as the first step in random access for LEO satellite Internet of Things. The proposed method allows for quick identification of collisions and accurate estimation of load while also being resistant to non-orthogonal interference. An analysis is performed on the probability of detecting the preamble, the probability of detecting collisions, and the accuracy of load monitoring. This analysis results in the development of an optimal set of probabilities for selecting preambles, which maximizes the precision of load monitoring.

To address the issue of data staleness caused by the slow sensor transmission rate and the high bit error rate of the satellite-to-ground link, ref. [110] introduces an age-optimal hybrid temporal-spatial generalized deduplication and automatic repeat request (HARQ-GD) protocol designed for high-sampling data collection in the satellite-integrated Internet of Things. By integrating temporal and spatial data correlations, encoding and decoding algorithms, and packet format optimization, the HARQ-GD protocol successfully mitigates data staleness. As a result, transmission times are reduced, and the compression rate is enhanced. The simulation results indicate that the HARQ-GD protocol reduces the Age of Information (AoI), a metric that quantifies the freshness of data, more effectively than conventional Generalized Deduplication (GD) and Hybrid Automatic Repeat Request With Chase Combining (HARQ-CC) schemes.

LEO satellite networks have emerged as a promising solution for ubiquitous IoT connectivity. A Grant-Free Random Access (GF-RA) mechanism is proposed in [111] to address the dynamic nature of IoT traffic and intermittent transmissions from randomly activated devices. This GF-RA system employs a modified Bernoulli-Rician Message Passing (BR-MP) technique with Expectation-Maximization (EM) for User Activity Detection (UAD) and Channel Estimation (CE). Extensive simulation results of the proposed BR-MP-EM algorithm validate its accuracy and show its resilience to channel impairments.

In [112], researchers propose an integrated satellite-terrestrial IoT network to extend coverage to areas not serviced by terrestrial cellular networks. The system employs Power Domain Non-Orthogonal Multiple Access for both LEO satellites and terrestrial base stations. However, interference arises between LEOs and base stations due to limited feedback line capacity. To mitigate this issue, researchers aim to minimize the overall network's transmission power while ensuring acceptable IoT device communication rates. To tackle this challenge, an Iterative Penalty Function (IPF)-based scheme is proposed for designing robust beamforming for satellites and base stations. Simulation results demonstrate the effectiveness of the proposed scheme in satellite-terrestrial IoT scenarios.

Table 7 summarizes the proposed schemes to enhance the reliability of IoT data transmission. The mitigation of interference, the development of novel modulation and signaling schemes, the use of RL-based resource allocation methods, HARQ feedback enhancement, the design of novel data encoding schemes and collision detection techniques, and robust beamforming schemes for NOMA-based networks are among the highlighted approaches.

**Table 7.** IoRT systems based on space-borne NTN networks: **enhanced data transmission reliability.**

| Refs. | NTN Architecture | Description |
|---|---|---|
| [103] | LEO | Mathematical derivation of an electromagnetic Power Flux Density (PFD) mask for mitigating satellite interference on terrestrial communication systems in the same frequency band. |
| [104,105] | GEO | Development of a novel modulation and signaling scheme based on Chirp-Spread Spectrum (CSS). |
| [106] | LEO-GEO | Development of a reinforcement learning framework to optimize resource allocation and enable robust data transmissions. |
| [107] | Generic | Use of Network Coding to enhance the feedback mechanism of the HARQ process. |
| [108] | LEO-LoRA | A system architecture is proposed, along with the development of a data encoding and packaging scheme for the reliable transmission of IoT data over low-speed satellite links. |
| [109] | Generic | Development of a novel collision detection scheme for satellite-based IoT that enables rapid collision detection and load estimation while being robust to non-orthogonal interference. |
| [110] | Generic | Use of data compression and transmission optimization to improve the HARQ process as well as the timeliness of data. |

Table 7. *Cont.*

| Refs. | NTN Architecture | Description |
|---|---|---|
| [111] | LEO | Development of a novel access scheme for terrestrial-satellite IoT communication that maintains low access delay, robustness to channel impairments, and enables reliable data transmission. |
| [112] | LEO | Development of a robust beamforming scheme for NOMA-based integrated satellite-terrestrial IoT networks to minimize power consumption and ensure reliable communication for a large number of IoT devices. |

### 5.6. Enhanced and Timely Data Acquisition

The study in [113] explores the use of CubeSats as low-cost satellite network building blocks for the Internet of Space Things. To enable seamless functionality for a variety of use cases and stakeholders, an automated network slicing methodology is proposed for space-ground integrated networks. Designed for extremely dense CubeSat networks, the system focuses on effective route computation and resource allocation while ensuring Service Level Agreement (SLA) adherence. Notably, it employs an SLA-based methodology, eliminating the need for prior knowledge of slice resource requirements. Furthermore, a case-driven evaluation scenario was employed to evaluate the adaptability and effectiveness of the proposed framework.

The study in [38] addresses the challenge of establishing IoT connectivity in remote areas with restricted communication infrastructure, such as the Arctic. The research presents an alternative solution involving a swarm of small, freely drifting satellites and industry-standard protocols. The authors validate the feasibility of this approach by simulating networking protocols and link characteristics across different satellite orbits and ground nodes. The findings indicate that small-satellite swarms can significantly minimize communication overhead and end-to-end request latency, enabling IoT deployment even in the most remote settings, like the Arctic.

The researchers in [67] explored the use of a satellite-enabled IoRT network to collect data in remote areas with no internet connectivity. Specifically, the Xingyun satellite constellation was employed to gather environmental data from the Tibetan Plateau. The monitoring system consisted of terrestrial, ground-based terminals equipped with satellite transceivers and environmental sensors. Five such terminals were deployed in challenging regions to monitor air temperature, relative humidity, precipitation, snow depth, land surface temperature, tree stemflow rate, and photosynthetically active radiation. Field experiments evaluated the performance of the proposed system, revealing its efficiency.

The research in [114] focuses on the problem of maintaining fast data updates in satellite-integrated IoT networks for applications like animal tracking and environmental monitoring. To address this, the authors propose Spatially Temporally Correlative Mutual Information (STI), a new metric that takes into account correlations between the most recent update message and the current state of the data source. They optimize channel slot allocation using a Markov decision process framework by maximizing the averaged STI over a specified updating period. The simulation results show that the proposed strategy outperforms traditional scheduled access schemes such as slotted ALOHA and Threshold-ALOHA.

In [115], a packet scheduling algorithm designed specifically for IoRT devices operating in 5G satellite networks is proposed. The algorithm achieves its objectives through the implementation of a cross-layer design approach. The principal objective of this packet scheduler is to optimize system throughput while ensuring a reasonable degree of fairness for both delay-sensitive and delay-tolerant IoRT services. The simulation results indicate that the proposed packet scheduler demonstrates better performance in terms of throughput, spectral efficiency, and fairness index when compared to previously proposed schemes.

The expanding M2M/IoT traffic and its potential for satellite delivery are examined in [116]. The emphasis is on comparing two major M2M/IoT protocol stacks, Constrained Application Protocol (CoAP) and Message Queuing Telemetry Transport (MQTT), on a satellite Random Access (RA) channel using the DVB-RCS2 specification. The completion time metric is critical in determining which protocol stack is preferable. The study emphasizes the advantages of the Publish/Subscribe (PUB/SUB) paradigm, particularly in satellite-based architectures, for its effectiveness in rapidly delivering new data to subscribers. The authors then propose integrating CoAP with the observer pattern and proxying functionality to optimize the PUB/SUB paradigm. CoAP outperforms MQTT on RA satellite channels, demonstrating the adaptability of application-layer tuning options.

A QoE-aware satellite constellation design scheme is proposed in [117] to improve user experience in satellite IoT applications. The satellite IoT network is composed of LEO satellites and ground-based IoT devices. Quality of Experience (QoE) factors are defined to evaluate coverage performance, communication efficiency, regional demand capacity, and profitability. A Multi-Layer Tabu Search (MLTS) optimization algorithm is then used to determine optimal satellite orbits that maximize QoE. The simulation results show that the proposed constellation is effective in optimizing multiple QoE factors, resulting in a better user experience.

Within the context of IoRT, the authors of [118] address the challenge of interconnecting widely dispersed IoT nodes. They propose a cooperative mechanism that employs feedback on link conditions and adaptive coding knowledge to optimize resource allocation and improve video transmission quality and fairness among IoRT nodes. The proposed methodology was validated by implementing it on an emulation platform, demonstrating its efficiency and advantages over existing methods. This cooperative mechanism has the potential to improve the performance and reliability of IoRT networks, especially in disaster recovery scenarios where seamless video transmissions from remote locations are critical.

A study in [119] proposes a joint optimization strategy for maximizing long-term network utility in a NOMA Satellite IoT downlink system. The proposed approach employs two virtual queues to manage both data queuing and power consumption. Lyapunov optimization is employed in order to attain network stability and optimize resource allocation, taking into account the constraints of onboard communication resources. Moreover, the authors propose a solution utilizing Successive Interference Cancellation (SIC) decoding in conjunction with Particle Swarm Optimization (PSO) to determine the optimal resource allocation strategy. The simulation results show that the joint optimization allocation operates efficiently in terms of long-term network utility, average data rate, and queuing delay.

Table 8 summarizes studies that have explored the potential of integrating space-borne NTNs into IoT networks to enhance data acquisition and improve timeliness. These studies have proposed innovative approaches for network slicing, small-satellite swarms, and QoE-aware constellation designs. They have also evaluated the performance of different protocols and resource allocation mechanisms and demonstrated that space-borne NTNs are a promising solution for enabling delay-sensitive IoRT applications in remote and challenging environments.

**Table 8.** IoRT systems based on space-borne NTN networks: **enhanced and timely data acquisition**.

| Ref. | NTN Architecture | Description |
|---|---|---|
| [113] | Generic | Development of an SLA-based automatic network slicing framework for ultra-dense SDN-based CubeSat networks. |
| [38] | LEO | Evaluating the feasibility of small-satellite swarms for IoT connectivity, focusing on reducing communication overhead and latency. |
| [67] | LEO | A study that demonstrates the feasibility of a satellite-enabled IoRT network for data retrieval in challenging Tibetan Plateau environments. |
| [114] | Generic | A metric is proposed to determine the optimal allocation of channel slots for data updates based on the timeliness of the information. |
| [115] | GEO | Development of a new packet scheduling algorithm for mixed traffic that improves throughput, ensures fairness, and maintains QoS for both delay-sensitive and delay-tolerant IoRT services. |

**Table 8.** *Cont.*

| Ref. | NTN Architecture | Description |
|---|---|---|
| [116] | Generic | Comparative performance evaluation of CoAP and MQTT protocols for rapid data delivery on satellite random access channels compliant with the DVB-RCS2 standard. |
| [117] | LEO | Development of a QoE-aware satellite constellation design scheme to enhance user experience in satellite IoT networks. |
| [118] | Generic | Development of a cooperative resource allocation mechanism that utilizes link condition feedback and adaptive coding to dynamically allocate resources and ensure consistent and reliable video streams. |
| [119] | LEO | Development of a joint optimization approach for successfully managing both data queuing delay and power consumption at the downlink of a NOMA-based satellite IoT system. |

### 5.7. Enhancement of Edge Computing Capabilities

Advancements in ultra-dense satellite communications, as explored in [120], could offer an alternative approach for expanding IoRT computational efficiency and resource utilization. IoRT devices can offload their processing tasks to satellite networks or relay them to terrestrial data centers via satellite communications. However, due to resource constraints, a single satellite cannot handle computation-intensive and complex operations. This paper proposes a collaborative task processing scheme that utilizes multiple satellites in a group to simultaneously process IoRT tasks, enhancing computational efficiency and resource utilization. To maximize completion delay reduction for all IoRT devices, the optimization problem is formulated as a Winner Determination Problem (WDP), and a low-complexity cooperative computation algorithm is presented as a solution.

Moreover, the growing importance of IoT in the information industry is emphasized in [121], highlighting the need to address network strain, particularly for remote IoT platforms. Edge computing is proposed as a solution for offloading processing closer to data sources. While satellite data transmission is crucial for remote IoT devices, conventional satellites lack versatility due to their specific application designs. To address this limitation, the authors propose transforming conventional satellites into adaptive space-edge computing nodes, enabling dynamic resource sharing and software management. Simulations demonstrate the efficiency of this approach compared to conventional satellite constellations, with the quality of service depending on the number of satellites, computational capacity, and task outsourcing strategies.

To address the challenge of managing outsourcing path selection and resource allocation for computation-intensive and delay-sensitive tasks in dynamic network environments, a novel Ka/Q-band satellite-terrestrial integrated network for IoT in remote locations is proposed in [122]. The problem is formulated as a Markov decision process, with the aim of maximizing offloaded tasks while minimizing LEO satellite power consumption and adhering to specific delay constraints. DRL techniques are applied in making optimal decisions by leveraging dynamic IoRT device queues and taking into account time-varying channel conditions and ground station computing capabilities. Simulation results demonstrate the effectiveness of the proposed scheme.

An integrated satellite-terrestrial network architecture is proposed in [123] to enhance delay-sensitive task outsourcing for IoRT applications. This architecture seamlessly integrates satellite and terrestrial networks, expanding communication resources, backhaul capacities, and coverage to address the challenges of IoRT deployment. A key focus of the proposed architecture is to optimize offloading link selection and bandwidth allocation for base stations and IoT users. To address the differentiated time granularities of various decision-making processes, a two-timescale stochastic optimization problem is formulated. The authors develop a Hybrid Proximal Policy Optimization (H-PPO)-based algorithm to efficiently address the subproblems involved in the optimization problem. Simulation results demonstrate the effectiveness of the proposed scheme, particularly in scenarios characterized by limited spectrum resources and heavy traffic volumes.

To enhance the performance of IoT applications in LEO satellite networks, the authors of [124] propose a hybrid offloading architecture that integrates satellite mobile edge computing, which enables IoT devices to directly access satellite-based computational resources, eliminating the need for distant cloud servers and minimizing transmission and computation energy consumption. This approach alleviates bandwidth constraints and minimizes transmission delays. Additionally, the authors present a multi-agent actor-critic RL algorithm to optimize outsourcing policy decisions while taking into account the limitations of satellite resources.

Table 9 summarizes the research efforts that have explored the use of space-borne NTN networks to enhance edge computing capabilities for IoRT applications. These efforts have demonstrated the potential of space-based NTNs to provide low-latency, high-bandwidth connectivity to remote IoRT devices, enabling the offloading of computation tasks to satellites for processing. This can significantly improve the performance and scalability of IoRT applications, particularly in resource-constrained environments.

**Table 9.** IoRT systems based on space-borne NTN networks: **enhancement of edge computing capabilities.**

| Ref. | NTN Architecture | Description |
|---|---|---|
| [120] | LEO | Development of a collaborative task processing scheme that leverages multiple satellites to process offloaded IoRT tasks. |
| [121] | LEO | Development of a hardware and software architecture that utilizes virtualization and flexible scheduling to transform traditional satellites into space edge computing nodes. |
| [122] | LEO | Modeling computation offloading as a Markov decision process and optimizing it using deep reinforcement learning to maximize offloaded tasks while ensuring delay performance. |
| [123] | LEO | Development of a policy optimization reinforced learning algorithm, which supports delay-sensitive task offloading to achieve better data transmission rates and reliability. |
| [124] | LEO | A hybrid computation offloading architecture is proposed, consisting of LEO satellites and cloud servers, to minimize task delay and reduce energy consumption utilizing multi-agent reinforcement learning. |

## 6. IoRT Systems Based on Hybrid NTN Networks

This category of IoRT systems seeks to exploit the combined strengths of satellites and UAVs to supplement the satellite backbone infrastructure with dynamic and mobile capabilities, enabling targeted coverage. However, this complexity translates into more sophisticated systems, resulting in fewer research works in this regard. Therefore, we identified a total of 11 related research works in this category, classified according to their core research objectives into the following subsections. As anticipated, the limited number of research works does not address the full range of research objectives identified in Section 4. Consequently, as we delve deeper into Section 8, a research gap is identified, emphasizing potential areas for future investigation.

### 6.1. *Enhancement of Energy Efficiency*

In [125], the authors investigate the feasibility of a satellite-aerial-terrestrial integrated network to effectively support the expanding IoT landscape. The study centers on cognitive radio-based networks, leveraging their capabilities to address the critical challenge of spectrum resource management in response to the increasing demands of IoT devices. The study emphasizes the significance of spectrum sharing across satellite, aerial, and terrestrial networks, highlighting the role of cognitive networks in fulfilling IoT requirements. The authors propose a cooperative beamforming scheme designed to optimize IoT communications' security and energy efficiency under constrained energy resources.

Leveraging UAV relays, ref. [126] explores energy-efficient resource allocation in a two-hop uplink communication for Space-Air-Ground IoRT networks. In remote areas where ground base stations lack reach, satellites and smart devices rely on UAV relays for seamless communication. By optimizing sub-channel selection, uplink transmission power control, and UAV relay placement, the proposed iterative technique maximizes system energy efficiency, addressing the challenges of IoRT networks. The Lagrangian dual decomposition approach is employed to determine the optimal sub-channel selection and power control policy based on UAV relay deployment positions. Successive convex approximations are then used to determine UAV relay placement. Numerical results demonstrate that the proposed scheme significantly improves system energy efficiency.

In [127], the authors propose a novel UAV-LEO integrated data collection system that addresses the challenges posed by the network's heterogeneity and the limited mobility of UAVs and LEOs. The system aims to maximize data gathering efficiency by optimizing IoT device bandwidth allocation and UAV trajectory optimization. Optimizing the transmission powers of UAVs and selecting LEO satellites are proposed to maximize data upload volume and minimize UAV energy consumption. Simultaneously optimizing IoT-UAV data collection and UAV-LEO data transmission is achieved by considering the relay role and cache capacity constraints of UAVs. This problem is efficiently solved using sequential convex approximation (SCA) and block coordinate descent (BCD). The proposed system is shown to perform efficiently in terms of energy consumption and total upload data volume.

Table 10 summarizes key research efforts in enhancing the energy efficiency of IoRT systems utilizing hybrid NTN architectures. Researchers have proposed various techniques, including cognitive network architectures, cooperative beamforming schemes, and iterative-based algorithms, to optimize energy consumption and maximize network performance under constrained energy resources. These advancements demonstrate the promising potential of hybrid NTN networks in addressing the energy efficiency challenges in IoRT applications.

**Table 10.** IoRT systems based on hybrid NTN networks: **enhancement of energy efficiency**.

| Ref. | NTN Architecture | Description |
|---|---|---|
| [125] | LEO-UAV | A cognitive network architecture together with a cooperative beamforming scheme aiming to optimize energy efficiency under constrained energy resources. |
| [126] | LEO-UAV | Iterative-based algorithms are proposed to maximize systems' energy efficiency in two-hop link communication for hybrid NTN-based IoRT networks. |
| [127] | LEO-UAV | Effective techniques have been developed to optimize UAV trajectories and transmit powers, aiming to maximize data upload volume while minimizing UAV energy consumption. |

### 6.2. Enhanced Throughput/Transmission Rate

In [128], the authors present a simulation of a scenario involving a large number of UAVs hovering in the sky and a GEO-deployed relay satellite. They delve into analyzing and optimizing the performance of IoRT networks with UAV access and GEO satellite backhaul. To address the modeling challenges posed by the complex two-tier network, they developed a two-level queue system composed of multiple UAV queues on the first level and the satellite queue on the second level. To design the two-level transmission service for each UAV group, they introduce the stochastic network calculus (SNC)-based min-plus convolution and the leftover service, which are used as mathematical tools to effectively address the complexities of the network structure. The simulation results demonstrate that the proposed approach is effective in maximizing throughput while guaranteeing delay bounds.

Satellite and UAV integrated networks have the potential to enable a variety of IoRT applications. With regard to IoRT applications, the authors of [64] investigate methods for enhancing downlink transmission. Radio frequency technology is used for satellite-to-UAV links, and free-space optical (FSO) technology is used for UAV-to-IoRT connectivity. To address the scarcity of statistical Channel State Information (CSI), the authors devise an optimization problem aimed at maximizing the system's Ergodic Sum Rate (ESR) while adhering to power and IoRT device rate constraints. To address this nonconvex problem, a beamforming scheme based on the Alternating Direction Method of Multipliers (ADMM) is proposed. To reduce implementation complexity, a suboptimal zero-forcing approach is introduced. For the proposed BF schemes, closed-form ESR expressions are derived, and simulations are employed to validate the theoretical analysis.

Addressing the critical absence of 5G coverage in maritime regions, ref. [129] proposes a nearshore network leveraging on-shore terrestrial base stations and tethered UAVs in virtual clusters. Employing non-orthogonal multiple access for efficient data transmission to maritime IoT devices, the network forms a hybrid satellite-UAV-terrestrial architecture. To mitigate interference, a joint power allocation problem is formulated to maximize network sum rates. Large-scale channel state information extracted from maritime IoT device locations is employed to optimize system overhead. An iterative power allocation algorithm is introduced, demonstrating the potential for enhanced coverage in maritime on-demand networks based on NOMA.

Due to the limited transmission power of smart devices in IoRT networks, UAVs can be used as relays to transmit data from smart devices to LEO satellites in Space-Air-Ground (SAG)-IoRT systems. To maximize system capacity, the authors of [130] propose a joint optimization approach that considers smart device connection scheduling, power control, and UAV trajectory. To achieve maximum system capacity, researchers alternately iterate smart device connection scheduling, power control, and UAV trajectory design. The proposed iterative solution outperforms both the static UAV scheme and the dynamic UAV scheme with a circular trajectory.

Table 11 summarizes research on IoRT systems based on Hybrid NTN Networks to enhance throughput and transmission rate. These studies propose novel frameworks and resource allocation strategies to address heterogeneity challenges and maximize network performance.

**Table 11.** IoRT systems based on hybrid NTN networks: **enhanced throughput/transmission rate.**

| Ref. | NTN Architecture | Description |
|---|---|---|
| [128] | GEO-UAV | Development of a framework for analyzing and optimizing hybrid UAV-GEO networks, addressing heterogeneity challenges and maximizing throughput while guaranteeing delay bounds. |
| [64] | GEO-UAV | Development of a beamforming scheme based on the alternating direction method of multipliers to maximize the ergodic sum rate of the system. |
| [129] | Generic Satellite-UAV | A study of a hybrid satellite-UAV-terrestrial network architecture for maritime on-demand coverage that employs NOMA to manage interference and maximize network sum rate. |
| [130] | LEO-UAV | Development of a resource allocation scheme that maximizes system capacity by jointly optimizing connection scheduling, power control, and UAV trajectory. |

### 6.3. *Enhanced and Timely Data Acquisition*

The study in [69] explores a satellite-based 5G scenario for massive machine-type communication, utilizing an intermediate layer of UAVs for efficient data collection from sensors dispersed across vast rural areas. Specifically, the application focuses on the timely detection of fire alarms. In the scenario under study, terrestrial internet connectivity is limited or unavailable, making UAVs the ideal solution for gathering sensor data and relaying it to a control station via satellite. The study proposes a system architecture tailored to this scenario and develops an analytical model to characterize the average delay involved in transmitting fire alarm notifications from the field to the control center. The mobile gateway framework provides an alternative for collecting data while also reducing the problem of network partition caused by static gateways. To validate the proposed approach, simulations were conducted, demonstrating its effectiveness.

To optimize data acquisition strategies in the IoRT, a novel integrated space–air–ground network employing unmanned aerial vehicles and low-earth orbit satellites is proposed [65].

The authors differentiate between delay-sensitive and delay-tolerant data, tailoring transmission methods accordingly. For delay-sensitive data, a real-time relay via LEO satellites ensures timely delivery, while delay-tolerant data are stored on UAVs for subsequent transmission to a data center. To address the complex optimization problem, a multidimensional algorithm, Rate Demand-Based Joint Optimization (RDJO), is introduced. RDJO optimizes UAV deployment, IoRT device bandwidth allocation, and UAV transmission power, significantly enhancing IoRT data acquisition efficiency. Simulations comparing RDJO to conventional algorithms validate its efficiency.

The research in [131] investigates the combined use of UAVs and LEO satellite networks to provide low-latency access to IoRT sensors. The authors evaluate LEO satellite-assisted UAV data collection, which utilizes two communication mechanisms: delay-tolerant data are transmitted to Earth via the carry-store mode of UAVs, while delay-sensitive data are transmitted via the UAV satellite network. Considering the limited payloads of UAVs, the authors aim to minimize the total energy cost of the UAV trajectory and transmission while meeting IoRT requirements. The efficiency of the proposed approach is proven by numerical results.

Table 12 summarizes the work on IoRT systems based on hybrid NTN networks for enhanced and timely data acquisition. Bandwidth allocation and UAV placement strategies are employed to optimize data transmission delay under delay sensitivity constraints.

**Table 12.** IoRT systems based on hybrid NTN networks: **enhanced and timely data acquisition.**

| Ref. | NTN Architecture | Description |
|---|---|---|
| [69] | LEO-UAV | A novel system architecture is proposed, along with an analytical model to quantify the mean delay in alarm notifications, with a particular emphasis on ensuring timely alarm detection. |
| [65] | LEO-UAV | Joint optimization of bandwidth allocation for IoRT devices and UAV location deployment and transmission power while considering data delay sensitivity. |
| [131] | LEO-UAV | Combined use of UAVs and LEO satellite networks to delay aware access to IoRT sensors while minimizing the overall energy cost of UAVs. |

### 6.4. Enhancement of Edge Computing Capabilities

The authors in [132] propose an MEC-enabled NTN for wide-area, time-sensitive IoT scenarios in remote regions. The proposed system utilizes a hierarchical architecture comprising GEO satellites, UAVs, HAPs, and LAPs, with a focus on a UAV swarm equipped with an MEC server and a satellite that relays data to the cloud server via a gateway. The NTN employs a cell-free architecture to address the complex propagation environment between communication and MEC and simultaneously coordinates resources to accommodate on-demand coverage and the uneven distribution of IoT devices. The problem is modeled using large-scale CSI to minimize computing latency in the NTN, while a power allocation and data stream scheduling system is also introduced.

Table 13 summarizes the research work on IoRT systems based on hybrid NTN networks for enhancement of edge computing capabilities. MEC-enabled framework along with large-scale CSI are employed to optimize latency and coverage for IoT devices.

**Table 13.** IoRT systems based on hybrid NTN networks: **enhancement of edge computing capabilities.**

| Ref. | NTN Architecture | Description |
|---|---|---|
| [132] | GEO | Development of a process-oriented framework that integrates NTNs with MEC to enable wide-area, time-sensitive IoT applications. By utilizing large-scale CSI, it optimizes latency while ensuring on-demand coverage for IoT devices. |

## 7. IoRT Systems Based on Aerial NTN Networks

The reliance on UAVs as the sole means of connectivity for isolated areas not covered by terrestrial wireless networks presents obvious limitations. As a result of these limitations, we identified a total of four research works in this category, classified according to their core research objectives into the following subsections. As expected, the limited number of research works covers only a subset of the research objectives identified in Section 4. Consequently, a research gap is evident, indicating potential areas that can be explored in future studies.

### 7.1. Enhancement of Energy Efficiency

This work introduces a drone-powered IoT relay system for efficient data collection in remote environmental monitoring applications [133]. The system utilizes 5-GHz communication technology for high-speed data transmission and low-power LoRa technology for energy optimization, enabling a sustained throughput of 3.5 MB/s for cached data collection at an altitude of 140 meters. This innovative approach offers a promising solution for intelligent environmental monitoring in regions lacking public networks, significantly enhancing data transmission from critical areas.

In this study, the authors examine the future of 6G wireless communication networks, seeking improved metrics to connect a large number of devices, which presents challenges for current IoT applications [134]. Integrating non-orthogonal multiple access and spatial modulation techniques, a novel method called Spatial NOMA (S-NOMA) is introduced to increase energy efficiency. Multiple-Input Multiple-Output (MIMO) enables spatial modulation to selectively activate transmission antennas per symbol interval, thereby optimizing data rates and minimizing interantenna interference. In addition, an energy efficiency-focused optimization method for power allocation is proposed for the S-NOMA scheme. This method improves energy efficiency for all users by incorporating antenna selection bits from all users, in contrast to traditional NOMA. Simulation results demonstrate the improved performance of S-NOMA with energy-efficient power allocation over conventional NOMA.

Table 14 summarizes the work on enhancing the energy efficiency of IoRT systems using aerial NTN Networks.

**Table 14.** IoRT systems based on aerial NTN networks: **enhancement of energy efficiency**.

| Ref. | NTN Architecture | Description |
|---|---|---|
| [133] | LAP-UAV | Development of a low-cost, UAV-enabled LoRa IoT relay system for enhanced energy efficiency and high-speed environmental data acquisition. |
| [134] | UAV | Development of a spatial NOMA scheme combined with multiple access and spatial modulation techniques to enhance energy efficiency in 6G wireless networks for massive IoT applications. |

### 7.2. Enhanced Throughput/Transmission Rate

As a solution for the 6G heterogeneous IoT, Clustered Non-Orthogonal Multiple Access (C-NOMA) is proposed in [135] in order to address spectrum efficiency, system throughput, and receiving-end complexity. It uses UAVs as Low-Altitude Platform Stations for localized communication reinforcement and high-altitude platform stations for broader coverage. The study aims to optimize four key factors: spectrum allocation, power control, horizontal coordinates, and hovering altitude for UAV 3D location planning and resource allocation. Simulations and theoretical analysis demonstrate that the proposed two-stage solution substantially improves the system uplink sum rate with relatively low complexity compared to previous methods.

The study in [136] proposes a Clustered NOMA (C-NOMA) system that utilizes UAVs as aerial base stations for Wireless-Powered Communication (WPC) to maximize the uplink average achievable sum rate of IoT terminals [136]. WPC enables IoT terminals to harvest energy from downlink radio frequency signals and subsequently use the harvested energy to transmit data to the uplink. By optimizing UAV trajectory and subslot allocation in a synergistic manner, the proposed system ensures that the uplink sum rate and UAV mobility constraints are met. The complex and nonconvex optimization problem is efficiently solved using an iterative algorithm. Numerical results validate the effectiveness of this scheme for a UAV-supported C-NOMA system.

Table 15 presents a concise summary of the information presented in this section.

**Table 15.** IoRT systems based on aerial NTN networks: **enhanced throughput/transmission rate.**

| Ref. | NTN Architecture | Description |
|---|---|---|
| [135] | HAP-UAV | A 6G clustered-NOMA HAP-UAV system architecture is proposed, and a UAV 3D location planning and resource allocation scheme is developed that enhances system throughput. |
| [136] | LAP UAV | A trajectory optimization and subslot allocation scheme is proposed for UAV-based clustered NOMA wireless-powered communication systems to improve the uplink average sum rate for IoT terminals. |

## 8. Recent Trends, Observations, and Future Research Challenges

This section analyzes current research trends observed in Sections 5–7 and identifies future research challenges related to integrating NTNs into the IoRT environment.

### *8.1. Research Trends and Observations*

IoRT presents unique challenges for connectivity and data gathering and processing, demanding innovative solutions that are cost-effective, scalable, and efficient. To address these challenges, we have identified key research trends that are shaping the future of IoRT. This subsection delves into promising technologies like CubeSats, NOMA, machine learning, and MEC. Examining these technologies provides valuable insights into how IoRT can overcome its hurdles and realize its full potential for revolutionizing remote IoT applications.

- *CubeSats: A Cost-Effective and Scalable Solution for Remote IoT Connectivity*

  Miniaturized satellites known as CubeSats are rapidly gaining traction as a cost-optimized solution for establishing communication links with geographically dispersed IoRT systems. Their miniature size and inherently low energy requirements make them ideal for constructing network constellations tailored to specific mission profiles and optimized for minimized energy expenditure. Furthermore, CubeSats exhibit exceptional scalability and adaptability, enabling the deployment of dense constellations with extensive coverage areas, particularly in regions beyond the reach of terrestrial infrastructure. This capability is further augmented by their proficiency in forming dynamic collaborative networks, which maximizes resource utilization and flexibility. Furthermore, CubeSats play a crucial role as excellent research platforms, promoting progress in compact electronics, energy-conserving techniques, and cost-efficient deployment strategies. These developments will have an immediate effect on the connection of IoRT systems because they lay the groundwork for future global connectivity that will be both more affordable and more accessible.

- *Softwarization Driving Agile and Scalable Remote IoT Networks*

  A key research path for NTN-based IoRT systems is centered around the softwarization of network infrastructure. The integration of SDN and NFV offers a flexible and dynamic architecture to achieve cost-efficient deployment in NTNs. Researchers are exploring the potential of software-defined radio for cost-effective gateway implementation, addressing interoperability challenges, and enhancing adaptability in diverse network environments. Additionally, the adoption of SLA-based automatic network slicing frameworks can also be noted as an emerging trend, enabling the efficient allocation of resources and customization of network slices tailored to the specific requirements of IoT devices. These trends collectively signify a shift towards more agile, cost-effective, and scalable solutions for the evolving landscape of remote IoT-based NTNs.

- *NOMA: A Key Enabler for Scalable and Efficient IoRT Connectivity*

  NOMA enhances spectral efficiency by enabling more devices to share the available frequency spectrum, thereby increasing capacity. This directly translates to increased network capacity, facilitating enhanced connectivity for remote IoT devices communicating with satellites and UAVs. NOMA, by leveraging a dynamic scheduling scheme, can dynamically adjust resource allocation based on the varying needs of IoT devices. This allows the system to handle a mix of devices with different data rates, latency sensitivities, and QoS demands by dynamically adapting the transmission parameters for each device. This adaptability further mitigates latency issues by prioritizing critical information transmission for real-time applications. NOMA's inherent scalability ensures effective connectivity in dynamic environments, accommodating a growing number of devices and users. Moreover, its compatibility with emerging technologies like edge computing and machine learning further enhances performance and facilitates efficient resource management and optimization.

- *Machine learning-enabled IoRT: Optimizing Network Performance*

  Machine learning is revolutionizing IoRT by unlocking the full potential of NTN systems. ML algorithms excel at optimizing resource allocation within the NTN network and for its users. They achieve this through efficient bandwidth distribution via techniques like predictive demand forecasting, dynamic traffic routing, and priority-based scheduling. Furthermore, reinforcement learning and deep RL can be effectively harnessed to mitigate the Doppler effect in satellite communication. These techniques accurately predict frequency changes and promptly adapt modulation techniques, leading to improved signal quality and data transmission. Beyond network optimization, ML plays a crucial role in data management. It helps identify anomalies, reduce data size, and extract relevant features, enabling efficient analysis and informed decision-making. This empowers the use of NTN for the IoRT, providing robust and effective connectivity solutions for diverse IoRT challenges.

- *MEC-enabled IoRT Task Offloading*

  Integrating MEC architecture with NTNs holds significant potential to enhance the performance of NTN-based IoRT systems. Depending on the NTN architecture, this integration can be achieved by incorporating computing capabilities into satellites or UAVs, essentially forming edge servers. These edge servers, positioned closer to IoT devices, can offload IoRT processing tasks, reducing reliance on distant cloud servers. Utilizing coordination techniques, often based on ML, enables the dynamic allocation of offloaded tasks to achieve various objectives, such as energy-efficient channel allocation, maximizing the number of offloaded tasks, or minimizing latency through delay-aware task offloading. Furthermore, real-time collaboration may enable edge nodes to exchange processed data and make collective decisions, further improving the effectiveness of MEC.

### 8.2. Future Research Challenges

Research on NTN-based IoRT systems continuously seeks to improve existing techniques, and new approaches and technologies offer further potential for advancement across all design objectives. However, despite significant research dedicated to IoRT systems based on space-borne NTN networks and their alignment with the design objectives outlined in Section 4, a notable gap exists for IoRT systems based on Hybrid and Aerial NTN networks. Research in these domains has not comprehensively addressed all identified design objectives, as will be further explored in the following.

- *Enhancing Coverage, Data Transmission Reliability and Timely Data Acquisition*

  For objectives related to coverage, data transmission reliability, and timely data acquisition in IoRT systems based on Hybrid and Aerial NTN networks, the identified research gap is partially attributed to the reliance on the mobility capabilities of UAVs, particularly LAPs, which can offer localized coverage, potentially improving both reliability and coverage and facilitating timely data acquisition. However, it is also crucial to acknowledge their inherent limitations in range, energy consumption, and integration with existing infrastructure. Dedicated research efforts are essential in these areas to develop effective schemes that comprehensively address all design objectives, unlocking the full potential of UAVs and bridging the gap towards robust and reliable IoRT networks.

- *Addressing Cost-Effective Deployment Challenges*

  Furthermore, a critical research gap persists in the area of cost-effective deployment and operation of IoRT systems based on Hybrid and Aerial NTN networks. Especially in the case of hybrid NTN networks, their inherently more complex architecture is expected to pose more challenges. Consequently, this presents fertile ground for exploration, where innovative approaches like CubeSats, which are explored for space-borne NTN networks, can be applied. However, it is important to note that while leveraging research from space-borne NTN networks to hybrid and aerial networks and vice versa can be promising, direct technology transfer is often not feasible. The research community needs to carefully assess the applicability of existing knowledge in each specific case, considering the distinct characteristics and challenges of each network type.
- *Enhancing Edge Computing Capabilities*

  Another significant research opportunity lies in enhancing the edge computing capabilities of IoRT systems based on aerial NTN networks. Exploring and unlocking the potential synergies between UAV swarms, edge computing, and AI holds significant promise for revolutionizing IoRT systems based on aerial NTN architectures. By fostering deeper research and development in this domain, we can pave the way for a future with intelligent, distributed networks that seamlessly connect and empower a vast array of remote devices. This holds the potential to revolutionize fields ranging from environmental monitoring and disaster response to precision agriculture.
- *Energy Efficiency of IoRT Systems*

  While enhancing edge computing capabilities presents a transformative opportunity for IoRT systems, another equally important research area lies in ensuring the energy efficiency of these networks. Although the research community has made efforts to address this challenge, further work remains to be conducted. Moreover, the ongoing drive to improve energy efficiency in NTNs, particularly for seamless integration with remote IoT systems, is expected to be incorporated into the upcoming 3GPP standards. This is crucial for the sustainability of NTNs, and therefore, the implementation of technologies like extended Discontinuous Reception (eDRX) and the adoption of energy-as-a-service criteria are expected to have a significant impact on the development of energy-efficient models for future NTNs, aligning with the recommendations outlined in the 3GPP standards.

## 9. Future Outlook toward 6G-Era

Despite the transformative impact of 5G networks on user experiences and IoT applications, their limitations, particularly in remote or isolated regions, are undeniable. These areas lack terrestrial network coverage, hindering seamless access and service provision for 5G-enabled IoT applications. In response to this challenge, ongoing research and scientific trends highlight the need for continuous 5G standard evolution, particularly in the realm of non-terrestrial networks. By seamlessly integrating these networks with their terrestrial counterparts, we pave the way for the emergence of futuristic 6G networks [137].

This shift in focus aims to address the limitations of current IoT connectivity and unlock new opportunities for remote applications [138,139]. As we look towards the 2030s, the advent of 6G networks is poised to overcome these obstacles, fulfilling the fundamental need for ubiquitous connectivity [140,141]. Research and industry are actively exploring the capabilities and requirements of 6G, with ultra-high frequency communications, artificial intelligence, edge computing, and non-terrestrial networks at the forefront [142–144]. These elements together will construct a fully networked 6G world, empowering remote IoT applications and emerging services. A 6G network provides the vision of a unified environment that seamlessly connects the globe, bridging the divide between remote and accessible areas[145–147].

### 9.1. NTNs Vision and The Forthcoming Energy-Awareness

Figure 4 illustrates the various stages of 5G network upgrades towards 6G-IoT as per the 3GPP Release cycles. Anticipating additional technical advancements, the evolution from 5G to 5G-Advanced standards is expected to address the limitations of non-terrestrial networks, particularly in areas with no terrestrial coverage. The ongoing development of 5G technology aims to enhance performance and cater to emerging use cases.

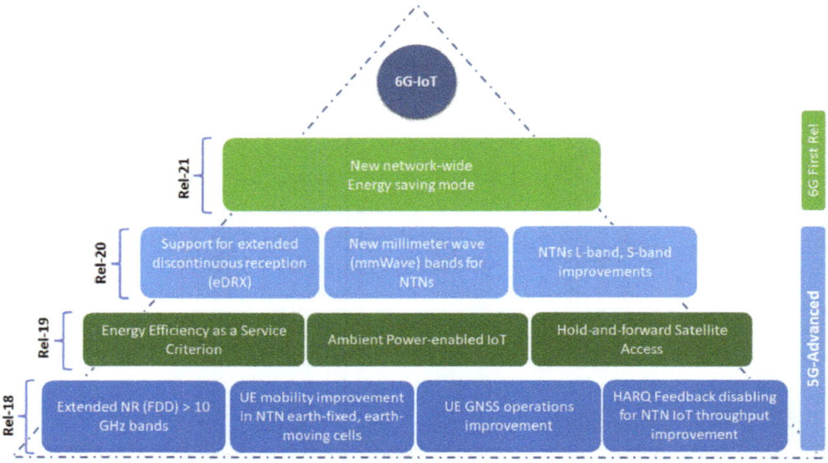


**Figure 4.** Overview of planned 3GPP standard releases on NTN.

The inherent versatility of 5G lays a robust foundation for enabling NTNs. Given the complexity of NTNs, especially satellite communication networks, a holistic approach is necessary for their design. Across multiple 3GPP releases, considerable attention has been dedicated to NTN design [72], signaling a commitment to making 5G from space a reality. Activities beyond standards are deemed essential for integrating non-terrestrial networks with IoT [41]. Subsequent 3GPP upgrades are poised to unlock possibilities for servicing remote locations.

### 9.2. NTNs in 3GPP Releases 18 and 19

Enhancements in 3GPP Release 18 (Rel-18) for NTNs encompass support for frequencies above 10 GHz, integration with 3GPP New Radio (NR), improved mobility management, handover procedures, power-saving features, and enhanced security and reliability [12,148]. These improvements are projected to facilitate diverse use cases, including global broadband connectivity, support for remote and underserved areas, disaster relief, connected vehicles, drones, and Industrial IoT. Furthermore, Rel-18 is expected to introduce the option to disable Hybrid Automatic Repeat Request (HARQ) feedback for NTN IoT by default, which can significantly improve throughput for devices with sporadic data traffic.

In the context of Release 19 (Rel-19), support for satellite-based 3GPP communication implies the availability of a backhaul network between the radio access node and the core network, as well as the wireless device's ability to utilize existing Global Navigation Satellite System (GNSS) functionality [149]. For the Rel-19 discussion, 3GPP will explore scenarios involving discontinuous backhaul connections. For example, a satellite could regularly orbit Earth and receive data from a location without direct transmission. In this scenario, it is necessary for satellite communication to have hold-and-forward capabilities.

While devices may use a 3GPP-based satellite communication architecture to determine their location in certain scenarios, the primary driver behind the upcoming 3GPP specifications for NTNs is the development of direct links with both user equipment (UE) and IoT devices.

In addition, Rel-19 is projected to integrate support for ambient power-enabled remote IoT devices, enabling them to harvest energy from their surroundings to power their operations. Apart from this, the Hold-and-Forward (HF) feature, which is also expected and being studied in Rel-19, is designed to overcome the issues of latency and reliability that are commonly encountered in satellite communications as it enables the satellite to temporarily hold data packets prior to their transmission to the ground station. Consequently, the hold-and-forward technique plays a crucial role in guaranteeing the integrity of data in challenging and remote conditions by storing packets that could potentially be lost as a result of interference or obstacles.

The scientific community has consistently emphasized that adherence to the 3GPP standards for NTNs is crucial to the successful implementation of future enterprise strategies [150]. This recognition of NTN's significance as a key work item and research area within the 3GPP has attracted a growing number of companies seeking to participate in this burgeoning field. However, the organization is actively exploring potential collaborations with emerging space companies and established corporate entities to harness the potential of 3GPP-based NTNs and expand their reach.

Energy Efficiency Challenges

In 3GPP Rel-18, significant focus has been directed towards enhancing the energy efficiency of NTNs [12]. These enhancements are intended to reduce power consumption in NTN base stations and user UEs while maintaining or even improving overall performance. One noteworthy improvement involves refining beamforming techniques, enabling more precise targeting of signals from NTN base stations towards UEs. This targeted signal delivery minimizes the power required for data transmission and reception. Furthermore, introducing new power-saving modes for NTN UEs is another crucial addition. These modes allow UEs to conserve power during periods of inactivity or when data transmission or reception is not actively in progress. In addition, network-assisted power management is incorporated to assist UEs in effectively managing power consumption. The network will provide valuable information about traffic conditions and available energy sources, empowering UEs to optimize their power usage efficiently [151].

Furthermore, leveraging AI and ML has emerged as a pivotal strategy for enhancing energy efficiency in NTNs. AI and ML algorithms are employed to dynamically adjust critical parameters such as beamforming, power levels, and network topology, leading to significant energy savings [152]. As recognized by 3GPP, energy efficiency will be a major focus of Rel-19, with its importance reflected in its consideration as a service criterion. Telecommunications operators are actively pursuing strategies to reduce energy usage in network infrastructure and wireless devices. Optimizing power consumption in the network equipment is paramount for achieving cost efficiency and reducing operating costs [153]. However, it is crucial to strike a balance between power reduction and maintaining QoS for end-users.

### *9.3. Future 3GPP Standard Releases*

Future versions beyond 3GPP Rel-18 and Rel-19 pave the way for future technical upgrades of 5G-NTN on a physical level, as well as a reconfiguration of 5G networks by 2030, when the first version for 6G networks will be released. This direction necessitates substantial investigation and is open to scientific testing and simulations by the scientific community. The need for improvements in energy efficiency and the limited availability of spectrum require further enhancements.

To tackle these challenges, 3GPP Release 20 (Rel-20) initiatives will address them by introducing innovative technologies such as extended discontinuous reception (eDRX) and millimeter wave (mmWave) bands [150,154–156]. The eDRX is a power-saving technology that enables NTN devices, like satellites, to enter periodic low-power sleep mode while maintaining continuous network communication. This is crucial for satellites that have limited energy resources, as it helps to increase their operational lifespans and decrease expenses. 3GPP specifications for mmWave bands for NTNs introduce novel channel models, modulation and coding schemes, and beamforming techniques. These advancements are specifically designed to optimize the propagation characteristics and interference environment of NTNs. As a result, they provide improved bandwidth, reduced susceptibility to interference, and the potential for high data rates and low latency [155].

In the direction of the space bands, specific advancements were made for L-band and S-band in the context of NTNs. For L-band, the standardization included Single Carrier Modulation and Coding Schemes (SC-MCS), providing enhanced spectral efficiency and lower latency compared to conventional multi-carrier modulation. Additionally, new channel models and interference mitigation techniques tailored for L-band NTN links were introduced [157]. In the case of S-band, the focus was on optimizing Dual Polarized (DP) and Quadrature Polarized (QP) antenna systems, improving resource allocation, and modulation schemes. The standardization also included beamforming techniques to concentrate transmitted signals, thereby extending the capacity and range of S-band NTN links. These standardized improvements in the L-band and S-band, driven by 3GPP Rel-20, are poised to significantly enhance the capabilities of NTNs, fostering innovations in areas like maritime connectivity, aerial broadband, and disaster relief [157,158].

In conclusion, 3GPP Rel-20 and Release 21 (Rel-21) mark a pivotal transition towards the next generation of the Internet of Things, commonly referred to as 6G-IoT. These releases signify significant progress in Non-Terrestrial Networks, introducing millimeter wave (mmWave) bands, enhancing L-band and S-band, and, notably, introducing network-wide energy-saving modes. Rel-20's mmWave bands provide expanded bandwidth, which is crucial for supporting emerging applications like video streaming, while advancements in L-band and S-band improve spectral efficiency and reduce latency, fostering wider coverage in remote areas. Building upon Rel-20, Rel-21 introduces energy-saving modes, enabling NTN devices to operate more sustainably.

The synergistic impact of these advancements positions 6G-IoT as a transformative force with the potential to revolutionize the connectivity and quality of experience (QoE) of users in remote areas and utilize services upon the remote IoT, including remote ultra-high precision agriculture, remote environmental condition monitoring, and remote supply chain optimization.

## 10. Conclusions and Future Work

In conclusion, this survey paper emphasizes the importance of non-terrestrial networks in the Internet of Remote Things ecosystem. These networks overcome the limitations of traditional terrestrial networks, enabling reliable connectivity in geographically isolated areas and challenging environments. The paper identifies seven key objectives for NTN research: energy efficiency, cost-effective deployment/operation, enhancement of coverage and availability, enhancement of throughput/transmission rate, enhancement of data transmission reliability, enhanced and timely data acquisition, and enhancement of edge computing capabilities. By addressing these objectives, NTNs for IoRT have the potential to revolutionize the way data are collected, processed, and utilized, enabling a wide range of applications in remote and isolated environments.

However, to fully unlock this potential, further research is necessary. Building upon the findings of this survey, our future work will focus on two key areas:

- Investigating the impact of distance, network resources, and aerial platform types on the feasibility and performance of remote IoT connectivity via NTNs. This will involve developing models to predict network performance and identify optimal configurations for different deployment scenarios.
- Exploring energy-efficient communication protocols and developing power-aware routing algorithms that leverage the unique characteristics of NTNs to minimize energy consump-

tion. This may involve integrating renewable energy sources into remote IoT devices and quantifying the environmental impact of different IoRT deployment scenarios.

By addressing these research areas, we aim to contribute to the development of a sustainable and efficient IoRT ecosystem, unlocking the full potential of NTNs.

**Author Contributions:** Investigation, S.P., D.T. and D.N.S.; Writing—original draft, S.P., D.T. and D.N.S.; Review & editing, D.N.S., A.R., G.K. and C.S.; Supervision, D.N.S., A.R., G.K. and C.S. All authors have read and agreed to the published version of the manuscript.

**Funding:** This research received no external funding.

**Informed Consent Statement:** Not applicable.

**Data Availability Statement:** Data sharing not applicable.

**Conflicts of Interest:** The authors declare no conflict of interest.

## Abbreviations

The following abbreviations are used in this manuscript:

| | |
|---|---|
| 3GPP | 3rd Generation Partnership Project |
| AI | Artificial Intelligence |
| AoI | Age of Information |
| B5G | Beyond-Fifth-Generation |
| BR-MP | Bernoulli-Rician Message Passing |
| CoAP | Constrained Application Protocol |
| CSI | Channel State Information |
| CSS | Chirp-Spread Spectrum |
| DL | Deep Learning |
| DRL | Deep Reinforcement Learning |
| DVB-RCS2 | Digital Video Broadcasting—Return Channel via Satellite 2 |
| DVB-RCS2 | Digital Video Broadcasting—Return Channel via Satellite 2 |
| eDRX | extended Discontinuous Reception |
| EM | Expectation-Maximization |
| FDD | Frequency-division Duplex |
| FL | Federated Learning |
| GEO | Geostationary Earth Orbit |
| GF-RA | Grant-Free Random Access |
| GNSS | Global Navigation Satellite System |
| HAPs | High-altitude platforms |
| HARQ | Hybrid Automatic Repeat Request |
| HSTN | Hybrid Satellite-Terrestrial Network |
| IIoT | Industrial Internet of Things |
| IMDs | IoT Mobile Devices |
| IoAT | Internet of Agricultural Things |
| IoBT | Internet of Battle Things |
| IoMT | Internet of Maritime Things |
| IoRT | Internet of Remote Things |
| IoST | Internet of Satellite/Space Things |
| IoT | Internet of Things |
| IoUT | Internet of Underground Things |
| IoUwT | Internet of Underwater Things |
| IPF | Iterative Penalty Function |
| LAPs | Low-altitude platforms |
| LEO | Low Earth Orbit |
| LPWAN | Low-Power Wide-Area Network |
| LTE-M | Long-Term Evolution Machine Type Communication |

| | |
|---|---|
| mMTC | Massive Machine Type Communication |
| MEC | Mobile Edge Computing |
| MEO | Medium Earth Orbit |
| ML | Machine Learning |
| MLTS | Multi-Layer Tabu Search |
| MQTT | Message Queuing Telemetry Transport |
| MTC | Machine Type Communication |
| NB-IoT | NarrowBand Internet of Things |
| NC | Network Coding |
| NOMA | Non-Orthogonal Multiple Access |
| NTN | Non-Terrestrial Network |
| PFD | Power Flux Density |
| PPP | Poisson Point Process |
| PSO | Particle Swarm Optimization |
| QoE | Quality of Experience |
| QoS | Quality of Service |
| RL | Reinforcement Learning |
| SBD | Short Burst Data |
| SGP4 | Simplified General Perturbation 4 |
| SIC | Successive Interference Cancellation |
| SINR | Signal-to-Interference-plus-Noise Ratio |
| S-IoT | Satellite Internet of Things |
| SLA | Service Level Agreement |
| SNR | Signal-to-Noise Ratio |
| STI | Spatially Temporally Correlative Mutual Information |
| UAV | Unmanned Aerial Vehicle |
| UCSS | Unipolar-coded CSS |
| UE | User Equipment |
| URLLC | Ultra-Reliable Low-Latency Communications |
| VLEO | Very Low Earth Orbit |
| WDP | Winner Determination Problem |
| WSN | Wireless Sensor Network |

## References


1. Bellini, P.; Nesi, P.; Pantaleo, G. IoT-Enabled Smart Cities: A Review of Concepts, Frameworks and Key Technologies. *Appl. Sci.* **2022**, *12*, 1607. [CrossRef]
2. Vailshery, L.S. Number of Internet of Things (IoT) Connected Devices Worldwide from 2019 to 2023, with Forecasts from 2022 to 2030. 2023. Available online: https://www.statista.com/statistics/1183457/iot-connected-devices-worldwide/ (accessed on 6 February 2024).
3. Shafique, K.; Khawaja, B.A.; Sabir, F.; Qazi, S.; Mustaqim, M. Internet of Things (IoT) for Next-Generation Smart Systems: A Review of Current Challenges, Future Trends and Prospects for Emerging 5G-IoT Scenarios. *IEEE Access* **2020**, *8*, 23022–23040. [CrossRef]
4. Gupta, B.; Quamara, M. An overview of Internet of Things (IoT): Architectural aspects, challenges, and protocols. *Concurr. Comput. Pract. Exp.* **2020**, *32*, e4946, [CrossRef]
5. Demertzis, K.; Tsiknas, K.; Taketzis, D.; Skoutas, D.N.; Skianis, C.; Iliadis, L.; Zoiros, K.E. Communication Network Standards for Smart Grid Infrastructures. *Network* **2021**, *1*, 132–145. [CrossRef]
6. Polymeni, S.; Skoutas, D.N.; Kormentzas, G.; Skianis, C. FINDEAS: A FinTech-Based Approach on Designing and Assessing IoT Systems. *IEEE Internet Things J.* **2022**, *9*, 25196–25206. [CrossRef]
7. Ojha, T.; Misra, S.; Raghuwanshi, N.S. Internet of Things for Agricultural Applications: The State of the Art. *IEEE Internet Things J.* **2021**, *8*, 10973–10997. [CrossRef]
8. Souri, A.; Hussien, A.; Hoseyninezhad, M.; Norouzi, M. A systematic review of IoT communication strategies for an efficient smart environment. *Trans. Emerg. Telecommun. Technol.* **2022**, *33*, e3736. [CrossRef]
9. Chaudhari, B.S.; Zennaro, M.; Borkar, S. LPWAN Technologies: Emerging Application Characteristics, Requirements, and Design Considerations. *Future Internet* **2020**, *12*, 46. [CrossRef]
10. Ogbodo, E.U.; Abu-Mahfouz, A.M.; Kurien, A.M. A Survey on 5G and LPWAN-IoT for Improved Smart Cities and Remote Area Applications: From the Aspect of Architecture and Security. *Sensors* **2022**, *22*, 6313. [CrossRef]
11. 3GPP. *Technical Specification Group Radio Access Network*; NR; Release 17; Technical Report TR 36.763; 3GPP: Sophia Antipolis, France, 2021.

2. 3GPP. *Technical Specification Group Radio Access Network*; NR; 5G System ; IoT and NTN Enhancements (Release 18); Technical Report TR 38.801; 3GPP: Sophia Antipolis, France, 2023.
3. Mahdi, M.N.; Ahmad, A.R.; Qassim, Q.S.; Natiq, H.; Subhi, M.A.; Mahmoud, M. From 5G to 6G Technology: Meets Energy, Internet-of-Things and Machine Learning: A Survey. *Appl. Sci.* **2021**, *11*, 8117. [CrossRef]
4. Qadir, Z.; Le, K.N.; Saeed, N.; Munawar, H.S. Towards 6G Internet of Things: Recent advances, use cases, and open challenges. *ICT Express* **2023**, *9*, 296–312. [CrossRef]
5. Mahmood, M.R.; Matin, M.A.; Sarigiannidis, P.; Goudos, S.K. A Comprehensive Review on Artificial Intelligence/Machine Learning Algorithms for Empowering the Future IoT Toward 6G Era. *IEEE Access* **2022**, *10*, 87535–87562. [CrossRef]
6. Varga, P.; Peto, J.; Franko, A.; Balla, D.; Haja, D.; Janky, F.; Soos, G.; Ficzere, D.; Maliosz, M.; Toka, L. 5G support for Industrial IoT Applications—Challenges, Solutions, and Research gaps. *Sensors* **2020**, *20*, 828. [CrossRef] [PubMed]
7. Chettri, L.; Bera, R. A Comprehensive Survey on Internet of Things (IoT) Toward 5G Wireless Systems. *IEEE Internet Things J.* **2020**, *7*, 16–32. [CrossRef]
8. Moges, T.H.; Lakew, D.S.; Nguyen, N.P.; Dao, N.N.; Cho, S. Cellular Internet of Things: Use cases, technologies, and future work. *Internet Things* **2023**, *24*, 100910. [CrossRef]
9. Fageda, X.; Suárez-Alemán, A.; Serebrisky, T.; Fioravanti, R. Air connectivity in remote regions: A comprehensive review of existing transport policies worldwide. *J. Air Transp. Manag.* **2018**, *66*, 65–75. [CrossRef]
10. Vannieuwenborg, F.; Verbrugge, S.; Colle, D. Choosing IoT-connectivity? A guiding methodology based on functional characteristics and economic considerations. *Trans. Emerg. Telecommun. Technol.* **2018**, *29*, e3308, [CrossRef]
11. Polymeni, S.; Plastras, S.; Skoutas, D.N.; Kormentzas, G.; Skianis, C. The Impact of 6G-IoT Technologies on the Development of Agriculture 5.0: A Review. *Electronics* **2023**, *12*, 2651. [CrossRef]
12. Plastras, S.; Polymeni, S.; Skoutas, D.N.; Kormentzas, G.; Skianis, C. Sustainable Networking Solutions in Remote IoT Environments: Use Cases, Challenges, and Solutions for Smart Agriculture. In *Re-Visioning Geography: Supporting the SDGs in the Post-COVID Era*; Klonari, A., De Lázaro y Torres, M.L., Kizos, A., Eds.; Springer International Publishing: Cham, Switzerland, 2023; pp. 307–325. [CrossRef]
13. Xia, T.; Wang, M.M.; Zhang, J.; Wang, L. Maritime Internet of Things: Challenges and Solutions. *IEEE Wirel. Commun.* **2020**, *27*, 188–196. [CrossRef]
14. Wei, T.; Feng, W.; Chen, Y.; Wang, C.X.; Ge, N.; Lu, J. Hybrid Satellite-Terrestrial Communication Networks for the Maritime Internet of Things: Key Technologies, Opportunities, and Challenges. *IEEE Internet Things J.* **2021**, *8*, 8910–8934. [CrossRef]
15. Sanchez-Gonzalez, P.L.; Díaz-Gutiérrez, D.; Leo, T.J.; Núñez-Rivas, L.R. Toward Digitalization of Maritime Transport? *Sensors* **2019**, *19*, 926. [CrossRef]
16. Saeed, N.; Alouini, M.S.; Al-Naffouri, T.Y. Toward the Internet of Underground Things: A Systematic Survey. *IEEE Commun. Surv. Tutor.* **2019**, *21*, 3443–3466. [CrossRef]
17. Salam, A.; Raza, U. Current Advances in Internet of Underground Things. In *Signals in the Soil: Developments in Internet of Underground Things*; Springer International Publishing: Cham, Switzerland, 2020; pp. 321–356. [CrossRef]
18. Cariou, C.; Moiroux-Arvis, L.; Pinet, F.; Chanet, J.P. Internet of Underground Things in Agriculture 4.0: Challenges, Applications and Perspectives. *Sensors* **2023**, *23*, 4058. [CrossRef]
19. Khalil, R.A.; Saeed, N.; Babar, M.I.; Jan, T. Toward the Internet of Underwater Things: Recent Developments and Future Challenges. *IEEE Consum. Electron. Mag.* **2021**, *10*, 32–37. [CrossRef]
20. Domingo, M.C. An overview of the internet of underwater things. *J. Netw. Comput. Appl.* **2012**, *35*, 1879–1890. [CrossRef]
21. Kao, C.C.; Lin, Y.S.; Wu, G.D.; Huang, C.J. A Comprehensive Study on the Internet of Underwater Things: Applications, Challenges, and Channel Models. *Sensors* **2017**, *17*, 1477. [CrossRef] [PubMed]
22. Mohsan, S.A.H.; Li, Y.; Sadiq, M.; Liang, J.; Khan, M.A. Recent Advances, Future Trends, Applications and Challenges of Internet of Underwater Things (IoUT): A Comprehensive Review. *J. Mar. Sci. Eng.* **2023**, *11*, 124. [CrossRef]
23. Akyildiz, I.F.; Kak, A. The Internet of Space Things/CubeSats. *IEEE Netw.* **2019**, *33*, 212–218. [CrossRef]
24. Jiao, J.; Wu, S.; Lu, R.; Zhang, Q. Massive Access in Space-Based Internet of Things: Challenges, Opportunities, and Future Directions. *IEEE Wirel. Commun.* **2021**, *28*, 118–125. [CrossRef]
25. Kott, A.; Swami, A.; West, B.J. The Internet of Battle Things. *Computer* **2016**, *49*, 70–75. [CrossRef]
26. Ang, K.L.M.; Seng, J.K.P. Application Specific Internet of Things (ASIoTs): Taxonomy, Applications, Use Case and Future Directions. *IEEE Access* **2019**, *7*, 56577–56590. [CrossRef]
27. Russell, S.; Abdelzaher, T. The Internet of Battlefield Things: The Next Generation of Command, Control, Communications and Intelligence (C3I) Decision-Making. In Proceedings of the MILCOM 2018 IEEE Military Communications Conference (MILCOM), Los Angeles, CA, USA, 29–31 October 2018; pp. 737–742. [CrossRef]
28. Palma, D.; Birkeland, R. Enabling the Internet of Arctic Things with Freely-Drifting Small-Satellite Swarms. *IEEE Access* **2018**, *6*, 71435–71443. [CrossRef]
29. De Sanctis, M.; Cianca, E.i.; Araniti, G.; Bisio, I.; Prasad, R. Satellite Communications Supporting Internet of Remote Things. *IEEE Internet Things J.* **2016**, *3*, 113–123. [CrossRef]

40. Qu, Z.; Zhang, G.; Cao, H.; Xie, J. LEO Satellite Constellation for Internet of Things. *IEEE Access* **2017**, *5*, 18391–18401. [CrossRe
41. Marchese, M.; Moheddine, A.; Patrone, F. IoT and UAV Integration in 5G Hybrid Terrestrial-Satellite Networks. *Senso* **2019**, *19*, 3704. [CrossRef]
42. Michailidis, E.T.; Potirakis, S.M.; Kanatas, A.G. AI-Inspired Non-Terrestrial Networks for IIoT: Review on Enabling Technologi and Applications. *IoT* **2020**, *1*, 21–48. [CrossRef]
43. Fang, X.; Feng, W.; Wei, T.; Chen, Y.; Ge, N.; Wang, C.X. 5G Embraces Satellites for 6G Ubiquitous IoT: Basic Models for Integrate Satellite Terrestrial Networks. *IEEE Internet Things J.* **2021**, *8*, 14399–14417. [CrossRef]
44. Centenaro, M.; Costa, C.E.; Granelli, F.; Sacchi, C.; Vangelista, L. A Survey on Technologies, Standards and Open Challenges i Satellite IoT. *IEEE Commun. Surv. Tutor.* **2021**, *23*, 1693–1720. [CrossRef]
45. Kua, J.; Loke, S.W.; Arora, C.; Fernando, N.; Ranaweera, C. Internet of Things in Space: A Review of Opportunities and Challenge from Satellite-Aided Computing to Digitally-Enhanced Space Living. *Sensors* **2021**, *21*, 8117. [CrossRef] [PubMed]
46. Li, J.; Kacimi, R.; Liu, T.; Ma, X.; Dhaou, R. Non-Terrestrial Networks-Enabled Internet of Things: UAV-Centric Architecture Applications, and Open Issues. *Drones* **2022**, *6*, 95. [CrossRef]
47. Azari, M.M.; Solanki, S.; Chatzinotas, S.; Kodheli, O.; Sallouha, H.; Colpaert, A.; Mendoza Montoya, J.F.; Pollin, S.; Haqiqatneja A.; Mostaani, A.; et al. Evolution of Non-Terrestrial Networks From 5G to 6G: A Survey. *IEEE Commun. Surv. Tut* **2022**, *24*, 2633–2672. [CrossRef]
48. Vaezi, M.; Azari, A.; Khosravirad, S.R.; Shirvanimoghaddam, M.; Azari, M.M.; Chasaki, D.; Popovski, P. Cellular, Wide-Are and Non-Terrestrial IoT: A Survey on 5G Advances and the Road Toward 6G. *IEEE Commun. Surv. Tutor.* **2022**, *24*, 1117–117 [CrossRef]
49. Sharif, S.; Zeadally, S.; Ejaz, W. Space-aerial-ground-sea integrated networks: Resource optimization and challenges in 6G. *J. Net Comput. Appl.* **2023**, *215*, 103647. [CrossRef]
50. Chen, X.; Xu, Z.; Shang, L. Satellite Internet of Things: challenges, solutions, and development trends. *Front. Inf. Techn Electron. Eng.* **2023**, *24*, 935–944. [CrossRef]
51. Saeed, N.; Almorad, H.; Dahrouj, H.; Al-Naffouri, T.Y.; Shamma, J.S.; Alouini, M.S. Point-to-Point Communication in Int grated Satellite-Aerial 6G Networks: State-of-the-Art and Future Challenges. *IEEE Open J. Commun. Soc.* **2021**, *2*, 1505–152 [CrossRef]
52. Saafi, S.; Vikhrova, O.; Fodor, G.; Hosek, J.; Andreev, S. AI-Aided Integrated Terrestrial and Non-Terrestrial 6G Solutions fc Sustainable Maritime Networking. *IEEE Netw.* **2022**, *36*, 183–190. [CrossRef]
53. Geraci, G.; López-Pérez, D.; Benzaghta, M.; Chatzinotas, S. Integrating Terrestrial and Non-Terrestrial Networks: 3D Opportunitie and Challenges. *IEEE Commun. Mag.* **2023**, *61*, 42–48. [CrossRef]
54. Hassan, S.S.; Kim, D.H.; Tun, Y.K.; Tran, N.H.; Saad, W.; Hong, C.S. Seamless and Energy-Efficient Maritime Coverage i Coordinated 6G Space–Air–Sea Non-Terrestrial Networks. *IEEE Internet Things J.* **2023**, *10*, 4749–4769. [CrossRef]
55. Leyva-Mayorga, I.; Soret, B.; Röper, M.; Wübben, D.; Matthiesen, B.; Dekorsy, A.; Popovski, P. LEO Small-Satellite Constellatior for 5G and Beyond-5G Communications. *IEEE Access* **2020**, *8*, 184955–184964. [CrossRef]
56. Wang, F.; Jiang, D.; Wang, Z.; Chen, J.; Quek, T.Q.S. Seamless Handover in LEO Based Non-Terrestrial Networks: Servic Continuity and Optimization. *IEEE Trans. Commun.* **2023**, *71*, 1008–1023. [CrossRef]
57. Xiao, Z.; Yang, J.; Mao, T.; Xu, C.; Zhang, R.; Han, Z.; Xia, X.G. LEO Satellite Access Network (LEO-SAN) towards 6G: Challenge and Approaches. *IEEE Wirel. Commun.* **2022**, 1 –8. [CrossRef]
58. Lin, X.; Cioni, S.; Charbit, G.; Chuberre, N.; Hellsten, S.; Boutillon, J.F. On the Path to 6G: Embracing the Next Wave of Low Eart Orbit Satellite Access. *IEEE Commun. Mag.* **2021**, *59*, 36–42. [CrossRef]
59. Xie, H.; Zhan, Y.; Zeng, G.; Pan, X. LEO Mega-Constellations for 6G Global Coverage: Challenges and Opportunities. *IEEE Acce* **2021**, *9*, 164223–164244. [CrossRef]
60. Alwis, C.D.; Kalla, A.; Pham, Q.V.; Kumar, P.; Dev, K.; Hwang, W.J.; Liyanage, M. Survey on 6G Frontiers: Trends, Application Requirements, Technologies and Future Research. *IEEE Open J. Commun. Soc.* **2021**, *2*, 836–886. [CrossRef]
61. Cao, X.; Yang, P.; Alzenad, M.; Xi, X.; Wu, D.; Yanikomeroglu, H. Airborne Communication Networks: A Survey. *IEEE J. Se Areas Commun.* **2018**, *36*, 1907–1926. [CrossRef]
62. Michailidis, E.T.; Maliatsos, K.; Skoutas, D.N.; Vouyioukas, D.; Skianis, C. Secure UAV-Aided Mobile Edge Computing for IoT: Review. *IEEE Access* **2022**, *10*, 86353–86383. [CrossRef]
63. Wang, X.; Guo, Y.; Gao, Y. Unmanned Autonomous Intelligent System in 6G Non-Terrestrial Network. *Information* **2024**, *15*, 3 [CrossRef]
64. Kong, H.; Lin, M.; Zhang, J.; Ouyang, J.; Zhu, W.P.; Alouini, M.S. Beamforming Design and Performance Analysis for Satelli and UAV Integrated Networks in IoRT Applications. *IEEE Internet Things J.* **2022**, *9*, 14965–14977. [CrossRef]
65. Yao, Y.; Dong, D.; Huang, S.; Pan, C.; Chen, S.; Li, X. Optimization of the Internet of remote things data acquisition based o satellite UAV integrated network. *China Commun.* **2023**, *20*, 15–28. [CrossRef]
66. Guo, H.; Li, J.; Liu, J.; Tian, N.; Kato, N. A Survey on Space-Air-Ground-Sea Integrated Network Security in 6G. *IEEE Commu Surv. Tutor.* **2022**, *24*, 53–87. [CrossRef]
67. Chen, Y.; Zhang, M.; Li, X.; Che, T.; Jin, R.; Guo, J.; Yang, W.; An, B.; Nie, X. Satellite-Enabled Internet of Remote Things Networ Transmits Field Data from the Most Remote Areas of the Tibetan Plateau. *Sensors* **2022**, *22*, 3713. [CrossRef]

8. Hamdan, S.; Ayyash, M.; Almajali, S. Edge-Computing Architectures for Internet of Things Applications: A Survey. *Sensors* **2020**, *20*, 6441. [CrossRef]
9. Giambene, G.; Addo, E.O.; Kota, S. 5G Aerial Component for IoT Support in Remote Rural Areas. In Proceedings of the 2019 IEEE 2nd 5G World Forum (5GWF), Dresden, Germany, 30 September–2 October 2019; pp. 572–577. [CrossRef]
0. Mwase, C.; Jin, Y.; Westerlund, T.; Tenhunen, H.; Zou, Z. Communication-efficient distributed AI strategies for the IoT edge. *Future Gener. Comput. Syst.* **2022**, *131*, 292–308. [CrossRef]
1. Callebaut, G.; Leenders, G.; Van Mulders, J.; Ottoy, G.; De Strycker, L.; Van der Perre, L. The Art of Designing Remote IoT Devices—Technologies and Strategies for a Long Battery Life. *Sensors* **2021**, *21*, 913. [CrossRef]
2. Zhu, X.; Jiang, C. Integrated Satellite-Terrestrial Networks Toward 6G: Architectures, Applications, and Challenges. *IEEE Internet Things J.* **2022**, *9*, 437–461. [CrossRef]
3. Chiti, F.; Fantacci, R.; Pierucci, L. Energy Efficient Communications for Reliable IoT Multicast 5G/Satellite Services. *Future Internet* **2019**, *11*, 164. [CrossRef]
4. Zhen, L.; Bashir, A.K.; Yu, K.; Al-Otaibi, Y.D.; Foh, C.H.; Xiao, P. Energy-Efficient Random Access for LEO Satellite-Assisted 6G Internet of Remote Things. *IEEE Internet Things J.* **2021**, *8*, 5114–5128. [CrossRef]
5. Boquet, G.; Martinez, B.; Adelantado, F.; Pages, J.; Ruiz-de Azua, J.A.; Vilajosana, X. Low-Power Satellite Access Time Estimation for Internet of Things Services over Non-Terrestrial Networks. *IEEE Internet Things J.* **2023**, *11*, 3206–3216. [CrossRef]
6. Yue, P.; Du, J.; Zhang, R.; Ding, H.; Wang, S.; An, J. Collaborative LEO Satellites for Secure and Green Internet of Remote Things. *IEEE Internet Things J.* **2023**, *10*, 9283–9294. [CrossRef]
7. Song, Z.; Hao, Y.; Liu, Y.; Sun, X. Energy-Efficient Multiaccess Edge Computing for Terrestrial-Satellite Internet of Things. *IEEE Internet Things J.* **2021**, *8*, 14202–14218. [CrossRef]
8. Zhao, B.; Liu, J.; Wei, Z.; You, I. A Deep Reinforcement Learning Based Approach for Energy-Efficient Channel Allocation in Satellite Internet of Things. *IEEE Access* **2020**, *8*, 62197–62206. [CrossRef]
9. Li, F.; Lam, K.Y.; Liu, X.; Wang, L. Resource Allocation in Satellite-Based Internet of Things Using Pattern Search Method. *IEEE Access* **2020**, *8*, 110908–110914. [CrossRef]
0. Dazhi, M.N.; Al-Hraishawi, H.; Shankar, B.; Chatzinotas, S. Terminal-Aware Multi-Connectivity Scheduler for Uplink Multi-Layer Non-Terrestrial Networks. In Proceedings of the 2022 IEEE Globecom Workshops (GC Wkshps), Rio de Janeiro, Brazil, 4–8 December 2022; pp. 1133–1139. [CrossRef]
1. Chu, J.; Chen, X.; Zhong, C.; Zhang, Z. Robust Design for NOMA-Based Multibeam LEO Satellite Internet of Things. *IEEE Internet Things J.* **2021**, *8*, 1959–1970. [CrossRef]
2. Tang, S.; Pan, Z.; Hu, G.; Wu, Y.; Li, Y. Deep Reinforcement Learning-Based Resource Allocation for Satellite Internet of Things with Diverse QoS Guarantee. *Sensors* **2022**, *22*, 2979. [CrossRef]
3. Kim, M.G.; Jo, H.S. Performance Analysis of NB-IoT Uplink in Low Earth Orbit Non-Terrestrial Networks. *Sensors* **2022**, *22*, 7097. [CrossRef]
4. Fraire, J.A.; Henn, S.; Dovis, F.; Garello, R.; Taricco, G. Sparse Satellite Constellation Design for LoRa-based Direct-to-Satellite Internet of Things. In Proceedings of the GLOBECOM 2020 IEEE Global Communications Conference, Taipei, Taiwan, 7–11 December 2020; pp. 1–6. [CrossRef]
5. Akyildiz, I.F.; Kak, A. The Internet of Space Things/CubeSats: A ubiquitous cyber-physical system for the connected world. *Comput. Netw.* **2019**, *150*, 134–149. [CrossRef]
6. Gavrilă, C.; Popescu, V.; Alexandru, M.; Murroni, M.; Sacchi, C. An SDR-Based Satellite Gateway for Internet of Remote Things (IoRT) Applications. *IEEE Access* **2020**, *8*, 115423–115436. [CrossRef]
7. Wang, L.; Liu, S.; Wang, W.; Fan, Z. Dynamic uplink transmission scheduling for satellite Internet of Things applications. *China Commun.* **2020**, *17*, 241–248. [CrossRef]
8. Almonacid, V.; Franck, L. Extending the coverage of the internet of things with low-cost nanosatellite networks. *Acta Astronaut.* **2017**, *138*, 95–101. [CrossRef]
9. Chen, Y.; Zhao, F.; Yang, R.; Kong, C.; Li, R.; Wang, J. Internet of Things over Beam-Hopping-Based Non-terrestrial Networks. In Proceedings of the 2023 26th Conference on Innovation in Clouds, Internet and Networks and Workshops (ICIN), Paris, France, 6–9 March 2023; pp. 74–81. [CrossRef]
0. Wong, A.; Chow, Y.T. Solar-Supplied Satellite Internet Access Point for the Internet of Things in Remote Areas. *Sensors* **2020**, *20*, 1409. [CrossRef] [PubMed]
1. Zhou, H.; Liu, L.; Ma, H. Coverage and Capacity Analysis of LEO Satellite Network Supporting Internet of Things. In Proceedings of the ICC 2019 IEEE International Conference on Communications (ICC), Shanghai, China, 20–24 May 2019; pp. 1–6. [CrossRef]
2. Zhou, D.; Sheng, M.; Wu, J.; Li, J.; Han, Z. Gateway Placement in Integrated Satellite–Terrestrial Networks: Supporting Communications and Internet of Remote Things. *IEEE Internet Things J.* **2022**, *9*, 4421–4434. [CrossRef]
3. Chan, C.C.; Al-Hourani, A.; Choi, J.; Gomez, K.M.; Kandeepan, S. Performance Modeling Framework for IoT-over-Satellite Using Shared Radio Spectrum. *Remote Sens.* **2020**, *12*, 1666. [CrossRef]
4. Liu, X.; Zhai, X.B.; Lu, W.; Wu, C. QoS-Guarantee Resource Allocation for Multibeam Satellite Industrial Internet of Things With NOMA. *IEEE Trans. Ind. Inform.* **2021**, *17*, 2052–2061. [CrossRef]

95. Aiyetoro, G.; Owolawi, P. Spectrum Management Schemes for Internet of Remote Things (IoRT) Devices in 5G Networks vi GEO Satellite. *Future Internet* **2019**, *11*, 257. [CrossRef]
96. Kodheli, O.; Andrenacci, S.; Maturo, N.; Chatzinotas, S.; Zimmer, F. Resource Allocation Approach for Differential Dopple Reduction in NB-IoT over LEO Satellite. In Proceedings of the 2018 9th Advanced Satellite Multimedia Systems Conference an the 15th Signal Processing for Space Communications Workshop (ASMS/SPSC), Berlin, Germany, 10–12 September 2018; pp. 1– [CrossRef]
97. Hashim, I.S.M.; Al-Hourani, A.; Wayne Rowe, S.T.; Scott, J.R. Adaptive X-Band Satellite Antenna for Internet-of-Things (IoT over Satellite Applications. In Proceedings of the 2019 13th International Conference on Signal Processing and Communicatio Systems (ICSPCS), Gold Coast, QLD, Australia, 16–18 December 2019; pp. 1–7. [CrossRef]
98. Yang, M.; Xue, G.; Liu, B.; Yang, Y.; Su, Y. Load Estimation Based Dynamic Access Protocol for Satellite Internet of Thing *Remote Sens.* **2022**, *14*, 6402. [CrossRef]
99. Zhu, Y.; Delamotte, T.; Knopp, A. Geographical NOMA-Beamforming in Multi-Beam Satellite-Based Internet of Things. I Proceedings of the 2019 IEEE Global Communications Conference (GLOBECOM), Waikoloa, HI, USA, 9–13 December 201 pp. 1–6. [CrossRef]
100. Ye, N.; Yu, J.; Wang, A.; Zhang, R. Help from space: Grant-free massive access for satellite-based IoT in the 6G era. *Digit. Com mun. Netw.* **2022**, *8*, 215–224. [CrossRef]
101. Yan, Y.; An, K.; Zhang, B.; Zhu, W.P.; Ding, G.; Guo, D. Outage-Constrained Robust Multigroup Multicast Beamform ing for Satellite-Based Internet of Things Coexisting With Terrestrial Networks. *IEEE Internet Things J.* **2021**, *8*, 8159–817 [CrossRef]
102. Kodheli, O.; Andrenacci, S.; Maturo, N.; Chatzinotas, S.; Zimmer, F. An Uplink UE Group-Based Scheduling Technique for 5G mMTC Systems Over LEO Satellite. *IEEE Access* **2019**, *7*, 67413–67427. [CrossRef]
103. Xia, T.; Wang, M.M.; You, X. Satellite Machine-Type Communication for Maritime Internet of Things: An Interference Perspectiv *IEEE Access* **2019**, *7*, 76404–76415. [CrossRef]
104. Hofmann, C.A.; Knopp, A. Ultranarrowband Waveform for IoT Direct Random Multiple Access to GEO Satellites. *IEEE Interne Things J.* **2019**, *6*, 10134–10149. [CrossRef]
105. Hofmann, C.A.; Knopp, A. Direct Access to GEO Satellites: An Internet of Remote Things Technology. In Proceed ings of the 2019 IEEE 2nd 5G World Forum (5GWF), Dresden, Germany, 30 September–2 October 2019; pp. 578–58 [CrossRef]
106. Zhou, D.; Sheng, M.; Wang, Y.; Li, J.; Han, Z. Machine Learning-Based Resource Allocation in Satellite Networks Supportin Internet of Remote Things. *IEEE Trans. Wirel. Commun.* **2021**, *20*, 6606–6621. [CrossRef]
107. Liu, S.; Jiao, J.; Ni, Z.; Wu, S.; Zhang, Q. Age-Optimal NC-HARQ Protocol for Multi-hop Satellite-based Internet of Things. I Proceedings of the 2021 IEEE Wireless Communications and Networking Conference (WCNC), Nanjing, China, 29 March–1 Apri 2021; pp. 1–6. [CrossRef]
108. Lysogor, I.; Voskov, L.; Rolich, A.; Efremov, S. Study of Data Transfer in a Heterogeneous LoRa-Satellite Network for the Interne of Remote Things. *Sensors* **2019**, *19*, 3384. [CrossRef]
109. Zhen, L.; Zhang, Y.; Yu, K.; Kumar, N.; Barnawi, A.; Xie, Y. Early Collision Detection for Massive Random Access in Satellite-Base Internet of Things. *IEEE Trans. Veh. Technol.* **2021**, *70*, 5184–5189. [CrossRef]
110. Xu, Y.; Li, Y.; Zhang, Q.; Yang, Z. Age-Optimal Hybrid Temporal-Spatial Generalized Deduplication and ARQ for Satellite Integrated Internet of Things. *IEEE Internet Things J.* **2022**, *9*, 24963–24979. [CrossRef]
111. Zhang, Z.; Li, Y.; Huang, C.; Guo, Q.; Liu, L.; Yuen, C.; Guan, Y.L. User Activity Detection and Channel Estimatio for Grant-Free Random Access in LEO Satellite-Enabled Internet of Things. *IEEE Internet Things J.* **2020**, *7*, 8811–882 [CrossRef]
112. Chu, J.; Chen, X. Robust Design for Integrated Satellite–Terrestrial Internet of Things. *IEEE Internet Things J.* **2021**, *8*, 9072–908 [CrossRef]
113. Kak, A.; Akyildiz, I.F. Towards Automatic Network Slicing for the Internet of Space Things. *IEEE Trans. Netw. Serv. Manag* **2022**, *19*, 392–412. [CrossRef]
114. Li, Y.; Xu, Y.; Zhang, Q.; Yang, Z. Age-Driven Spatially Temporally Correlative Updating in the Satellite-Integrated Internet o Things via Markov Decision Process. *IEEE Internet Things J.* **2022**, *9*, 13612–13625. [CrossRef]
115. Aiyetoro, G.; Owolawi, P. Dynamic Packet Scheduling for Internet of Remote Things (IoRT) devices in 5G Satellite Network In Proceedings of the 2020 International Conference on Advances in Computing and Communication Engineering (ICACCE Las Vegas, NV, USA, 22–24 June 2020; pp. 1–6. [CrossRef]
116. Bacco, M.; Colucci, M.; Gotta, A. Application protocols enabling internet of remote things via random access satellite channel In Proceedings of the 2017 IEEE International Conference on Communications (ICC), Paris, France, 21–25 May 2017; pp. 1– [CrossRef]
117. Dai, C.Q.; Zhang, M.; Li, C.; Zhao, J.; Chen, Q. QoE-Aware Intelligent Satellite Constellation Design in Satellite Internet of Thing *IEEE Internet Things J.* **2021**, *8*, 4855–4867. [CrossRef]
118. Bisio, I.; Lavagetto, F.; Luzzati, G. Cooperative Application Layer Joint Video Coding in the Internet of Remote Thing *IEEE Internet Things J.* **2016**, *3*, 1418–1426. [CrossRef]

19. Jiao, J.; Sun, Y.; Wu, S.; Wang, Y.; Zhang, Q. Network Utility Maximization Resource Allocation for NOMA in Satellite-Based Internet of Things. *IEEE Internet Things J.* **2020**, *7*, 3230–3242. [CrossRef]
20. Chen, T.; Liu, J.; Tang, Q.; Huang, T.; Liu, Y. Cooperative Task Processing for the Internet of Remote Things through Ultra-Dense Satellite Systems. In Proceedings of the 2021 International Conference on Space-Air-Ground Computing (SAGC), Huizhou, China, 23–25 October 2021; pp. 83–88. [CrossRef]
21. Wang, Y.; Yang, J.; Guo, X.; Qu, Z. Satellite Edge Computing for the Internet of Things in Aerospace. *Sensors* **2019**, *19*, 4375. [CrossRef]
22. Chen, T.; Liu, J.; Ye, Q.; Zhuang, W.; Zhang, W.; Huang, T.; Liu, Y. Learning-Based Computation Offloading for IoRT Through Ka/Q-Band Satellite–Terrestrial Integrated Networks. *IEEE Internet Things J.* **2022**, *9*, 12056–12070. [CrossRef]
23. Han, D.; Ye, Q.; Peng, H.; Wu, W.; Wu, H.; Liao, W.; Shen, X. Two-Timescale Learning-Based Task Offloading for Remote IoT in Integrated Satellite–Terrestrial Networks. *IEEE Internet Things J.* **2023**, *10*, 10131–10145. [CrossRef]
24. Qin, Z.; Yao, H.; Mai, T.; Wu, D.; Zhang, N.; Guo, S. Multi-Agent Reinforcement Learning Aided Computation Offloading in Aerial Computing for the Internet-of-Things. *IEEE Trans. Serv. Comput.* **2023**, *16*, 1976–1986. [CrossRef]
25. Ruan, Y.; Li, Y.; Zhang, R.; Cheng, W.; Liu, C. Cooperative Resource Management for Cognitive Satellite-Aerial-Terrestrial Integrated Networks Towards IoT. *IEEE Access* **2020**, *8*, 35759–35769. [CrossRef]
26. Li, Z.; Wang, Y.; Liu, M.; Sun, R.; Chen, Y.; Yuan, J.; Li, J. Energy Efficient Resource Allocation for UAV-Assisted Space-Air-Ground Internet of Remote Things Networks. *IEEE Access* **2019**, *7*, 145348–145362. [CrossRef]
27. Ma, T.; Zhou, H.; Qian, B.; Cheng, N.; Shen, X.; Chen, X.; Bai, B. UAV-LEO Integrated Backbone: A Ubiquitous Data Collection Approach for B5G Internet of Remote Things Networks. *IEEE J. Sel. Areas Commun.* **2021**, *39*, 3491–3505. [CrossRef]
28. Zhu, Y.; Bai, W.; Sheng, M.; Li, J.; Zhou, D.; Han, Z. Joint UAV Access and GEO Satellite Backhaul in IoRT Networks: Performance Analysis and Optimization. *IEEE Internet Things J.* **2021**, *8*, 7126–7139. [CrossRef]
29. Fang, X.; Feng, W.; Wang, Y.; Chen, Y.; Ge, N.; Ding, Z.; Zhu, H. NOMA-Based Hybrid Satellite-UAV-Terrestrial Networks for 6G Maritime Coverage. *IEEE Trans. Wirel. Commun.* **2023**, *22*, 138–152. [CrossRef]
30. Wang, Y.; Li, Z.; Chen, Y.; Liu, M.; Lyu, X.; Hou, X.; Wang, J. Joint Resource Allocation and UAV Trajectory Optimization for Space–Air–Ground Internet of Remote Things Networks. *IEEE Syst. J.* **2021**, *15*, 4745–4755. [CrossRef]
31. Jia, Z.; Sheng, M.; Li, J.; Niyato, D.; Han, Z. LEO-Satellite-Assisted UAV: Joint Trajectory and Data Collection for Internet of Remote Things in 6G Aerial Access Networks. *IEEE Internet Things J.* **2021**, *8*, 9814–9826. [CrossRef]
32. Liu, C.; Feng, W.; Tao, X.; Ge, N. MEC-Empowered Non-Terrestrial Network for 6G Wide-Area Time-Sensitive Internet of Things. *Engineering* **2022**, *8*, 96–107. [CrossRef]
33. Zhang, M.; Li, X. Drone-Enabled Internet-of-Things Relay for Environmental Monitoring in Remote Areas Without Public Networks. *IEEE Internet Things J.* **2020**, *7*, 7648–7662. [CrossRef]
34. Jia, M.; Gao, Q.; Guo, Q.; Gu, X. Energy-Efficiency Power Allocation Design for UAV-Assisted Spatial NOMA. *IEEE Internet Things J.* **2021**, *8*, 15205–15215. [CrossRef]
35. Qin, P.; Zhu, Y.; Zhao, X.; Feng, X.; Liu, J.; Zhou, Z. Joint 3D-Location Planning and Resource Allocation for XAPS-Enabled C-NOMA in 6G Heterogeneous Internet of Things. *IEEE Trans. Veh. Technol.* **2021**, *70*, 10594–10609. [CrossRef]
36. Na, Z.; Liu, Y.; Shi, J.; Liu, C.; Gao, Z. UAV-Supported Clustered NOMA for 6G-Enabled Internet of Things: Trajectory Planning and Resource Allocation. *IEEE Internet Things J.* **2021**, *8*, 15041–15048. [CrossRef]
37. Ghildiyal, Y.; Singh, R.; Alkhayyat, A.; Gehlot, A.; Malik, P.; Sharma, R.; Akram, S.V.; Alkwai, L.M. An imperative role of 6G communication with perspective of industry 4.0: Challenges and research directions. *Sustain. Energy Technol. Assessments* **2023**, *56*, 103047. [CrossRef]
38. Pattnaik, S.K.; Samal, S.R.; Bandopadhaya, S.; Swain, K.; Choudhury, S.; Das, J.K.; Mihovska, A.; Poulkov, V. Future Wireless Communication Technology towards 6G IoT: An Application-Based Analysis of IoT in Real-Time Location Monitoring of Employees Inside Underground Mines by Using BLE. *Sensors* **2022**, *22*, 3438. [CrossRef]
39. Imoize, A.L.; Adedeji, O.; Tandiya, N.; Shetty, S. 6G Enabled Smart Infrastructure for Sustainable Society: Opportunities, Challenges, and Research Roadmap. *Sensors* **2021**, *21*, 1709. [CrossRef]
40. Alraih, S.; Shayea, I.; Behjati, M.; Nordin, R.; Abdullah, N.F.; Abu-Samah, A.; Nandi, D. Revolution or Evolution? Technical Requirements and Considerations towards 6G Mobile Communications. *Sensors* **2022**, *22*, 762. [CrossRef]
41. Di, B.; Song, L.; Li, Y.; Poor, H.V. Ultra-Dense LEO: Integration of Satellite Access Networks into 5G and Beyond. *IEEE Wirel. Commun.* **2019**, *26*, 62–69. [CrossRef]
42. Yap, K.Y.; Chin, H.H.; Klemeš, J.J. Future outlook on 6G technology for renewable energy sources (RES). *Renew. Sustain. Energy Rev.* **2022**, *167*, 112722. [CrossRef]
43. Charef, N.; Ben Mnaouer, A.; Aloqaily, M.; Bouachir, O.; Guizani, M. Artificial intelligence implication on energy sustainability in Internet of Things: A survey. *Inf. Process. Manag.* **2023**, *60*, 103212. [CrossRef]
44. Laroui, M.; Nour, B.; Moungla, H.; Cherif, M.A.; Afifi, H.; Guizani, M. Edge and fog computing for IoT: A survey on current research activities & future directions. *Comput. Commun.* **2021**, *180*, 210–231. [CrossRef]
45. Esmat, H.H.; Lorenzo, B.; Shi, W. Toward Resilient Network Slicing for Satellite–Terrestrial Edge Computing IoT. *IEEE Internet Things J.* **2023**, *10*, 14621–14645. [CrossRef]

146. Cui, H.; Zhang, J.; Geng, Y.; Xiao, Z.; Sun, T.; Zhang, N.; Liu, J.; Wu, Q.; Cao, X. Space-air-ground integrated network (SAGIN) fo 6G: Requirements, architecture and challenges. *China Commun.* **2022**, *19*, 90–108. [CrossRef]
147. López, O.L.A.; Alves, H.; Souza, R.D.; Montejo-Sánchez, S.; Fernández, E.M.G.; Latva-Aho, M. Massive Wireless Energy Transfe Enabling Sustainable IoT Toward 6G Era. *IEEE Internet Things J.* **2021**, *8*, 8816–8835. [CrossRef]
148. 3GPP. *Technical Specification Group Core Network and Terminals; Study on Enhancements for Non-Terrestrial Networks (NTNs)*; Release 1 Technical Report TR 23.756; 3GPP: Sophia Antipolis, France, 2023.
149. Chun, S. *3GPP Rel-19 toward 6G*; Technical Report, Ofinno: Reston, VA, USA, 2022.
150. 3GPP. *Study on Non-Terrestrial Networks (NTNs) for 5G and Beyond*; Technical Report R4-2220239; 3GPP: Sophia Antipoli France, 2023.
151. Höyhtyä, M.; Boumard, S.; Yastrebova, A.; Järvensivu, P.; Kiviranta, M.; Anttonen, A. Sustainable Satellite Commun cations in the 6G Era: A European View for Multilayer Systems and Space Safety. *IEEE Access* **2022**, *10*, 99973–10000 [CrossRef]
152. Fraga-Lamas, P.; Lopes, S.I.; Fernández-Caramés, T.M. Green IoT and Edge AI as Key Technological Enablers for a Sustainab Digital Transition towards a Smart Circular Economy: An Industry 5.0 Use Case. *Sensors* **2021**, *21*, 5745. [CrossRef]
153. Allam, Z.; Bibri, S.E.; Jones, D.S.; Chabaud, D.; Moreno, C. Unpacking the '15-Minute City' via 6G, IoT, and Digita Twins: Towards a New Narrative for Increasing Urban Efficiency, Resilience, and Sustainability. *Sensors* **2022**, *22*, 136 [CrossRef] [PubMed]
154. 3GPP. *Survey of UAVs, Satellites, Hybrid NTNs for IoRT Scenarios with Respect to Energy Efficiency (EE)*; Technical Report RP-22351 3GPP: Sophia Antipolis, France, 2023.
155. 3GPP. *Study on Non-Terrestrial Networks (NTNs) for 5G and Beyond—OAM for NTNs*; Technical Report RP-23080 3GPP: Sophia Antipolis, France, 2023.
156. 3GPP. *Satellite Node Radio Transmission and Reception*; 5G; NR; Technical Report TS 38.108 v17.4.0; 3GPP: Sophia Antipoli France, 2023.
157. Pastukh, A.; Tikhvinskiy, V.; Dymkova, S.; Varlamov, O. Challenges of Using the L-Band and S-Band for Direct-to-Cellula Satellite 5G-6G NTN Systems. *Technologies* **2023**, *11*, 110. [CrossRef]
158. Dicandia, F.A.; Fonseca, N.J.G.; Bacco, M.; Mugnaini, S.; Genovesi, S. Space-Air-Ground Integrated 6G Wireless Communicatio Networks: A Review of Antenna Technologies and Application Scenarios. *Sensors* **2022**, *22*, 3136. [CrossRef] [PubMed]

MDPI
St. Alban-Anlage 66
4052 Basel
Switzerland
www.mdpi.com

*Sensors* Editorial Office
E-mail: sensors@mdpi.com
www.mdpi.com/journal/sensors

www.ingramcontent.com/pod-product-compliance
Lightning Source LLC
Chambersburg PA
CBHW061831210926
51850CB00006B/855